Delta Modulation Systems

DELTA MODULATION SYSTEMS

R. Steele, B.Sc., C.Eng., M.I.E.E.
*Senior Lecturer, Electronics and
Electrical Engineering Department,
Loughborough University of Technology*

A HALSTED PRESS BOOK

JOHN WILEY & SONS
New York – Toronto

© Raymond Steele, 1975

First Published, 1975

Published in the U.S.A. and Canada
by Halsted Press, a Division
of John Wiley & Sons, Inc.
New York.

Library of Congress Cataloging in Publication Data
Steele, Raymond
   Delta modulation systems.

  "A Halsted Press book."
  1. Delta modulation. I. Title.
TK5102.5.S7   1975   621.38   74-11299
ISBN 0-470-82104-3

Printed in England by
J. W. Arrowsmith Ltd., Bristol, England

CONTENTS

Preface

1. **LINEAR DELTA MODULATION** — 1
   - 1.1. Introduction — 1
   - 1.2. Linear d.m. system — 4
   - 1.3. Delayed encoding — 14
   - 1.4. Delta-sigma modulation — 17
   - 1.5. Encoding d.c. signals — 20
   - 1.6. Models for linear delta modulation systems — 20
   - 1.7. Linear delta modulator using a sample and hold circuit — 28
   - 1.8. Time-division multiplexing linear d.m. signals — 29

2. **DOUBLE INTEGRATION AND EXPONENTIAL DELTA MODULATION** — 35
   - 2.1. Introduction — 35
   - 2.2. Double integration delta modulation — 35
   - 2.3. Exponential delta modulation systems — 57
   - 2.4. Exponential delta-sigma modulation — 64
   - 2.5. Pulse delta modulation — 66

3. **CALCULATION OF QUANTIZATION NOISE** — 78
   - 3.1. Introduction — 78
   - 3.2. Brief history — 78
   - 3.3. Representation of the delta modulator — 79
   - 3.4. Normal spectrum technique — 81
   - 3.5. nth order autoconvolution of low-pass white noise — 85
   - 3.6. nth order autoconvolution of band-pass white noise — 88
   - 3.7. Autocorrelation function of sampled waveforms — 91
   - 3.8. Autocorrelation function of the signal at the output of a sample and hold — 94
   - 3.9. The signal from the local decoder — 98
   - 3.10. Noise power at the decoder output — 105
   - 3.11. Performance of the delta modulator — 110

4. **SLOPE OVERLOAD NOISE IN LINEAR D.M. SYSTEMS** — 118
   - 4.1. Introduction — 118
   - 4.2. Slope overload with sinusoidal input signals — 118

4.3.	Slope overload with Gaussian signals	126
4.4.	Slope limiter model	143
4.5.	Hermite polynomial model	150

5. IDLE CHANNEL NOISE AND TRANSMISSION ERRORS IN LINEAR D.M. SYSTEMS — **152**
5.1.	Idle channel noise in d.m. codecs	152
5.2.	Transmission errors in d.m. systems	158

6. ASYNCHRONOUS DELTA MODULATION SYSTEMS — **165**
6.1.	Asynchronous delta modulation	165
6.2.	Asynchronous pulse delta modulation	172
6.3.	Rectangular wave modulation	178
6.4.	Discussion	181

7. SYLLABICALLY COMPANDED DELTA MODULATION — **183**
7.1.	Introduction	183
7.2.	Syllabically companded delta-sigma modulation systems	184
7.3.	Continuous d.m.	200
7.4.	Externally companded d.m.	204
7.5.	SCALE	206
7.6.	Digitally controlled d.m.	213
7.7.	Speech-reiteration d.m.	214
7.8.	Compression and expansion circuits	219

8. INSTANTANEOUSLY COMPANDED DELTA MODULATION SYSTEMS — **223**
8.1.	Introduction	223
8.2.	Statistical d.m.	224
8.3.	Discrete adaptive d.m.	227
8.4.	High information d.m.	228
8.5.	Bosworth–Candy delta-sigma modulation	234
8.6.	Pulse group d.m.	238
8.7.	Companded double integration d.m.	241
8.8.	Digital delta modulation	243
8.9.	First order constant factor d.m.	243
8.10.	Second order constant factor d.m.	252
8.11.	Robust d.m.	260
8.12.	Similarity between the instantaneously adaptive d.m. systems	261
8.13.	Television performance	262

9. DELTA AND PULSE CODE MODULATION — **266**
9.1.	Introduction	266

Contents

9.2. Principle of linear p.c.m.	266
9.3. Linear p.c.m. by a d.m. technique	268
9.4. A-law p.c.m. by d.m. technique	270
9.5. Comparisons between the performance of p.c.m. and d.m.	283
9.6. Dither	297
9.7. Effect of successive modulations on signal-to-noise ratio	298

10. ENCODERS WITH MULTI-LEVEL QUANTIZERS — **301**
- 10.1. Introduction — 301
- 10.2. Multi-level d.m. and differential p.c.m. — 301
- 10.3. Multi-level d.s.m. — 323
- 10.4. Sliding scale system — 327
- 10.5. Syllabically companded two-bit p.c.m. — 330

11. DELTA MODULATION DIGITAL FILTERS — **337**
- 11.1. Introduction — 337
- 11.2. D.M. non-recursive filter with analogue coefficients — 338
- 11.3. D.M. recursive filter with analogue coefficients — 342
- 11.4. Discussion — 345
- 11.5. Programmable binary transversal filter — 347
- 11.6. Recursive filter using recirculating shift registers — 351
- 11.7. Elementary filter section — 352
- 11.8. Digital implementation of d.m. filters — 352

12. INSTRUMENTATION USING D.M. — **354**
- 12.1. Introduction — 354
- 12.2. Measurement of noise in a d.m. codec — 354
- 12.3. D.M. time scaler — 358
- 12.4. Distance domain d.m. — 362
- 12.5. Delta-sigma wattmeter — 365
- 12.6. Speed control of an induction motor — 368

APPENDIX Definitions of quantization noise — 372

Index — 373

PREFACE

There have been many developments to the basic delta modulation system since its inception in 1946. The original systems had an ability to encode continuous signals, such as speech and television, into binary signals and to subsequently decode them but these early systems had some serious disadvantages. The present generation of delta modulation systems, however, have a vastly improved performance and their added complexity has been compensated by the impact of large scale integration techniques. This improvement has arrived too late for some speech networks where the already large investment in pulse code modulation systems will ensure its entrenchment. Delta modulation systems, however, have a role to play in local subscriber networks.

In certain types of military communication systems delta modulation has been used for some years in preference to pulse code modulation. There are situations where delta modulation is preferable to pulse code modulation, and vice versa, but the position is fluid because of the many new types of delta modulation systems continually being invented. It is hoped that the fluidity of the art will become apparent on reading the text.

The delta modulation systems described are all of the waveform tracking kind, that is the decoded signal closely resembles the original encoded signal. Consequently subjective systems where the recovered signal at the receiver is not the original, but which is subjectively acceptable, are not dealt with here. Emphasis is placed on the way delta modulation systems behave for such definable signals as sinewaves, steps, band-limited white noise, etc. The subjective performance in speech and television systems although considered is not the prime purpose of this book, rather the aim is to describe how delta modulation systems function and in so doing to give an insight into their performance when encoding the many possible types of signals which are encountered in practice.

The basic principles of delta modulation are described in detail and much of the experimental and theoretical research in this field is reviewed. The work is therefore directed to practising engineers, research workers and post-graduate students, although it will also be of value to final year undergraduates.

Chapter 1 describes the characteristics and parameters of linear delta modulation without recourse to much mathematical analysis. It is therefore advisable that this chapter should be studied carefully by the reader who is making his first acquaintance with the subject.

Preface

Chapter 2 considers an improvement to the delta modulation system requiring the presence of a predictor, and continues with a description of exponential delta modulation.

An important parameter in describing the performance of the delta modulation system is the decoded signal-to-noise ratio. Noise in delta modulation systems is difficult to quantify because the binary encoder, that is the delta modulator, has an amplitude quantizer, sampler and feedback loop. This problem is examined in Chapters 3 and 4 and because of the complexities the mathematical content is more extensive than in the previous two chapters. Chapters 3 and 4 may be omitted on a first reading as the essential features of the signal-to-noise ratio are contained in the first two chapters.

Chapter 5 considers the effects of asymmetry in the delta modulator and the effects of transmission errors, while in Chapter 6 asynchronous delta modulators which transmit binary signals that are quantized in amplitude but not in time are investigated.

Having presented the basic delta modulation system in some detail in the first six chapters the foundations have been laid which enable the progression to more complex systems.

Chapter 7 is concerned with syllabically companded delta modulation for speech signals, while Chapter 8 is essentially a review of the many different types of instantaneously companded delta modulation systems. These types of systems are suitable for encoding signals which change rapidly, such as occur in television. The surprising variety and number of delta modulation systems which have been designed makes the task of providing a comprehensive answer to the inevitable question 'How does the performance of delta modulation compare to that of pulse code modulation?' an extremely difficult one. Nevertheless, an answer, if somewhat inadequate, is given in Chapter 9. In this chapter the segmented A-law code for the 30/32 channel p.c.m. system is discussed, and it is shown how delta modulators can be arranged to produce pulse code modulators. Chapter 10 describes a type of differential pulse code modulation which is called here multi-level modulation. Also included in this chapter is syllabically companded pulse code modulation.

The last two chapters unlike the others are not concerned exclusively with telecommunications. Chapter 11 describes how delta modulators can be used to create digital filters which can be implemented in hardware. Finally, Chapter 12 describes how delta modulation can be utilised in the field of instrumentation. Examples are given of time scaling transients, automatic trace recorder, a wattmeter, etc.

The material for the text originates from the course of lectures the Author gave while working at the Royal Naval College, Greenwich, and subsequently at the Electronic and Electrical Engineering Department of Loughborough University of Technology, and more particularly from the research programme at the latter Establishment. The Author thanks Professor J. W. R. Griffiths for providing the research facilities at Loughborough, and acknowledges those students with whom he had the privilege of working on research projects over the last six years. He thanks the British Post Office for introducing him to the Normal Spectrum technique used in Chapter 3 and to his former student M. Passot for his valuable assistance in connection with this chapter. He appreciates the continuous encouragement and support of his wife, Brenda, throughout the preparation of the manuscript, and not least for typing it. Lastly the Author thanks those colleagues, students and friends who kindly read and commented on parts of the text.

Raymond Steele

Electronic and Electrical Engineering Dept.,
Loughborough University of Technology,
Leicestershire

Chapter 1

LINEAR DELTA MODULATION

1.1. INTRODUCTION

The main objectives of this chapter are to describe the behaviour of the linear delta modulator system and to discuss its important characteristics and limitations.

It will be shown that a delta modulator is a non-linear sampled-data closed loop control system which accepts a band-limited analogue signal and encodes it into binary form for transmission through a telecommunication channel. At the receiver the binary signal is decoded into a close replica of the original signal.

Before enlarging on the principles of delta modulation it is reasonable to ask if there is any necessity to go through the processes of encoding and decoding, instead of just transmitting the analogue signal directly. The answer to this question lies in the imperfections of telecommunication channels and their associated terminal equipments. When band-limited continuously varying signals such as speech or television are transmitted over these channels they are degraded by the effect of noise, dispersion and non-linearities. This degradation is increased if signals are conveyed over long distances necessitating many stages of amplification.

A method of overcoming the deterioration in the recovered signal is to encode the continuous message signal at the transmitter into signals which are quantized in both time and amplitude. Suppose that the amplitude of the encoded signal is allowed only two values, i.e. a simple binary code is used. Some error is involved in the encoding process due to the quantization of the continuous signal, but under certain conditions this can be made very small. Having transformed the continuous signal into a binary one and transmitted it through the channel it arrives at its destination in a distorted form. The receiver observes this distorted binary signal at each sampling instant and answers a simple question: 'Which binary level does this sample represent?' This question is repeatedly asked at every sampling instant and the answer manifests itself as a reformed binary waveform which is then decoded into a continuous signal. If the answers to the above questions are all correct this continuous signal is a close facsimile of the original signal encoded at the transmitter, where the only errors which occur are due to the encoding and decoding processes.

In the case of long distance communications repeaters are used. Each repeater reforms the incoming signal into a binary one before transmitting it to the next repeater. Providing the distortion in the channel, terminal equipment and repeaters is insufficient to cause an error, long distance communication can be achieved without any increase in the overall error when compared to short distance transmissions.

It should be emphasised that for many signals which are subjectively interpreted by the observer, such as speech and television, perfect transmission is unnecessary. The binary transmission system makes it possible to transmit signals which when decoded are of an acceptable quality although the distortion level would make analogue transmission unacceptable.

If extraneous distortion becomes excessive such that the binary levels are frequently wrongly interpreted by the receiver causing 'digital errors', then the digital systems will become unworkable unless redundancies are built into the binary signals enabling error detection or error detection and correction techniques to be used. Unless there are other reasons, like encription, for using a digital transmission system it may be better to revert to an analogue one in the presence of substantial channel distortion.

Suppose a situation exists, as it often does in practice, where digital transmission is preferable to analogue transmission. Having optimised the channel for the transmission of say binary signals, an important question to be answered is 'What is the best form of encoding to use?' For many practical situations the answer is likely to be a form of pulse code modulation, or one of the numerous forms of differential modulation such as delta modulation, which is of course the subject to be considered here. Some comparisons are made (Section 9.5) between pulse code modulation and delta modulation but because of the large variety of possible situations this is of limited value. The approach used here is to present a comprehensive account of delta modulation and to leave it to the reader to decide if it is applicable to his particular problem.

It will be shown that some types of delta modulation systems are particularly suited for operating in situations of high error rate while others can accommodate the rapid changes such as those found in television signals, etc. Although these are often complex systems, examined in detail in subsequent chapters, they are, however, derived from the basic delta modulation system called linear delta modulation, abbreviated linear d.m., which is the subject of this chapter.

Linear delta modulation was first described by Deloraine, van Miero and Derjavitch in a French patent[1] taken out in August 1946. By the early 1950's detailed descriptions of linear delta modulation had been made by de Jager,[2] Libois,[3] van de Weg,[4] Zetterberg[5] and others.[6-9] Theoretical and practical contributions are still being made to the subject, which is indicative of its hidden complexities.

Linear delta modulation

Before attempting any detailed descriptions it is first necessary to define a linear delta modulation system. It is simply an encoder and a decoder. However, because an encoder is often called a coder, the coder and decoder pair are frequently abbreviated in the literature to codec. The linear delta modulation system is synonymous, therefore, with codec.

The encoder is accommodated at the transmitter and receives a band-limited analogue input signal and produces a binary output signal. The decoder located in the receiver accepts this binary signal and decodes it back into a close replica of the original analogue signal. The terminal equipment at both the transmitter and receiver need not be considered for the purposes of this book.

The principle of linear d.m. is now described. An analogue input signal is encoded by the delta modulator into binary pulses which are conveyed to the terminal equipment for transmission. These pulses are also locally decoded back into an analogue waveform by an integrator in the feedback loop and subtracted from the input signal to form an error which is quantized to one of two possible levels depending on its polarity. The output of the quantizer is sampled periodically to produce the output binary pulses. The closed-loop arrangement of the delta modulator ensures that the polarity of the pulses are adjusted by the sign of the error signal which causes the local decoded waveform to 'track' the input signal. Stated another way, the delta modulator produces binary pulses at its output which represent *the sign of the difference* between the input and feedback signals–hence the prefix 'delta'. The modulation process is linear because the local decoder, i.e. an integrator, is a linear network. This terminology is used due to the existence of another class of delta modulators having non-linear local decoders, discussed in Chapters 7, 8, and 9.

The decoding of the binary pulses is readily effected. For errorless transmission the binary pulses are recovered at the receiver and are passed through the local decoder, an integrator, to produce a waveform which differs from the original signal by the error signal in the encoder. The final decoded signal is obtained by low-pass filtering the waveform at the output of the local decoder to remove any noise due to quantization effects in the encoder which lie outside the message band. Section 1.2 enlarges on the encoding behaviour with the aid of Figures 1.1 and 1.2.

As delta modulation is conceptually simple it is therefore all the more surprising to find that the system is difficult to analyse. The encoder prohibits simple analysis–there is the quantizer, which is of course a non-linear device, the time sampler which causes the output pulses to be both time and amplitude quantized, and the feedback loop. If a random input is applied, having at best known stationary statistical properties, then the evaluation of the noise in the decoded output signal becomes very difficult.

1.2. LINEAR D.M. SYSTEM

The schematic diagram of the basic delta modulator system is shown in Figure 1.1. An analogue signal is band-limited by a band-pass filter F_i having critical frequencies f_{c1} and f_{c2}, where $f_{c2} > f_{c1}$, to form the input signal $x(t)$. The waveform $L(t)$ at the output of the delta modulator consists of pulses of duration τ seconds spaced T seconds apart ($T \gg \tau$) having amplitudes of either $\pm V$ volts. These pulses occur at clock times, i.e. at a rate $f_p = 1/T$ where f_p is considerably greater than the Nyquist rate $2f_{c2}$. The delta modulator acts as an analogue to digital converter having an analogue input signal $x(t)$ and a binary output signal $L(t)$. The relationship between $x(t)$ and $L(t)$ is such that $L(t)$ is a binary representation of $x(t)$, where the rate of occurrence of each binary pulse is directly proportional to the instantaneous slope of $x(t)$. If the slope of the input signal $x(t)$ is positive then while this condition exists the output waveform $L(t)$ has more positive pulses than negative ones. The situation is reversed when $x(t)$ has a negative slope.

Figure 1.2 shows how the voltage waveform of a delta modulator varies with time when the input signal is a sinusoid. The pulses composing the binary output waveform are drawn for convenience having negligible width τ. When these pulses are integrated by the integrator[10,11,12] in the feedback network the resulting waveform $y(t)$ consists of steps having magnitude $\pm \gamma$ volts and duration T seconds, which oscillate about the

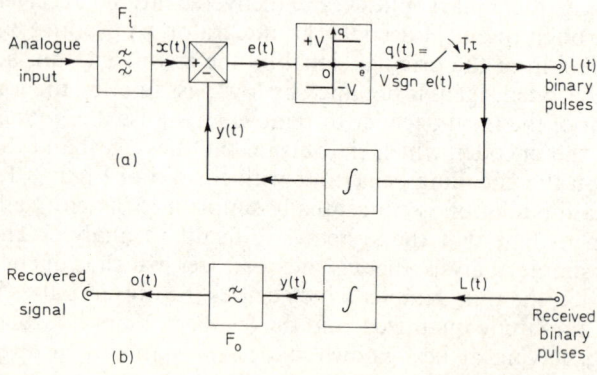

Fig. 1.1. Linear delta modulation system: (a) encoder, (b) decoder

Linear delta modulation

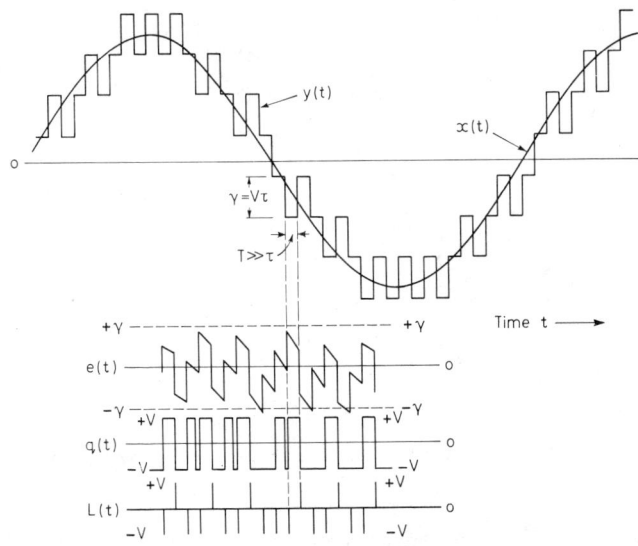

Fig. 1.2. Linear d.m. waveforms when the encoder is correctly tracking the input signal

analogue input signal $x(t)$. The difference between $x(t)$ and $y(t)$ is the error signal $e(t)$, which is analysed in detail in Chapter 3. This error signal is quantized to limits $\pm V$, which means the sign and not the magnitude of the error is quantized. The output of the quantizer is sampled every T seconds to produce the $L(t)$ pulses.

If $e(t) \geq 0$ at a clock instant, a positive pulse will be produced at the output of the encoder. When this pulse is integrated $y(t)$ is increased by a positive step. This increase in $y(t)$ will be subtracted from $x(t)$ and a change in the magnitude of the error signal occurs. If the error has not become negative by the next clock instant the output of the encoder will again be a positive pulse. As long as $e(t) \geq 0$ at successive clock instants a sequence of positive pulses will be produced. Eventually $y(t)$ will become greater than $x(t)$, at a clock instant $e(t) < 0$, and a negative pulse will occur at the output of the encoder resulting in a diminution in the $y(t)$ waveform by an amount γ. Thus the encoder attempts to minimise the error waveform when an input is present by varying the polarity of the pulses at the output of the modulator at clock instants. An inspection of Figure 1.2 and, more dramatically, Figure 1.4 shows that when the slope of the sinusoid is large and negative there are more negative than positive pulses generated. The situation is reversed when the slope is large and positive. At the maxima and minima of the waveform where the slope is close to zero there are approximately equal numbers of positive and

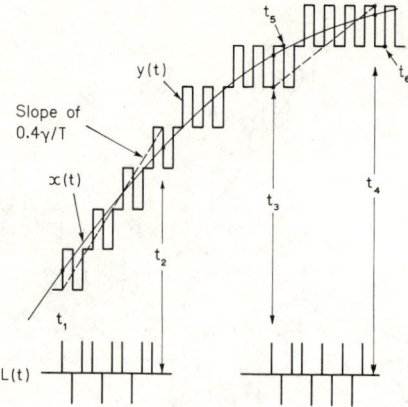

Fig. 1.3. *Mean value of L(t) approximates to mean value of the slope of x(t)*

negative $L(t)$ pulses. The encoder attempts to generate an $L(t)$ pattern whose mean value approximates to the mean value of the slope of the sinusoid over a short period of time. This is illustrated in Figure 1.3. From period t_1 to t_2 there are 10 clock periods and the $L(t)$ waveform generates seven positive pulses and three negative ones. The average value of $L(t)$ over this period is $4V\tau/10T$. An ideal integrator has an impulse response $h_i(t)$ which equals unity for $t \geq 0$ and zero for $t < 0$. A solitary $L(t)$ pulse generates a step $\gamma = V\tau$ volts at the output of the integrator. Observe that γ is in volts because the impulse response has dimensions $(s)^{-1}$. The average value of $L(t)$ can now be written as $0.4\gamma/T$. The change in $x(t)$ over the same time interval t_1 to t_2 is 3γ, and this corresponds to a mean slope of $0.3\gamma/T$ which is an approximation to the mean value of $L(t)$. If γ is small and f_p is large this approximation is improved. For a period of $10T$, from time t_3 to t_4, the slope of $x(t) = 0.1\gamma/T$ and the average value of $L(t) = 0.2\gamma/T$. However, if the average of $L(t)$ is taken from time t_5 to t_6 the slope of $x(t)$ is still $0.1\gamma/T$ but the average value of $L(t)$ is now zero. The average value of $L(t)$ now oscillates about the average slope of $x(t)$. The wide deviations of the average of $L(t)$ about the average of $x(t)$ demonstrate the desirability of minimising γ provided tracking can be maintained.

1.2.1. Decoder

The decoder in linear d.m. consists of an integrator and a simple filter. Assuming errorless transmission the $L(t)$ signal is recovered and integrated to give $y(t)$. This $y(t)$ signal is identical to the feedback to the error point in the encoder. As $y(t)$ only differs from the input signal $x(t)$ by a relatively small error signal $e(t)$ it follows that the signal at the output of the integrator in the decoder is a good reproduction of the original input signal. The step-like nature of $y(t)$ is removed by passing this signal through a filter whose frequency pass band is the same as the message band f_{c1} to f_{c2}, i.e. filters F_i and F_0 may be considered to be identical. Further simplifications may be added to the decoder by making the filter F_0 a low pass[13] rather than a band-pass[14] one. This is because the noise below f_{c1} is generally not troublesome.

The simplicity of the decoder in linear d.m. is one of its attractions, particularly when the integrator can be produced with just a resistor and capacitor.

1.2.2. Local decoder

Local decoder is used throughout the book to mean the system which is placed in the feedback path of the encoder. In the linear d.m. it is just an integrator, but in other encoders it can become rather complex, as in Chapter 8.

The decoder is the local decoder followed by a band limiting filter whose main function is to exclude unwanted noise.

1.2.3. Zero order hold

A practical encoder will have a zero order hold circuit following the sampler. This circuit when in receipt of a sample causes its value to be held at a constant amplitude for one clock period. The result is that the $L(t)$ waveform now consists of binary levels which may or may not change at clock instants, rather than narrow pulses of duration τ and amplitude $\pm V$. The hold circuit must be used in practice because narrow pulses with their large bandwidths cannot be countenanced for transmission purposes; in addition it is easier to produce binary levels than narrow pulses. The zero-order hold circuit can be placed outside or inside the closed loop. Because a sample and zero order hold circuit can be implemented with a D-type flip-flop it is usual to place the zero order hold circuit in the feed-back loop. This results in $y(t)$ changing from a waveform with steps to one with ramps and this might intuitively lead one to expect a significant reduction in the error signal. A detailed discussion of the effect on the noise at the output of the decoder by the inclusion in the delta modulator of a sample and hold circuit is given in Section 1.7. It is shown there that although the presence of the sample and hold circuit does effect the shape of the spectrum of the distortion components in the signal $y(t)$ at the output of the local decoder it only has a negligible effect on the distortion components which reside in the message band and are passed by the final filter F_0. Consequently the signal-to-noise ratio at the output of the decoder is, to a close approximation, the same irrespective of whether the linear d.m. codec generates narrow binary pulses or contains a sample and hold circuit. Generally, the discussion of the various delta modulation systems will be made assuming that the encoder generates narrow binary pulses or impulses, although the presence of a sample and hold circuit in the encoder will be considered occasionally, as in the case of the exponential d.m. codecs described in Section 2.3. Note that irrespective of whether $L(t)$, the output waveform from the delta modulator, consists of binary pulses or levels it will be processed by the transmitter terminal equipment to a form suitable for transmission. For example, the $L(t)$ waveform may be filtered, amplified and perhaps then used to modulate a carrier waveform.

1.2.4. Idling behaviour

The linear d.m. system (codec) is in its idling state when there is no analogue input signal. In the steady state the binary output signal $L(t)$ is composed of alternate positive and negative pulses resulting in a feedback signal $y(t)$ which is nearly a square wave having a period $2/f_p$ and amplitudes $\pm \gamma/2$. The error waveform is equal to $-y(t)$ as there is no input signal. The result is that the waveform at the output of the quantizer is also a square waveform, having the same frequency as that of the error signal, but with amplitudes $\pm V$.

The idling pattern of the linear d.m. system is 1 0 1 0 1 0..., where the logical ones and zeros refer to the positive and negative pulses of the $L(t)$ waveform respectively.

A feature of this idling pattern is that when an analogue signal is applied, the slope required to produce an all ones' pattern has the same magnitude and opposite polarity as the slope which will produce an all zeros' pattern. The encoder is therefore symmetrical. The behaviour of an asymmetrical encoder which does not have a ...1 0 1 0 1 0... pattern is discussed in Chapter 5.

Because the $y(t)$ waveform when idling is as described above its frequency spectrum is composed of lines where the fundamental and therefore the lowest frequency is half the clock rate.

The final filter F_0 in the decoder rejects frequencies above the highest frequencies which are expected in $x(t)$ when the latter is applied to the encoder. As the clock rate is greatly in excess of these expected frequencies in $x(t)$, generally by a factor of 10 or more, the waveform $y(t)$ in the idling state is completely rejected by the filter F_0. The result is that when the coder is idling no signal emerges from the output of the decoder.

When no input is presented to the encoder it has an oscillatory idling pattern; this does not imply an unstable situation, on the contrary, it must be emphasised that the linear delta modulator is very stable. The idling feedback waveform $y(t)$ never increases its magnitude or changes from a square waveform. If the encoder does have some asymmetry then the waveform $y(t)$ only changes in prescribed limits as discussed in Section 5.1. Similarly when an input signal is present the delta modulator always attempts to track this signal, even if the encoder is asymmetrical. This inherent stability of the linear delta modulator has been theoretically verified by Gersho.[15] The presence of a second integrator in the feedback loop does cause instability and this condition is investigated in Section 2.2.

1.2.5. Slope overload

When the step size γ in the waveform $y(t)$ is reduced compared to that in Figure 1.2, without changing the clock rate, a condition occurs when the feedback voltage $y(t)$ is not always varying about the input signal $x(t)$. This is shown in Figure 1.4 where there are sequences of positive and negative steps in $y(t)$ and the error is now in excess of γ. By doubling the amplitude and at the same time halving the frequency of the input sinusoid the relationship between $x(t)$ and $y(t)$ is unchanged. Similar sequences of positive and negative steps in $y(t)$ occur because the slope of the sinewave is unaltered. The slope of $x(t)$ is simply related (see Section 4.2) to the length of the sequence of similar polarity pulses in the waveform $L(t)$, and

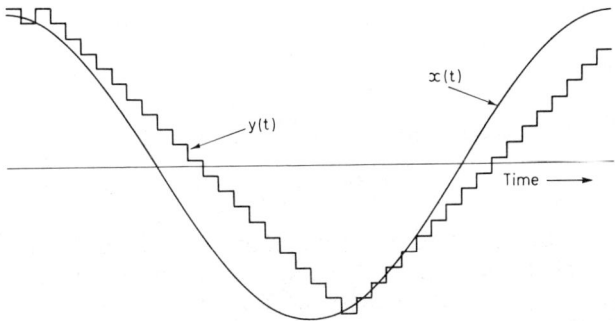

Fig. 1.4. $y(t)$ waveform when encoder is substantially slope overloaded by a sinusoidal input signal $x(t)$

Linear delta modulation

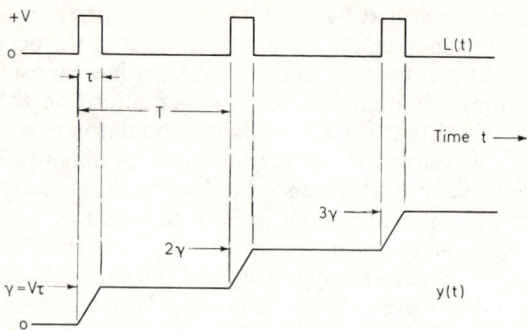

Fig. 1.5. *y(t) in response to a sequence of three positive pulses at the output of the encoder*

during these sequences the mean slope of $y(t)$ is a constant. This condition, where the feedback waveform $y(t)$ does not always vary about the input signal, i.e. track it, is referred to as 'slope overload'. Generally, slope overload starts when the slope of $x(t)$ is greater than the maximum slope associated with $y(t)$. However, during the condition of slope overload the slope of $x(t)$ may be zero, as is apparent from Figure 1.4.

The limitations imposed on the sinusoid to prevent slope overload will now be examined. When the overload condition occurs a sequence of identical polarity pulses occur at the output of the linear d.m. encoder. The situation is illustrated in Figure 1.5 for a sequence of positive output pulses. Assuming perfect integration, the feedback waveform $y(t)$ increases by $\gamma = V\tau$ volts per clock period. The maximum rate of increase of $y(t)$ is

$$\zeta = \gamma/T = \gamma f_p \tag{1.1}$$

For an input waveform

$$x(t) = E_s \sin 2\pi f_s t \tag{1.2}$$

$$x'(t) = E_s 2\pi f_s \cos 2\pi f_s t \tag{1.3}$$

The condition of slope overload is generally avoided if

$$E_s 2\pi f_s \leq \gamma f_p \tag{1.4}$$

The overload characteristic is a graph of the maximum value E_{sm} of E_s in decibels against $\log_{10} f_s$, which avoids overloading the delta modulator. From Equation 1.4 it follows that a delta modulator cannot encode high frequency sinewaves without evoking a slope overload condition unless the amplitude is restricted. High frequency signals can of course be accommodated by increasing the product γf_p.

A comprehensive discussion of overload characteristics is given in Section 2.3.2.

1.2.6. Amplitude range

The maximum amplitude E_{sm} of the input sinusoid which does not overload the encoder is given by the equality of Equation 1.4

$$E_{sm} = \frac{\gamma f_p}{2\pi f_s} \qquad (1.5)$$

When idling the waveform $y(t)$ at the output of the integrator is square with a peak-to-peak value of γ. In order to disturb this idling square wave the input sinusoid must have an amplitude $\geq \gamma/2$. When E_s is only just greater than $\gamma/2$, the encoder is virtually incapable of tracking the input signal, and the decoded output has severe distortion. The greater E_s becomes relative to $\gamma/2$ the better the tracking; the minimum value of E_s which gives an acceptable decoded signal-to-noise ratio is clearly a subjective choice. The ratio of E_{sm} to this value of E_s is referred to as the dynamic range, DR, of the input.

However, an amplitude range, AR, will be defined[16] which is the ratio of the value of E_s which overloads the encoder, i.e. E_{sm}, to the value of E_s which just disturbs the idling pattern, i.e. $\gamma/2$. Hence

$$AR = \frac{f_p}{\pi f_s} \qquad (1.6)$$

AR is maximised by ensuring that at f_{c2}, the highest frequency to be encoded, $f_p \gg f_{c2}$. This illustrates that the bit rate for a delta modulator must be in excess of the Nyquist rate for a tolerable AR.

1.2.7. Quantization noise

Quantization noise occurs when the encoder is tracking correctly and arises from the voltage and time quantization used in the encoding process. Applying definition (b) in the Appendix, it is the signal obtained by filtering $e(t)$ by F_0 in the decoder of Figure 1.1. It must be emphasised that quantization noise if often referred to as granular noise, but throughout this text the former name will be used.

A distinction must be made between slope overload and quantization noise. The latter occurs, as described above, when the encoder is correctly tracking the input signal, whereas slope overload noise occurs when the slope of the input signal is too great to be tracked by the encoder, resulting in a considerable error signal as described in Section 1.2.5. Slope overload noise is discussed in detail in Chapter 4.

A simple and useful formula for quantization noise which is valid for a variety of situations will be intuitively argued here. This formula is related to the more general and precise results of quantization noise which are calculated in Chapter 3.

An important factor in establishing this approximate formula is that for a wide variety of input signals, particularly band-limited random ones, the quantization noise is substantially uncorrelated with the input signals. This enables the formula for quantization noise to be specified without reference to the signal being encoded, i.e. to be applicable for a wide variety of input signals.

Now the decoded signal $o(t)$ is obtained by band-pass filtering the recovered $y(t)$. As $y(t)$ only differs from $x(t)$ by the error signal $e(t)$ the quantization noise as stated above is the part of the error signal which is passed by the final filter. Consequently it is necessary to comment upon the way in which that part of the spectral density function of $e(t)$ in the message band of $x(t)$ varies with the parameters of the system.

The formula assumes that $S_{nn}(f)$, the spectral density of $e(t)$, is substantially flat over the message band as in pulse code modulation. Computer simulation shows that this is often the case if the message signal is random, or if it is periodic, it does not have a stable harmonic relationship to the clock, i.e. the usual condition.

Because $S_{nn}(f)$ is flat the quantization noise N_q^2 is proportional to f_c, where f_c is equal to $f_{c2} - f_{c1}$.

As the clock rate is increased the inband components of $S_{nn}(f)$ fall linearly with clock rate f_p. There is a definite limit to this relationship as can be seen from Figure 3.17 in Chapter 3. Therefore, within limitations, $N_q^2 \propto 1/f_p$.

Raising the step height γ (proportional to V) increases the amplitudes in $e(t)$ and consequently $N_q^2 \propto \gamma^2$.

From the above remarks[2]

$$N_q^2 = K_q \cdot \frac{f_c \gamma^2}{f_p} \qquad (1.7)$$

where K_q is a constant of proportionality. The limitations on the range of f_c, f_p and γ for which Equation 1.7 is valid are deducible from Figures 3.16 to 3.18.

The values for K_q for good conditions of encoding have been quoted approximately as $\frac{1}{3}$. Unfortunately the range where Equation 1.7 is valid is limited because K_q is only constant over a restricted range of the variables f_c, f_p and γ. For example, the lack of constancy of K_q when a low-pass white noise signal, having an r.s.m. value of unity, is applied to the input of the encoder can be shown as follows:

For a constant ratio $f_p/f_{c2} = 32$ and γ between 0·4 and 1·0, K_q is reasonably constant at 0.45; when γ is increased to 1·8, $K_q = 0.9$. For $f_p/f_{c2} = 8$ or 64 and γ held constant at 0·65, $K_q \simeq 0.6$; for the intermediate ratios $f_p/f_{c2} = 16$ or 32, $K_q \simeq 0.35$.

For normal encoding conditions Equation 1.7 can be applied providing the restrictions on K_q are appreciated. The accuracy of the formula deteriorates as the bandwidth of the white noise input reduces and $S_{nn}(f)$ is no longer flat in the message band, see Figure 3.16.

1.2.8. Signal-to-quantization noise ratio

Generally the signal-to-quantization noise ratio is defined in linear d.m. as the ratio of the input signal power S^2 to the decoded noise power N_q^2 in the presence of the decoded signal. If the input signal is low-pass white noise of power σ^2 then the signal-to-quantization noise ratio, abbreviated s.q.n.r., is

$$\text{s.q.n.r.} = \frac{1}{K_q} \cdot \frac{f_p}{f_c} \left(\frac{\sigma}{\gamma}\right)^2 \qquad (1.8)$$

This equation assumes that no slope overloading has occurred. The maximum value of s.q.n.r. for this type of input signal is presented in Section 9.5.1.3.

For sinusoidal inputs there is a continuous spectrum which is approximately flat over the message band, often with an additional line spectrum. It has been found[2] that Equation 1.7 is valid if K_q is of the order $\frac{1}{3}$.

If E_s is the amplitude of the sinusoid, the input power is $E_s^2/2$, and the s.q.n.r. is

$$\text{s.q.n.r.} = \frac{1}{2K_q} \left(\frac{f_p}{f_c}\right) \left(\frac{E_s}{\gamma}\right)^2 \qquad (1.9)$$

As the noise power is substantially independent of the signal power S^2 the s.q.n.r. increases linearly with increasing S^2 and has a peak value when

$$S^2 \simeq E_{sm}^2/2$$

From Equations 1.5 and 1.9 the peak s.q.n.r. is represented by the simple formula

$$\widehat{\text{s.q.n.r.}} = \frac{1}{8\pi^2 K_q} \cdot \frac{f_p^3}{f_c f_s^2} \qquad (1.10)$$

When E_s exceeds E_{sm} the encoder becomes overloaded and the noise power increases by a greater amount than the increase in signal power, leading to a decrease in the signal-to-noise ratio.

Observe that there is only one value of E_s for which the s.q.n.r. is a maximum. If the input signal is too large, as shown in Figure 1.4, slope

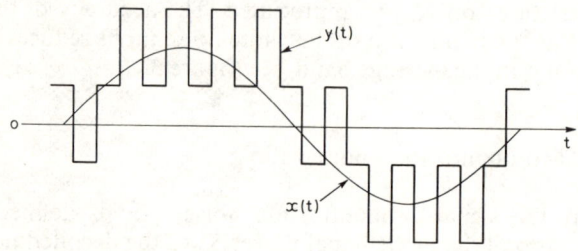

Fig. 1.6. Tracking a low amplitude sinusoidal input signal

overloading occurs, whereas Figure 1.6 illustrates the existence of a large quantization error due to a too small input amplitude. Consequently one input amplitude exists which minimises the error signal, and if the phase of $x(t)$ and the clock waveform are suitably related the maximum s.q.n.r. occurs when $E_s \simeq E_{sm}$. For simplicity it will be assumed that for sinusoidal input signals s.q.n.r. occurs when $E_s = E_{sm}$.

When the input signal $x(t)$ is low-pass white noise, the s.q.n.r. arises when some slope overloading occurs, but the r.m.s. value of the slope of $x(t)$ is well below the maximum rate of increase of the feed-back waveform $y(t)$, defined by Equation 1.1. The r.m.s. values of $x(t)$ which give the s.q.n.r. for various encoding conditions are deducible from Figures 3.18 and 3.19.

1.3. DELAYED ENCODING

If the input signal $x(t)$ is delayed by kT seconds, where k is a constant, and the delayed signal $x_d(t)$ is applied to the encoder then the tracking capability of the encoder can be improved if it is able to use the knowledge about its future signal. Newton[17] used this concept to improve the performance of the encoder in the vicinity of overload. The principle of the system will now be described with reference to Figure 1.7. The difference signal Δx and $L(t)$ are fed to the logic whose function is to decide the polarity of the feedback pulse at each clock instant. Now because Δx is measured across a fixed time delay kT it provides an estimate $\pm \Delta x/kT$

Linear delta modulation

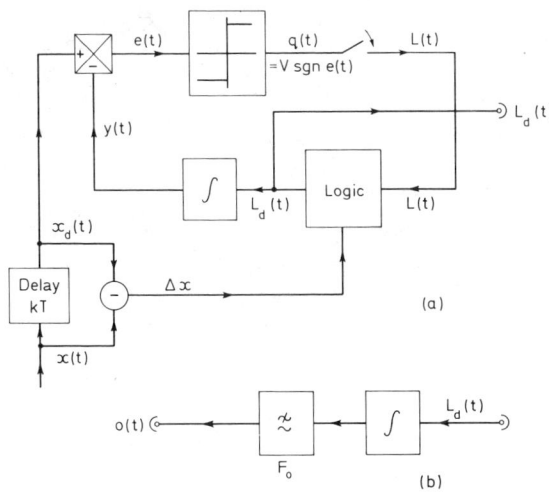

Fig. 1.7. Delayed encoding applied to a linear d.m. codec; (a) encoder, (b) decoder. (After Newton[17])

of the gradient of the signal which is shortly to be encoded. If this estimate is more positive than the maximum positive gradient γf_p of the feedback signal $y(t)$, then the logic instructs positive pulses to be fed to the integrator irrespective of $L(t)$. If an estimate of the gradient suggests that $x_d(t)$ will soon have a negative gradient in excess of $-\gamma f_p$ the logic overrides $L(t)$ and $L_d(t)$ reverts to a series of negative pulses until the estimate of the slope is less than $-\gamma f_p$. When

$$\frac{\Delta x}{kT} < |\gamma f_p|$$

the encoder does not anticipate overloading in the immediate future and the logic connects $L(t)$ to the integrator to produce the linear d.m. in Figure 1.1.

The encoder remains a linear delta modulator at all times, unlike the adaptive encoders discussed in Chapters 7, 8 and 9. The magnitude of the $L_d(t)$ pulses are the same as $L(t)$ but the value of a particular $L_d(t)$ pulse is independent of the sign of the error when the encoder anticipates future slope overloading. This feature is demonstrated in Figure 1.8 for $k = 5$. At time t_r an encoder without slope overload prediction will feed a negative pulse to the integrator and as a result a considerable discrepancy soon results between $x_d(t)$ and $y(t)$. With delayed encoding the value of the $x_d(t)$ signal, kT bits in the future, is already known and as the

Linear delta modulation

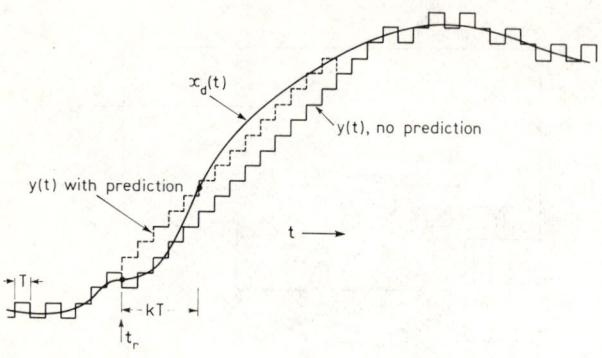

Fig. 1.8. Effect of slope overload prediction on the feedback waveform

average slope of $x_d(t)$ will soon be in excess of γ/T the sign of the pulse fed to the integrator at t_r is reversed. In this way the mean square error between $y(t)$ and $x_d(t)$ is reduced.

The decoder is identical to that of the linear d.m. shown in Figure 1.1. The improvement in the performance of the system is therefore achieved by increasing the complexity of the encoder.

The choice of k depends on the characteristics of the signal being encoded. For speech signals k is approximately 10. The signal-to-noise ratio of the decoded signal is improved with delayed encoding by some two to three dB. The reduction in the decoded noise is due not only to the total mean square error being smaller but by the increase in the number of zero crossings in the error signal which decreases the ratio of noise in the message band to the noise above this band.

Delayed encoding has not been widely implemented in practical systems because the performance of the delta modulator can generally be improved, beyond that of the encoder of Figure 1.7 by simply using a double integrator in the feedback loop as described in Section 2.2. The latter arrangement also has the advantage that it is easier to implement. Of course delayed encoding can be used in conjunction with double integration to get a further improvement in the decoded signal-to-noise ratio. This argument can be extended further, and it is found that a type of delayed encoding[18,19] is of value in adaptive d.m. encoders which vary the magnitude of the pulses fed back to the error point in a manner dependent on the slope of the input signal in order to achieve good tracking. Some adaptive strategies may cause the encoder to become unstable. Delayed encoding has a stabilising influence on these types of encoders and means that encoders can be produced with high signal-to-noise ratios.

1.4. DELTA-SIGMA MODULATION

The linear delta-sigma modulator[20,21] is composed of a delta modulator preceded by an integrator, i.e. the delta modulator encodes the integral of the input signal $x(t)$. A better name would be sigma-delta modulation. The waveform at the output of the integrator in the decoder shown in Figure 1.1 tracks the integral of $x(t)$ with some error $e(t)$. The recovery of $x(t)$ therefore involves the addition of a differentiator to the decoder to compensate for the extra integrator placed before the input to the encoder, and a low pass filter to reduce the effects of quantization. The arrangement is shown in Figure 1.9. As the integrator and differentiator in the decoder are complimentary they can be removed to produce a decoder which is simply a sharp cut-off low pass (sometimes band-pass) filter. The two integrators in the encoder can be replaced by one integrator placed after the error signal because

$$\int x(t)\,dt - \int L(t)\,dt = \int e(t)\,dt$$

The delta-sigma modulation system is shown in Figure 1.9 is exactly equivalent to the one depicted in Figure 1.10.

From an implementation point of view the encoder has the same system components as a delta modulator but they are slightly rearranged. The decoder is simpler without the integrator.

The repositioning of the integrator in the linear delta-sigma modulation encoder (linear d.s.m. encoder) compared to the linear d.m. encoder results in some performance changes. In a delta modulator the overload characteristic given by Equation 1.5 applies for an input signal $E_{sm} \sin 2\pi f_{sm} t$. If an input

$$x(t) = E_{sm} \sin 2\pi f_s t$$

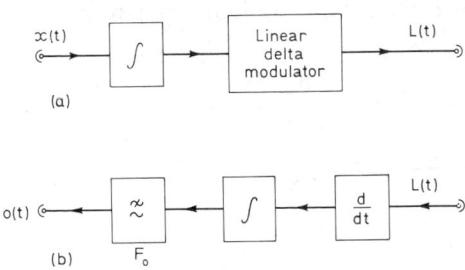

Fig. 1.9. *Delta-sigma modulation codec:* (a) *encoder,* (b) *decoder*

is applied to the delta modulator where $f_s > f_{sm}$ the encoder will experience slope overload. However, $x(t)$ is integrated in the linear d.s.m. encoder shown in Figure 1.9(a), i.e.

$$\int x(t)\,dt = -(E_{sm}/2\pi f_s)\cos 2\pi f_s t$$

after setting the constant to zero. This signal is then applied to the delta modulator and has a maximum slope of

$$\left|\frac{d}{dt}\left\{-\frac{E_{sm}}{2\pi f_s}\cdot\cos 2\pi f_s t\right\}\right|_{max} = |E_{sm}\sin 2\pi f_s t|_{max} = E_{sm}$$

The maximum slope in the feedback signal $y(t)$ in the delta modulator, produced by a sequence of ones or zeros at the output of the encoder, is γf_p as seen from Equation 1.1. Equating this maximum slope with that of the sinusoid at the output of the integrator the overload characteristic of the delta-sigma modulator is

$$E_{sm} = \gamma f_p \qquad (1.11)$$

Observe that the overload characteristic is independent of the frequency of the input signal and consequently this encoder is suitable for a wide range of applications.

The placing of the integrator in front of the delta modulator means that the quantization effects are still produced in the delta modulator. The noise spectral density function of the error signal in a delta modulator is assumed to be flat over the message band, as was described in Section 1.2.7. The decoder in a delta-sigma modulator differs from a delta modulator decoder in that there is no integrator. The result is that instead of the spectral density of the noise at the decoder output being flat with frequency it is modified by the reciprocal of the power transfer function of the missing integrator. From Equation 1.7 the spectral density of the quantization noise of a delta modulator over the message band is

$$S_{nn}(f) = \frac{K_q \gamma^2}{f_p} \qquad (1.12)$$

This becomes for a delta-sigma modulator

$$S_{nn}(f) = \frac{K_q \gamma^2}{f_p}\cdot(2\pi f)^2 \qquad (1.13)$$

because the power transfer function of an integrator is $1/(2\pi f)^2$. Hence

Linear delta modulation

the quantization noise at the output of the decoder is

$$N_q^2 = 4\pi^2 \frac{K_q \gamma^2}{f_p} \int_0^{f_{c2}} f^2 \, df = \frac{4\pi^2}{3} \cdot K_q \gamma^2 \frac{f_{c2}^3}{f_p} \qquad (1.14)$$

where 0 to f_{c2} is the pass-band of the decoding filter F_0.

1.4.1. Signal-to-quantization-noise ratio

For a sinusoidal input having a power $E_s^2/2$ the signal-to-quantization-noise ratio is found from Equation 1.14 after applying the same arguments as in Section 1.2.8, i.e.

$$\text{s.q.n.r.} = \frac{3}{8\pi^2 K_q} \left(\frac{E_s}{\gamma}\right)^2 \frac{f_p}{f_{c2}^3} \qquad (1.15)$$

$$\widehat{\text{s.q.n.r.}} = \frac{3}{8\pi^2 K_q} \cdot \left(\frac{f_p}{f_{c2}}\right)^3 \qquad (1.16)$$

Observe that the s.q.n.r. for both linear d.m. and linear d.s.m. depends on the cube of the clock rate. The minimum theoretical channel bandwidth is $(f_p/2)$—in practical systems it is often equal to f_p. The s.q.n.r. in linear d.s.m. is independent of the input frequency, unlike linear d.m. where it is inversely proportional to the square of the frequency of the input sinusoid. Comparisons between linear d.s.m. and d.m. are made in Section 9.5.

1.4.2. Amplitude range

Amplitude range is defined in Section 1.2.6. If $\gamma/2$ is the amplitude of a sinewave which just disturbs the idling pattern of a delta modulator, then the corresponding input for a delta-sigma modulator is

$$(\gamma/2) \sin 2\pi f_s t = \int E_{st} \sin 2\pi f_s t \, dt = \frac{E_{st}}{2\pi f_s} \sin\left(2\pi f_s t - \frac{\pi}{2}\right)$$

after setting the constant of integration to zero.

A sinusoid having an amplitude $E_{st}/2\pi f_s$ when applied to the input of a delta-sigma modulator will just disturb the idling condition. Consequently

$$E_{st} = \pi \gamma f_s \qquad (1.17)$$

From Equation 1.11 the maximum signal amplitude $E_{sm} = \gamma f_p$. The amplitude range is therefore

$$\text{AR} = f_p / \pi f_s \qquad (1.18)$$

the same as for a delta modulator. Thus, although the values of E_{sm} and E_{st} are different for a delta modulator and a delta-sigma modulator their ratios are identical.

1.5. ENCODING D.C. SIGNALS

Both linear d.m. and linear d.s.m. systems can be used to convey d.c. information. A linear d.m. encoder when initially confronted with say a positive d.c. signal X has a large positive error. Successive clock periods generate a sequence of positive pulses which 'pump-up' the integrator until $y(t)$ becomes greater than X. The error now changes sign at half the clock rate and the waveform $L(t)$ is identical to its idling pattern. The waveform $y(t)$ oscillates about X and its amplitudes are alternately $+\gamma$ and $-\gamma$.

For errorless transmission and no d.c. drift in the quantizer characteristic, this $y(t)$ waveform is generated at the output of the integrator in the decoder. However d.c. transmission is extremely vulnerable to transmission errors, each one of which causes an error in $y(t)$ of $\pm 2\gamma$. These errors are accumulative, although some errors have a nullifying effect. The decoder output has the appearance of a d.c. signal added to a random noise voltage. In order to mitigate the accumulative effect of transmission errors in the decoded signal it is advisable to use integrators having time constants as small as is compatible with the requirements for encoding higher frequency input signals.

The linear d.s.m. encoder generates a binary pattern $L(t)$ whose average value is equal to the applied d.c. The decoder, which is just a low-pass filter, averages $L(t)$ to recover the d.c. signal. If the linear d.s.m. codec contains leaky integrators then the applied d.c. input signal manifests itself at the output of the integrator (shown in Figure 1.9(a)), after a time equal to many integrator time constants, as another d.c. signal. As a consequence the linear d.s.m. and the linear d.m. codecs produce the same binary output waveforms consisting of alternate positive and negative pulses.

1.6. MODELS FOR LINEAR DELTA MODULATION SYSTEMS

The calculation of the spectral density function of the error signal in the delta modulator of Figure 1.1 is difficult. The task is simplified if the delta modulator is represented by an exact model which is of the open loop variety.

Two models will be discussed here which are only applicable when the encoder is not overloaded. Other models which are applicable when the encoder is overloaded are described in Chapter 4.

Linear delta modulation

1.6.1. Discrete pulse-phase modulation model

In pulse-phase modulation, abbreviated p.p.m., the rate of output pulses is proportional to the slope of the message signal. In a delta modulation process the rate of positive binary pulses (relative to the rate of negative binary pulses) is also proportional to the slope of the input signal. However, unlike delta modulation p.p.m. pulses are not confined to fixed time locations. If an open-loop model of a delta modulator is to be established with p.p.m. then the pulses from the p.p.m. must be time quantized such that these pulses coincide with the clock pulses of the delta modulator.

The discrete pulse-phase modulation model (abbreviated d.p.p.m.) of the delta modulator has been described by Flood and Hawksford[22] and is shown in Figure 1.11. The analogue input signal $x(t)$ phase modulates a carrier whose frequency is half the clock frequency. The maximum frequency deviation of this carrier is equal to $f_p/2$. This means that the instantaneous value of the carrier varies from zero to the clock rate f_p. When the input signal $x(t)$ is zero the output $m(t)$ from the phase modulator is a sinewave whose frequency is $f_p/2$. This waveform is passed through a zero-crossing detector to produce a square-wave $q(t)$ where the duration of each positive and negative level is $1/f_p$. $q(t)$ is connected to the pulse stretcher. The function of this circuit is to extend any $q(t)$ pulse by a duration of $1/f_p$ measured from the instant when $x(t)$ passed through a zero-crossing with a *positive* slope. Hence for the case when $x(t)$ is zero $q(t)$ is unaffected by the pulse stretcher, i.e. $q(t) = p(t)$ and $p(t)$ is sampled at a rate f_p with the result that the line waveform $L_M(t)$ is a square waveform having a pulse repetition rate of $f_p/2$. Thus, $L_M(t) = q(t)$ when $x(t)$ is zero, and $L_M(t)$ is identical to the idling waveform observed in a delta modulator.

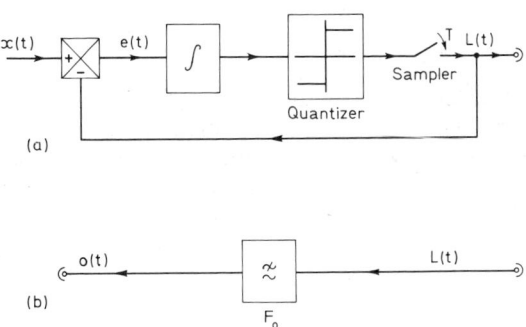

Fig. 1.10. Alternative representation of the delta-sigma modulation codec; (a) encoder, (b) decoder

Linear delta modulation

When an input signal $x(t)$ is applied to the model the waveform at the output from the phase modulator is

$$m(t) = M \cos \left(2\pi \frac{f_p}{2} \cdot t + \phi(t)\right) \quad (1.19)$$

where $\phi(t)$ is the instantaneous value of the phase angle and is linearly proportional to the input signal

$$\phi(t) = K_d x(t) \quad (1.20)$$

where K_d is constant for the modulator.

If the instantaneous frequency of $m(t)$ is f_i

$$2\pi f_i t = 2\pi \frac{f_p}{2} t + \phi(t)$$

$$f_i = \frac{f_p}{2} + \frac{1}{2\pi} \frac{d\phi(t)}{dt}$$

and from Equation 1.20

$$f_i = \frac{f_p}{2} + \frac{K_d}{2\pi} \cdot \frac{dx(t)}{dt} \quad (1.21)$$

From Equation 1.21 the frequency deviation, Δf, from the unmodulated carrier $f_p/2$ is

$$\Delta f = \left| \frac{K_d}{2\pi} \frac{dx(t)}{dt} \right|_{max} \quad (1.22)$$

The input signal $x(t)$ must be restricted so that Δf does not exceed $\pm f_p/2$, thereby preventing overmodulation of $m(t)$. Hence $\Delta f = f_p/2$

$$\frac{K_d}{2\pi} \left| \frac{dx(t)}{dt} \right|_{max} = \frac{f_p}{2} \quad (1.23)$$

For the actual delta modulator the overload condition occurs if the maximum slope of $x(t)$ is equal to the maximum slope of the waveform $y(t)$ at the output of the integrator in the feedback loop. The equation relating to this condition has been established in Section 1.2.5 (see Equation 1.1)

$$\left| \frac{dx(t)}{dt} \right|_{max} = \gamma f_p$$

where γ is the step height in the $y(t)$ waveform. Thus, from Equation 1.23

$$K_d = \pi/\gamma \quad (1.24)$$

Linear delta modulation

This result enables $m(t)$ to be expressed as

$$m(t) = M \cos[\pi\{f_p t + (x(t)/\gamma)\}] \quad (1.25)$$

Let the number of complete positive rotations of the phase of $m(t)$ at time t be R ($R \geq 1$). Then if ψ is the excess phase, i.e. $0 \leq \psi < 2\pi$,

$$\psi + \left(2\pi R - \frac{\pi}{2}\right) = \pi\left(f_p t + \frac{x(t)}{\gamma}\right) \quad (1.26)$$

The first positive zero-crossing occurs when the phase of $m(t)$ is $(3\pi/2)$, i.e. when $R = 1$, $\psi = 0$. The subsequent zero crossings occur when $\psi = 0$ and $R = 2, 3, 4, 5, \ldots$, i.e. when the phase of $m(t)$ is $7\pi/2$, $11\pi/2$, $15\pi/2$, $19\pi/2 \ldots$, respectively.

Suppose at time t, in Equation 1.26, N sampling periods have occurred, i.e. $t = N/f_p$. The Equation 1.26 becomes

$$\frac{\psi}{\pi} + 2R - \frac{1}{2} = N + \frac{x(N/f_p)}{\gamma} \quad (1.27)$$

The delta modulator is arranged so that the initial condition of the signal $y(t)$ at the output of the integrator in the feedback loop is $\gamma/2$. This enables $y(t)$ to oscillate symmetrically about zero if the input signal $x(t)$ is zero and creates the correct idling pattern at the output of the coder. Thus when $N = 1$, $y(t) = \gamma/2$. Hence after time $t = N/f_p$

$$y\left(\frac{N}{f_p}\right) = \gamma N_p - \gamma N_n + \frac{\gamma}{2}$$
$$= (2N_p - N + \tfrac{1}{2})\gamma \quad (1.28)$$

and $N = N_p + N_n$ where N_p and N_n are the number of positive and negative pulses at the output of the delta modulator.

To make the model compatible with the actual encoder the waveform $y_M(t)$ at the output of the integrator in the model (see Figure 1.11) is given an initial condition $\gamma/2$ such that

$$y_M\left(\frac{N}{f_p}\right) = (2N_p - N + \tfrac{1}{2})\gamma \quad (1.29)$$

In the model each complete rotation of the phase angle of $m(t)$ results in a positive pulse appearing at the output. Hence

$$R = N_p \quad (1.30)$$

Substituting this value of R in Equation 1.27 and using Equation 1.29

$$y_M\left(\frac{N}{f_p}\right) = x\left(\frac{N}{f_p}\right) + \gamma\left(1 - \frac{\psi}{\pi}\right) \quad (1.31)$$

Linear delta modulation

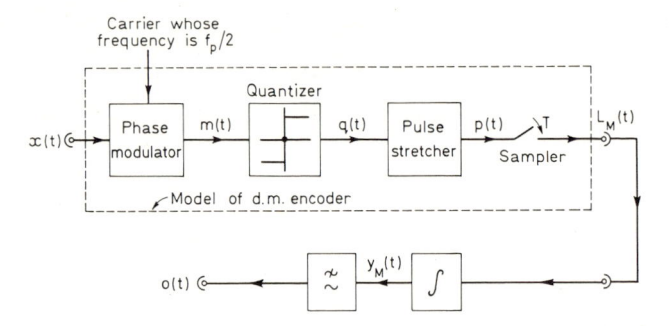

Fig. 1.11. Discrete-pulse-phase modulation model. (After Flood and Hawksford[22])

The error term in Equation 1.31 is

$$e\left(\frac{N}{f_p}\right) = \gamma\left(1 - \frac{\psi}{\pi}\right) \tag{1.32}$$

As

$$0 \leq \psi < 2\pi$$

$$0 \leq \frac{\psi}{\pi} < 2$$

Then

$$-\gamma < e\left(\frac{N}{f_p}\right) \leq \gamma \tag{1.33}$$

Equations 1.31 and 1.33 shows that the output $y(t)$ differs from the input by a magnitude which is not greater than the step height γ. This is the situation in the actual delta modulator and confirms the accuracy of the model. The implication is that the pulse patterns of the model and the coder are identical for a particular input signal when the initial conditions of both encoder and model are selected in the manner described above.

The voltage at the output of the integrators in the model and encoder are identical and consequently the noise characteristics are identical. This means that this open-loop model can be used to calculate the noise spectral density of a delta modulator.

Figure 1.12 shows typical waveforms in the model. Notice that positive pulses of length $1/f_p$ are produced whenever the $m(t)$ waveform has a positive slope passing through zero. Consequently $p(t)$ (see Figure 1.11) consists of positive pulses of length $1/f_p$ and negative pulses of variable length. The magnitude of these pulses is V. The waveform $p(t)$ is sampled

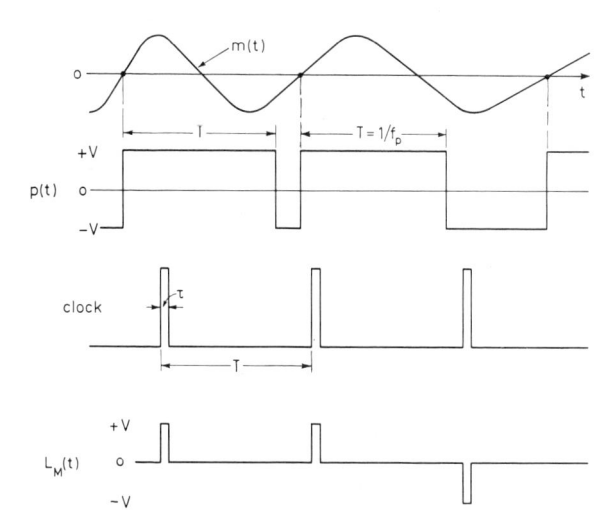

Fig. 1.12. *Typical waveforms in the discrete-pulse-phase modulation model*

at the clock rate of the delta modulator, i.e. at f_p, and pulses are produced which are positive if $p(t)$ is $+V$ or negative if $p(t)$ is $-V$. These pulses are of duration τ and when they are integrated they produce a voltage step γ equal to $V\tau$. If instantaneous sampling is considered then the magnitude of the $p(t)$ binary levels is γ, as an impulse of strength γ produces a step at the output of a perfect integrator of γ volts.

The waveform $L_M(t)$ is identical to the corresponding waveform $L(t)$ in a delta modulator for any $x(t)$ which does not overload the encoder. This has been verified by computer simulation and experimentally.

1.6.2. Infinite staircase model

The infinite staircase model was proposed by Van de Wegg in his classic paper.[4] It is a simpler model than the discrete pulse-phase modulation model and is both accurate and obvious.

The waveform $y(t)$ at the output of the integrator located in the feedback loop of the delta modulator has a magnitude which differs from the input signal $x(t)$ at any sampling instant by an amount which is less than $\pm \gamma$, where γ is the step-size in the waveform $y(t)$. This situation is illustrated by the waveform in Figure 1.13(b) for an arbitrary input signal. In this figure odd and even sampling periods have been designated by O and E respectively. If the magnitude of $y(t)$ is observed at any two adjacent E or

26					Linear delta modulation

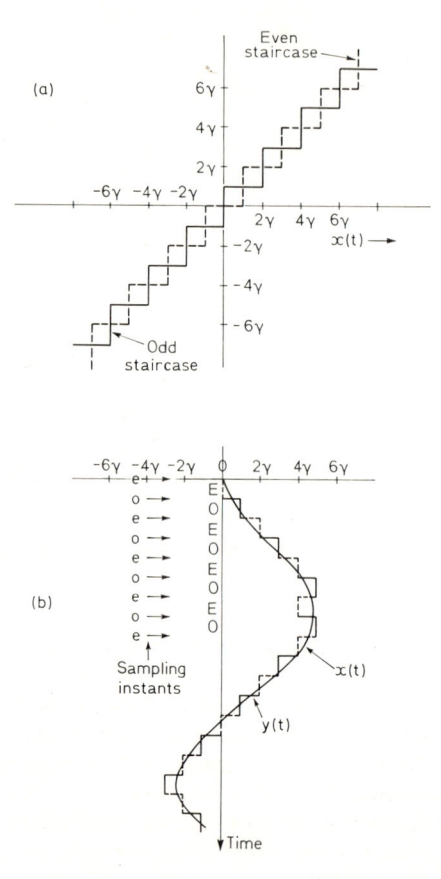

Fig. 1.13. (a) Infinite staircase functions, (b) y(t) waveform at odd and even sampling periods

O levels then it is found to be either increased by 2γ, decreased by 2γ or unchanged. This is to be expected as changes in $y(t)$ at sampling instant are either $+\gamma$ or $-\gamma$. The odd and even sampling instants in Figure 1.13(b) have been arranged such that the values of $y(t)$ at even times are even multiples of γ while at odd times they are odd multiples of γ.

These observations lead to the consideration of two infinite staircase functions having uniform steps of magnitude 2γ and uniform 'treads' of 2γ. The even staircase function will be used to establish the values of $y(t)$ at even sampling times while the odd staircase function will establish $y(t)$ at odd sampling times. These staircase functions are displayed in Figure 1.13(a). They have identical shapes but are displaced from each other by γ in both cartesian coordinates.

Linear delta modulation

The infinite staircase model of the delta modulator is formed with these two staircase functions, i.e., quantizers, two switches S_1 and S_2 to achieve instantaneous sampling, an adder, and a zero-order hold circuit. The arrangement is shown in Figure 1.14. The waveforms $y_e(t)$ and $y_o(t)$ at the output of the quantizers are sampled at even and odd sampling times respectively, i.e., at a rate of half the delta modulator clock rate. Denoting the instantaneously sampled signals by a raised asterisk, the signal at the output of the adder is

$$y^*(t) = y_e^*(t) + y_o^*(t)$$

Thus $y^*(t)$ is an impulse sequence where the impulses occur at the delta modulator clock rate f_p and the impulses from the odd and even staircase functions are interleaved. The $y^*(t)$ impulse sequence is passed to a zero-order hold circuit which holds each impulse for $1/f_p$ seconds. The result is that $y(t)$ changes its magnitude by γ each clock period. This is because successive $y_e^*(t)$ and $y_o^*(t)$ impulses differ from each other by magnitude γ only. The result is that $y(t)$ is identical to the $y(t)$ of the actual delta modulator shown in Figure 1.1.

An example of the response of the model to an input $x(t)$ is shown in Figure 1.13. Initially when t is zero, $y(t)$ is zero. Consequently the first staircase function to be sampled must be the even one as $y_e(t)$ is zero when $x(t)$ is in the range between $-\gamma$ and $+\gamma$. If the odd sampling instant is designated 'o', the value of $y_o(t)$ corresponding to $x(t)$ is sampled and conveyed via the adder to the hold circuit. At the next sampling instant, which is an even one and is denoted by 'e', the value of $y_e(t)$ corresponding to $x(t)$ is sampled and held. At the following sampling instant the odd staircase function is sampled, and so on. The result is that the output from the hold circuit is as shown in Figure 1.13(b) and is *identical* to the output from an actual delta modulator when the same input signal is applied.

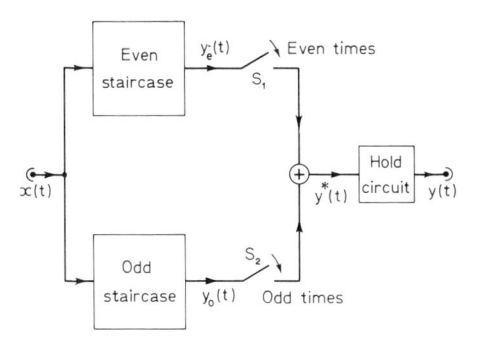

Fig. 1.14. *Infinite staircase model of linear delta modulator*

1.7. LINEAR DELTA MODULATOR USING A SAMPLE AND HOLD CIRCUIT

The effect of introducing a zero-order hold circuit immediately after the sampler was briefly discussed in Section 1.2.3. Having described the infinite staircase model of the delta modulator a more quantitative investigation into the effect of the hold circuit on the performance of this encoder can now be made.

The purpose of the hold circuit is to extend the duration of the $L(t)$ pulses (shown in Figure 1.5) from τ to T seconds. The $L(t)$ waveform now consists of binary levels rather than binary pulses. In order to maintain the same step-size $\pm \gamma$ in the waveform $y(t)$, V must be changed to

$$V\tau/T = V_h$$

Figure 1.15 shows the feedback waveforms $y(t)$ and $y_h(t)$ corresponding to the absence and insertion of the hold circuit respectively, for an output logical sequence of 1 1 1 1 0 1 0 1 1 0.

Whether the introduction of the zero-order hold reduces the quantization noise is a question best answered by noting that in the infinite staircase model of a delta modulator (Figure 1.14) the sampled $y(t)$ signal is passed through a zero-order hold circuit to give $y(t)$. In order to produce the waveform $y_h(t)$ when a zero order hold circuit is in the encoder the signal $y^*(t)$ shown in the model must be passed through a network having an impulse response $h_2(t)$. This network must be such that the convolution of $y^*(t)$ with $h_2(t)$ produces $y_h(t)$. This requires that $h_2(t)$ be of the form shown in Figure 1.16.

It will be shown in Section 3.10 that $y^*(t)$ has a spectral density function $S^*_{yy}(f)$ which consists of two components. One component is linearly related to the input spectral density function while the other component, $S^*_{nn}(f)$, is due to distortion noise.

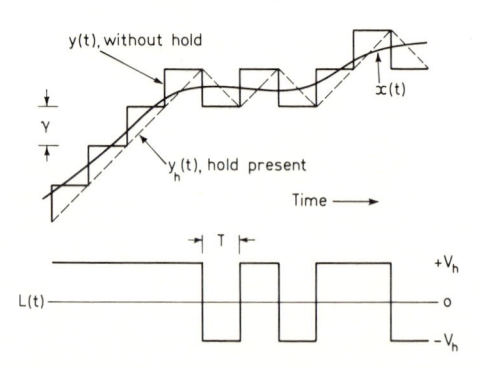

Fig. 1.15. *The effect of the hold circuit on the feedback waveforms*

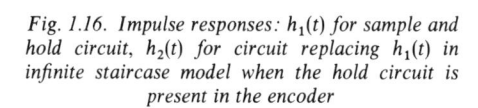

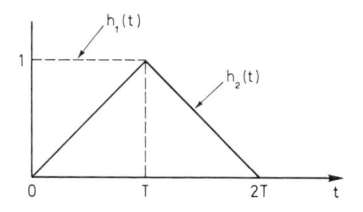

Linear delta modulation

Fig. 1.16. *Impulse responses: $h_1(t)$ for sample and hold circuit, $h_2(t)$ for circuit replacing $h_1(t)$ in infinite staircase model when the hold circuit is present in the encoder*

The distortion spectral density function at the output of the decoder is

$$S_{nn}(f) = S_{nn}^*(f)|H(j2\pi f)|^2$$

Without the zero-order hold in the encoder $H(j2\pi f)$ is a zero-order hold network, as shown in Figure 1.14, whose impulse function $h_1(t)$ is shown in Figure 1.16. The Fourier transform of $h_1(t)$ is $H(j2\pi f)_1$.

$$|H(j2\pi f)_1| = \frac{1}{f_p} \cdot \frac{\sin(\pi f/f_p)}{(\pi f/f_p)}$$

When the zero-order hold is included in the encoder $H(j2\pi f)$ is the Fourier transform of the impulse function of $h_2(t)$, namely

$$|H(j2\pi f)_2| = \frac{1}{f_p} \cdot \left\{ \frac{\sin(\pi f/f_p)}{\pi f/f_p} \right\}^2$$

As the clock frequency f_p is always much greater than the highest frequency f_{c2} in the input signal, the values of $|H(j2\pi f)_1|$ and $|H(j2\pi f)_2|$ are approximately the same over the frequency range 0 to f_{c2}.

The conclusion is that the introduction of the zero-order hold circuit in the encoder has negligible effect on the distortion noise.

1.8. TIME-DIVISION MULTIPLEXING LINEAR D.M. SIGNALS

The presence of the feedback loop in a delta modulator causes it to be a tracking encoder. It therefore has to track its analogue input signal at all times which prohibits it from being shared simultaneously between a number of analogue input signals.

For the purpose of illustration Figure 1.17 shows a number of analogue inputs $a_1, a_2, a_3, \ldots, a_k$ being encoded by k delta modulators. Suppose that the output of the encoders are sampled once every d.m. period T. Then, if an n-bit synchronisation code is added every T seconds the bit rate conveyed to the terminal equipment is $(k+n)f_p$. This is a delta modulation–time-division multiplexed signal, abbreviated d.m./t.d.m., and is filtered and perhaps used to modulate a carrier in the terminal equipment prior to transmission. When this signal arrives at the receiver it is filtered,

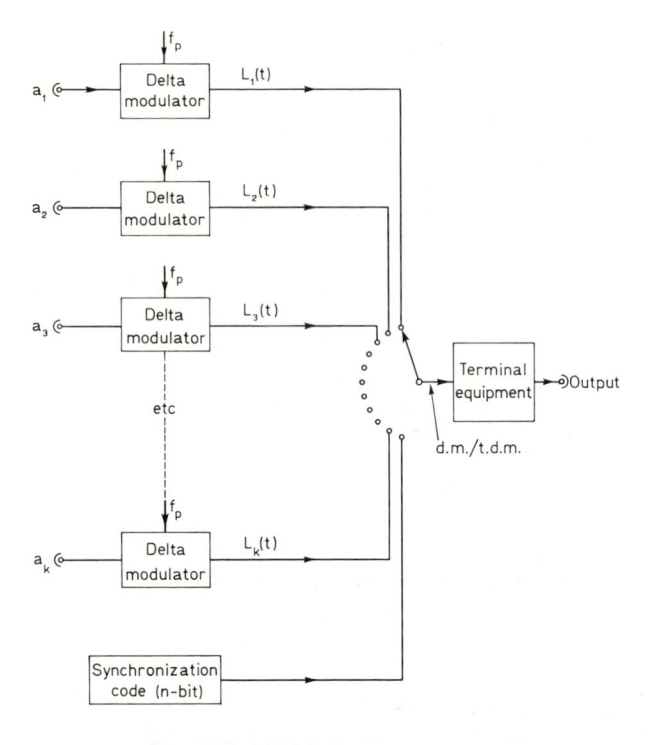

Fig. 1.17. Multiplexing arrangement

demodulated, amplified etc., by the receiver's terminal equipment in order to recover the d.m./t.d.m. signal. The *n*-bit synchronization code is extracted in order to ensure that each signal is directed to the appropriate decoder. Figure 1.18 shows the waveforms for one delta modulation period when $k = 3$ and $n = 1$.

It may be that a form of pulse compression multiplexing is used where instead of sending one sample of $L_1(t)$ followed by one sample of $L_2(t)$, etc. in every T seconds, N bits of $L_1(t)$ are transmitted followed by N bits of $L_2(t)$, etc. This requires that the encoding of each analogue signal occurs at f_p bits/s and the binary data is put into a buffer store from which it is removed when the N bit slot associated with each channel becomes available. The receiver de-multiplexes the t.d.m. signal and puts each N-bit d.m. 'word' into a buffer associated with each decoder. The data is inserted into each receiver buffer once every word slot, i.e. the data enters the buffer at the high bit rate of the t.d.m. signal, infrequently. However, it is extracted continuously from the buffer at the lower bit rate f_p in order to decode the d.m. binary signal. Such an arrangement[24] is suggested for

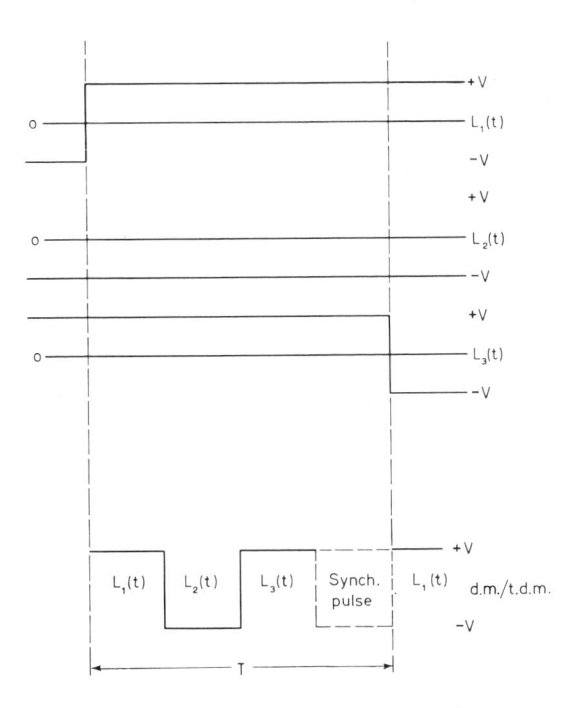

Fig. 1.18. *Waveforms in one d.m. clock period*

use in a local subscriber's network where the encoding and decoding of the voice signals occur at the hand-set.

The synchronisation code may take many forms, but if some channels are likely to be occasionally overloaded or forced to idle then synchronisation codes of continuous sequences of all ones or all zeros, or alternate polarity sequences, should be avoided. If all the data channels presented with signals band-limited to f_{c2} the highest frequency which a decoded sinusoid can have is just less than f_{c2}. By arranging for the synchronisation code to have a periodic sequence which when decoded will produce a sinusoid whose frequency is in excess of f_{c2} the receiver can readily identify the synchronisation channel from the t.d.m. input signal with the aid of a band-pass filter. If the t.d.m. signal has the data and synchronisation channels in a given sequence, then once the receiver has identified the synchronisation channel it knows the time location of each data channel and is able to distribute these data channels to their appropriate decoder.

For examples on time division multiplex systems, see References 23 to 37.

REFERENCES

1. Deloraine, E. M., Van Miero, S. and Derjavitch, B., French Pat. 932 140
2. De Jager, F., 'Delta modulation—a new method of p.c.m. transmission using the 1 unit code.' *Philips Research Report*, **7**, 442–466, December (1952)
3. Libois, L. J., 'Un nouveau procédé de modulation codée la modulation en delta', *L'Onde Electronique*, **32**, 26–31, January (1952)
4. van de Weg, N., 'Quantizing noise of a single integration delta modulation system with an N-digit code,' *Philips Research Report*, 367–385, October 8 (1953)
5. Zetterberg, L. H., 'A comparison between delta modulation and pulse code modulation', *Ericsson Technics*, **I**, 95–154 (1955)
6. French Pat. 987 238, (22.5.48)
7. Cutler, C. C., U.S. Pat. 2605361, (29.7.52)
8. Schouten, J. F., de Jager, F. and Greefkes, J. A., 'Delta modulation—a new modulation system for telecommunications', *Communication News*, 115–124, June 12 (1952)
9. Vasseur, J. P., 'Comparison of code modulation and classical modulation', *Ann. Radioelect.*, **9**, 137–149, April (1954)
10. Stata, R., 'Minimizing integrator errors', *Electrotechnology*, **82**, No. 4, 46–50, (1968)
11. Effenburger, J. A. and Steele, R., 'A transistorised integrator', *Electronic Engineering*, 230–234, April (1965)
12. Clayton, G. B., *'Operational Amplifiers'*, Butterworths, London (1971)
13. Fulford, J. F., 'Design considerations for active low-pass filters', *S.R.D.E. Christchurch*, Report No. TEL 1/66, February (1966)
14. Ackroyd, M. H., *Digital Filters*, Buttèrworths, London (1973)
15. Gersho, A., 'Stochastic stability of delta modulation', *Bell Systems Tech. J.*, **51**, No. 4, 821–841, April (1972)
16. Johnson, F. B., 'Calculating delta modulator performance', *I.E.E.E. Trans. Audio and Electroacoustics*, **AU-16**, No. 1 121–129, March (1968)
17. Newton, C. M. B., 'Delta modulation with slope-overload prediction', *Electronic Letters*, **7**, No. 9, 272–274, April 30 (1970)
18. Cutler, C. C., 'Delayed encoding; stabilizer for adaptive coders', *I.E.E.E. Trans. Communication Technology*, **COM-19**, No. 6, 898–907, December (1971)
19. Aughenbaugh, G. W., Irwin, J. D. and O'Neal, J. B., 'Delayed differential pulse code modulation', *Proceedings 2nd Annual Princeton Conference on Information Sciences and Systems*, Princeton, N.J., 124–130, March (1968). New York; I.E.E.E.
20. Inose, H., Yasuda, Y. and Murskani, J., 'A communication system by code modulation delta-sigma', *J. Inst. Elect. Eng.* Japan, **44**, No. 11, 1775–1780, November 1961
21. Inose, H. and Yasuda Y., 'A unity bit coding by negative feedback', *Proc. I.E.E.E.*, 1524–1535, November (1963)
22. Flood, J. E. and Hawksford, M. J., 'Exact model for delta modulation process', *Proc. I.E.E.* **118**, No. 9, 1155–1161, September (1971)
23. Hoare, D. W., Ivison, J. M. and Qazi, S., 'A multichannel biomedical telemetry system using delta modulation', *8th International Conference on Medical and Biological Engineering*, Session 30-8, Chicago, July (1969)
24. Flood, J. E. and Shurrock, C. R. J., 'Application of delta modulation in the telephone subscribers network', *International Zurich seminar on integrated systems for speech, video and data communications*, 1–9, March (1972)
25. Wing, P. A., 'Wisdom-self-synchronising delta-modulation orthogonal-channel modulator', *Electronic Letters*, **4**, No. 11, 211–212, May 31 (1968)
26. Inose, H., Yasuda, Y., Murakami, J. and Fujita, H., 'New modulation technique simplifies circuits', *Electronics*, **36**, No. 4, 52–55, January (1963)
27. Sharma, P. D., 'Multiplexing of r.w.m. signals' (Part 1), *Electronic Eng.*, 159–162, March (1968)

28. Sharma, P. D., 'Multiplexing of r.w.m. signals', Part 2 *Electronic Eng.* 215–220, April (1968)
29. Enomoto, O., Tomozawa, A., Katayama, H., Kaneko, H. and Sekimoto, T., 'An experimental high speed electronic switching system using delta modulation', *Nippon Electric Co. Res. Dev.*, No. 8, 126–138, October (1966)
30. Watson, R. B., Hudson, O. K., 'Transmitting system using delta modulation, *Electronics*, **29**, No. 10, 164, October (1965)
31. Magnauki, H., 'Wideband channel for emergency communication', *I.R.E. Trans. Vehicular Comm.* (U.S.A.), **VC-10**, No. 2, 40–45, August (1961)
32. Magnauki, H., "Wideband channel for emergency communication', *I.R.E. Internat. Conf. Rec.* (U.S.A.), **9**, 80–84, Pt. 8, (1961)
33. Inose, H., Yasuda, Y. and Marakaima, J. 'A telemetering system by code modulation, delta-sigma modulation, *I.R.E. Trans. Space Elect. Tele.* (U.S.A.), **SET 8**, No. 3, 204–209, September (1962)
34. Inose, H., Yasuda, Y., Kawai, and Tarkagi, M., 'The subscriber-line circuit and the signalling and line system for an experimental time division exchange featuring delta modulation techniques', *I.R.E. Trans. Comm. Syst.* (U.S.A.), C8–10, No. 4, 377–407, December (1962)
35. Daguet, J., Breant, P., Gaguere, C., Bellanger, M. and Lamaine, P., 'Multiplex telephonique a 60 voies, a modulation en delta a pente asservie', *Cables et Transmission*, **23**, No. 2, 178–197 (1969)
36. Snowden, D. R. C., 'A small station satellite system using delta modulation'. *Intelstat, I.E.E. International Conference on digital Satellite Communication*, 26, November (1969). London; I.E.E.
37. Enomoto, O., Goto, H. and Ichijo, K., 'A high speed switching network for p.c.m. switching systems,' *N.E.C. Research and Development*, No. 18, 16 July (1970)

Extended Bibliography of Linear Delta modulation.

38. Williams, B. H., 'Delta modulation system using junction transistors', *Electronic Engineering*, **34**, 674–680, November (1959)
39. Debart, H., 'A study of delta modulation systems from the point of view of information theory, *Cables et Transmission*, **15**, No. 3, 118–204, July (1961)
40. Dase, H., 'Delta modulation', *Elektrotek Tideskrift*, **74**, No. 4, 51–55, February (1961)
41. Arter, F. W., 'Simple delta modulation system', *Proc. I.R.E.*, (Australia) **9**, 517–523, September (1962)
42. Nalijak, C. A. and Tripp, J. S., 'A deterministic study of delta modulation', *I.E.E.E. Internat. Conf. Rec.* **2**, Pt. 8, 247–259, (1963)
43. Masckloff, R. H., 'Delta modulation', *Electro-technology*, **13**, No. 1, 91–97, January (1964)
44. Habara, K., 'Delta modulation using Esaki diodes', *J. Inst. Elect. Comm. Engrs. Japan*, **46**, No. 6835, June (1963)
45. Schilling, D. L., Clarke, K. K. and Pickholtz, R. L., *A space communications study— Status Report*, NASA-CR-101126, March (1969)
46. Balder, J. C. and Kramer, C., 'Analog-to-digital conversion by means of delta modulation', *I.E.E.E. Trans. Space Electronics and Telemetry*, **SET-10**, No. 3, 87–90, September (1968)
47. Biorci, G. and Maestrini, R., 'Intelligibility of delta-modulated signals in presence of noise', *Elettrotecnica*, **42**, No. 2, 1967. (*Electrical and Electronics Abstracts*, **72**, 24083, (1969))
48. Lender, A. and Kozuch, M., 'Single-bit delta modulating systems', *Electronics*, **34**, 125–129, November 17 (1961)
49. Murphy, G. J. and Shuraym, G. 'Delta modulation in feedback control system', *Proceedings National Electronic Conference*, **21**, 609–612, October 25–27 (1965)

50. Sekimoto, T., and Kaneko, H., 'A method of delta modulation', Record National Convention *Inst. Elec. Comm. Eng. Japan*, 317–319, 1958
51. Nakamaru, Y., and Kaneko, H., 'Delta Modulation encoder', *Nippon Electric Co. Research and Development*, No. 1, 46–54, October (1960)
52. Rao, G. V. S., 'Delta modulation', J. Instn. Engrs. (India) Electronics Engng. Div., **47**, No. 12, Pt. ET3 SP, 354–362 (1967)
53. Debart, H., 'Coded transmission with reduced bandwidth'. *Revue Cethedec* (France), **3**, No. 12, 51–57, (1967)
54. McKibbin, J. and Stradling, J. B. M. 'The use of delta modulation in pulse transmission systems', *Inst. Engrs. Australia Elec. Eng. Trans.*, **EE6**, No. 1, 35–39, March (1970)
55. Loc, F. J. and Arndt, G. D., 'A delta modulation threshold extension technique for frequency modulation systems', *S.W.I.E.E.C.O. Record of Technical Papers of the 22nd Southwestern I.E.E.E. Conference and Exhibition*, Dallas, 33–37, April (1970). New York; I.E.E.E.
56. Schemer, R. A., 'A survey of delta modulation systems', *Trans. S. African Inst. Elect. Egnrs*, **59**, No. 9, 203–213, September (1968)
57. Cermak, J., 'Delta modulation and its application'. *Sdelovaci Tech.* (Czechoslovakia), **16**, No. 12, 426–429, (1968). (See *Electrical Electronics Abstracts*, No. 20877, (1969))
58. Anon. 'Delta modulation', *Electronic Products*, **12**, No. 11, 41, March (1970)
59. Jones, P. S., 'Direct digital code conversion in large-scale voice communication systems'. Proceedings 3rd Hawaii International Conference on System Sciences, Honolulu, 395–398, January (1970). Hollywood Western Periodicals Co.
60. Nitadori, K., 'Statistical analysis of delta p.c.m.', *Electronics and Communications in Japan*, **41**, No. 48, 17–26, February (1965)
61. Cotton, R. V., Remm, R. L., and Strohmeyer, G. R., 'Analaysis of delta modulation encoding using computer simulation', *Aerospace Systems Conference*, Seattle, Wash., July 11–15 (1966) and *I.E.E.E. Trans. Aerospace and Electronic Systems*, Supplement, **AES-2**, 380–387, July (1966)

Other references to linear delta modulation are to be found at the end of Chapters 2 to 6 inclusive.

Chapter 2

DOUBLE INTEGRATION AND EXPONENTIAL DELTA MODULATION

2.1. INTRODUCTION

This chapter considers double integration and exponential delta modulation. These modulation systems are linear in that the signals fed back to the error point in the encoders are linearly related to the output binary signals, as described in Section 1.1. The encoders in these systems are non-linear while the decoders are linear.

The exponential d.m. systems are of the type which are realised in hardware form, i.e. they are to some extent the practical versions of the linear d.m. described in Chapter 1.

Double integration d.m. introduces problems of stability which are discussed in detail. It will be shown to have a superior signal-to-noise-ratio when compared to that of the single integration and exponential d.m., provided that it is made stable by a prediction technique. Although double integration d.m. is a linear d.m. system it will be known throughout this book by the former name, and the title linear d.m. will be confined to the d.m. system which has a single integrator in the local decoder.

2.2. DOUBLE INTEGRATION DELTA MODULATION

Suppose a second integrator is introduced into the feedback loop of the linear delta modulator, as shown in Figure 2.1, with the intuitive expectation that the feedback signal will be a closer representation of the input signal than is obtained with the single integration encoder.[1] It will be shown that the encoder will have oscillatory characteristics and an overload characteristic which falls off at 12 dB/octave. However, the oscillations in the encoder can be satisfactorily damped by applying a simple predictor in the feedback loop.

For convenience, consider the narrow $L(t)$ pulses at the output of the encoder to be impulses of strength $\pm V\tau$. These impulses produce a waveform $y_1(t)$ at the output of the integrator which is composed of steps of magnitude $\pm \gamma$ volts. These steps are integrated by the second integrator to produce the feedback waveform $y_2(t)$. Each step in $y_1(t)$ produces a

Double integration and exponential delta modulation

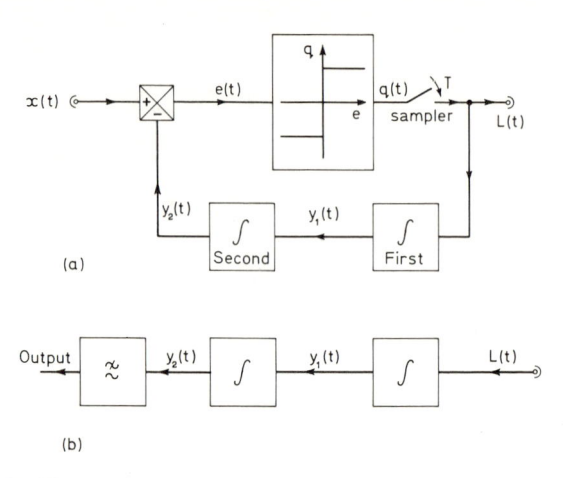

Fig. 2.1. Double integration delta modulation system, (a) encoder, (b) decoder

ramp in $y_2(t)$ which has a value γT after a clock period of T seconds, i.e. the ramps have a slope of γ. Both of the integrators are assumed to be ideal having an impulse response which is unity when $t \geq 0$ and zero when $t < 0$. The dimension of the impulse function is s^{-1}.

The impulse at the output of the encoder cause changes in the slope of the waveform $y_2(t)$. This means that the difference between the rate of occurrence of each type of binary impulse is proportional to the second derivative of the input signal $x(t)$. For a sinusoidal input having an amplitude E_s and a frequency f_s the maximum rate of change of slope is $E_s(2\pi f_s)^2$. Figure 2.2 shows the output pulses $L(t)$ drawn as impulses of strength $V\tau$(Vs) and the corresponding $y_1(t)$ and $y_2(t)$ waveforms. The slope of $y_1(t)$ is γ/T which gives the overload condition of the single integration linear delta modulator, see Equation 1.4. The maximum increase in the slope of $y_2(t)$ is γ/T(Vs^{-2}). The overload condition can now be specified by the equation

$$E_{sm}(2\pi f_s)^2 = (\gamma/T) \qquad (2.1)$$

where E_{sm} is the maximum amplitude of the input sinusoid which does not overload the encoder.

2.2.1. Idling modes

Double integration linear d.m. is not stable and tends to have oscillatory properties. Figure 2.3 shows the behaviour of the encoder in an idling mode where the input signal is zero. Instead of the binary pattern

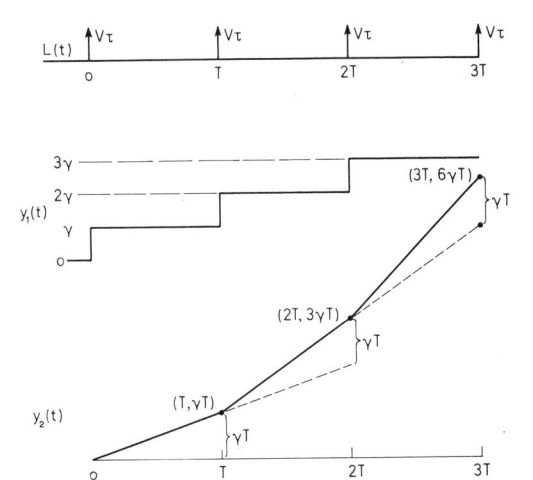

Fig. 2.2. $y_1(t)$ and $y_2(t)$ waveforms when the binary output is a series of positive impulses

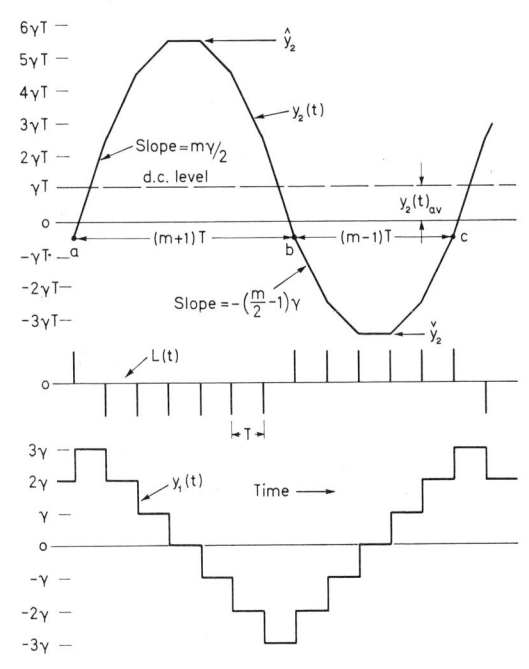

Fig. 2.3. Oscillatory behaviour in the idling mode, $m = 6$, $\mu = 0.833$

1 0 1 0 1 0..., as described previously for the single integration delta modulator, it is composed of alternate sequences of m ones and m zeros; the feedback waveform is oscillatory and generally has a d.c. component. The value of m determines the order of the idling mode. In Figure 2.3, $m = 6$.

Once the waveform $y_2(t)$ has changed polarity from negative to positive the error signal becomes negative and a negative impulse is generated at the output at the next clock instant. Although $y_1(t)$ is reduced by this impulse it is still positive with the result that $y_2(t)$ continues to increase but at a reduced rate. The error signal continues to remain negative and at each sampling instant $y_1(t)$ is decreased by γ and the slope of $y_2(t)$ is changed by γ/T. After m negative impulses have been generated $y_2(t)$ passes through zero and the error changes sign. However, the slope of $y_2(t)$ is now at its maximum in the negative direction and a further m positive impulse must be generated at the output of the encoder in order to again reverse the sign of the error signal. The oscillatory characteristic of the encoder is due to the slope of $y_2(t)$ having its maximum slope when its magnitude is zero.

Case when m is even

Consider the behaviour of the encoder starting at point a in Figure 2.3. A positive impulse is generated because $y_2(t) < 0$, and $y_2(t)$ increases its slope by γ to $+m\gamma/2$ say, as $y_2(t)$ passes through its zero crossing. This choice of slope results in the period of the oscillation being the sum of m

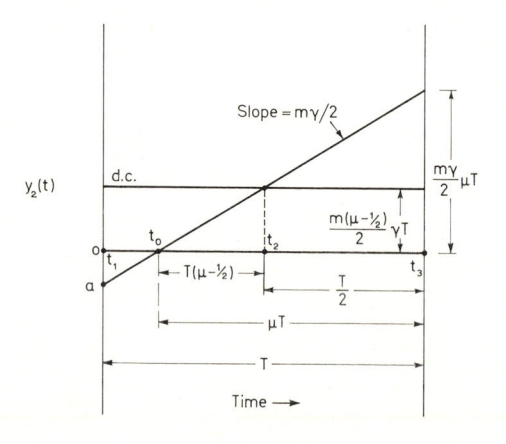

Fig. 2.4. Variation of $y_2(t)$ during the clock period in which it passes through zero with a positive slope

Double integration and exponential delta modulation

zeros followed by m ones. The inclined line in Figure 2.4 shows the waveform $y_2(t)$ during the time slot, or sampling period, at a. The zero crossing is at time t_0 which is μT seconds from the end of the slot at time t_3, where $0 < \mu \leq 1$. At time t_3, $y_2(t)$ has positive magnitude $m\gamma T\mu/2$ which results in the error changing sign and the encoder generating a negative impulse at its output causing the slope of $y_2(t)$ to decrease by γ. Successive negative impulses in the waveform $L(t)$ cause the slope to decrease until it becomes zero, when $y_2(t)$ has its maximum value $\hat{y}_2$ after $m/2$ clock periods.

$$\hat{y}_2 = \frac{m\gamma}{2}\mu T + \left\{\frac{m}{2}\gamma - \gamma\right\}T + \left\{\frac{m\gamma}{2} - 2\gamma\right\}T + \cdots + \gamma T$$

$$= \frac{\gamma T}{2}\{m\mu + (m-2) + (m-4) + \cdots + 2\}$$

Writing $m\mu$ as $(m\mu - m) + m$

$$\hat{y}_2 = \frac{\gamma T}{2}\left\{m(\mu - 1) + \frac{m}{2}\left(\frac{m}{2} + 1\right)\right\}$$

$$= \frac{\gamma T m}{2}\left(\mu - \frac{1}{2} + \frac{m}{4}\right) \tag{2.2}$$

The $y_1(t)$ waveform continues to decrease and $y_2(t)$ reduces from $\hat{y}_2$ with an initial slope of $-\gamma$ which increases to $-(m\gamma/2)$ after $mT/2$ seconds. When the first positive impulse occurs in the $L(t)$ waveform, i.e. at the first clock instant after $y_2(t)$ changes sign (point 'b' in Figure 2.3) the time taken for $y_2(t)$ to go from a to b is

$$(mT/2) + T + (mT/2) = (m+1)T$$

At b

$$y_2(t) = m\gamma T(1-\mu)/2$$

and at successive clock instants the slope of $y_2(t)$ changes by γ from $-m\gamma/2$ at the zero crossing until following $m/2$ pulses the slope is zero. The negative peak value of $y_2(t)$ is

$$\check{y}_2 = -\frac{m\gamma}{2}T(1-\mu) - \left\{\left(\frac{m\gamma}{2} - \gamma\right)T + \left(\frac{m}{2}\gamma - 2\gamma\right)T + \cdots + \gamma T\right\}$$

$$= -\frac{m\gamma T}{2}(1-\mu) - \frac{m}{4}\gamma T\left(\frac{m}{2} - 1\right)$$

$$= -\frac{m\gamma T}{2}\left(\frac{1}{2} - \mu + \frac{m}{4}\right) \tag{2.3}$$

The peak to peak value of y_2 is

$$\hat{y}_2 - \check{y}_2 = (m^2\gamma T/4) \tag{2.4}$$

From point b onwards the $L(t)$ waveform is a sequence of positive impulses which attempt to minimise the magnitude of $y(t)$. By the time $\check{y}_2$ occurs $(m/2)-1$ clock periods have elapsed. $y_2(t)$ remains at $\check{y}_2$ for one clock period and then decreases, with an increased slope per clock period of γ; after $(m/2)-1$ periods point c is reached. The time duration between b and c is

$$[(m/2)-1+1+(m/2-1)] = (m-1)T$$

As the magnitude and slope of $y_2(t)$ at a and c are identical the waveform is oscillatory with a frequency

$$f_m = \frac{1}{ac} = \frac{1}{(m+1)T+(m-1)T} = \frac{1}{2mT} = \frac{f_p}{2m} \tag{2.5}$$

The d.c. component of $y_2(t)$ is the average value of $y_2(t)$ in the time slot associated with the zero crossing. From Figure 2.4 this value is

$$y_2(t)_{av} = m(\mu-\tfrac{1}{2})\gamma T/2 \tag{2.6}$$

Figure 2.5 shows the $y_2(t)$ waveform for $m = 6$ and 3 values of μ.

Case when m *is odd*

When m is even the peak values of $y_2(t)$ are flat for one clock period, as shown in Figure 2.5. When m is odd the peak values are pointed, as

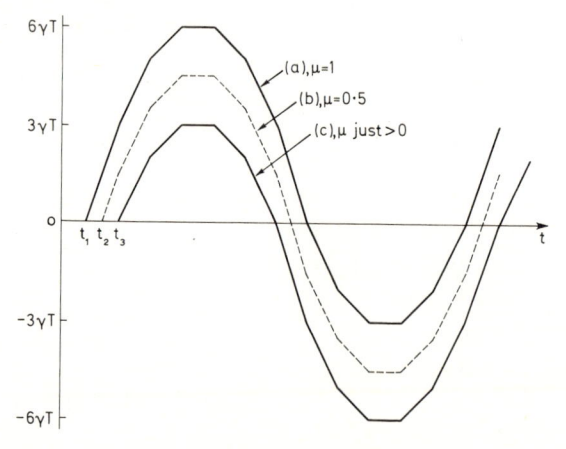

Fig. 2.5. *Feedback waveform* $y_2(t)$ *as a function of* μ *for* $m = 6$

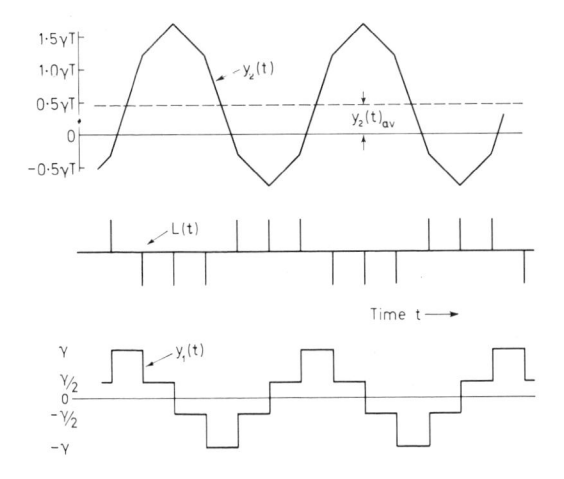

Double integration and exponential delta modulation 41

Fig. 2.6. Idling waveforms for $m = 3$, $\mu = 0.8$

shown in Figure 2.6, and this slight change in the characteristic of $y_2(t)$ results in the above equations being amended to

$$\hat{y}_2 = \frac{m\gamma}{2}\mu T + \left\{\left(\frac{m\gamma}{2}-\gamma\right)T + \left(\frac{m\gamma}{2}-2\gamma\right)T + \cdots + \frac{\gamma T}{2}\right\}$$

which reduces to

$$\hat{y}_2 = \frac{\gamma T}{2}\left[m\left(\mu-\frac{1}{2}\right)+\frac{m^2+1}{4}\right] \tag{2.7}$$

and

$$\check{y}_2 = -\frac{m\gamma T}{2}(1-\mu) - \left\{\left(\frac{m}{2}\gamma-\gamma\right)T + \left(\frac{m\gamma}{2}-2\gamma\right)T + \cdots + \frac{\gamma T}{2}\right\}$$

which when simplified becomes

$$\check{y}_2 = -\frac{\gamma T}{2}\left[m\left(\frac{1}{2}-\mu\right)+\frac{m^2+1}{4}\right] \tag{2.8}$$

The peak to peak value of $y_2(t)$ is

$$\hat{y}_2 - \check{y}_2 = (m^2+1)(\gamma T/4) \tag{2.9}$$

The d.c. value for m having odd integral values is given by Equation 2.6.

2.2.2. Encoding an input signal

When an input signal is applied with a slow rate of change of slope the encoder may be stable. However any rapid changes in the input will force the $y_2(t)$ signal to become oscillatory, as shown in Figure 2.7.

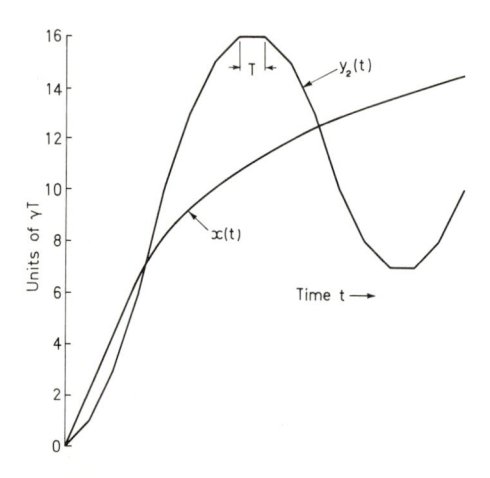

Fig. 2.7. Oscillatory behaviour of $y_2(t)$ when tracking an arbitrary input signal $x(t)$

Clearly the encoder is of little value unless the oscillatory behaviour can be damped. Fortunately this can be achieved by means of a predictor placed in the feedback loop.

2.2.3. Stabilising the encoder's response by means of prediction

A fault with the double integration linear d.m. is that the feedback signal $y_2(t)$ often crosses the input signal $x(t)$ with a slope much greater than that of $x(t)$. This is demonstrated in Figure 2.7. When idling, many different oscillatory waveforms are possible as described in the previous section.

In order to minimise the oscillatory nature of $y_2(t)$ a simple prediction system is used[2] as shown in Figure 2.8. The first integrator produces a step-like waveform $y_1(t)$ where the size of each step is γ. The feedback waveform is not $y_2(t)$ as in the previous case but $y_3(t)$ where

$$y_3(t) = \int y_1(t)\,dt + PTy_1(t) = y_2(t) + PT\frac{dy_2(t)}{dt} \qquad (2.10)$$

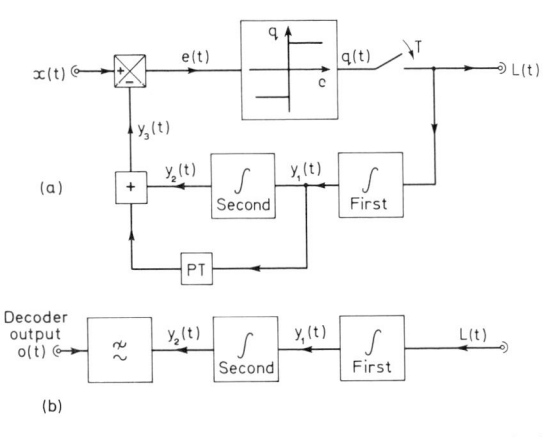

Double integration and exponential delta modulation 43

Fig. 2.8. Double integration d.m. codec with prediction: (a) encoder, (b) decoder

where P is a prediction constant. Figure 2.9 shows the behaviour of $y_3(t)$ for a ...1 1 0 0... output pattern. In this figure $P = 0.5$. At a clock instant $y_3(t)$ is immediately changed by γPT in the same direction as that of the impulse at the encoder output. This ensures that $y_3(t)$ has a step change in the right direction to minimise the error signal $e(t)$. In the figure it can be seen that at time $2T$ the $y_3(t)$ waveform jumps to N. This is the same as if the encoder had prior knowledge of the required slope at M, i.e. the predictor effectively arranges for the correct slope at M even before the encoding occurs at time $2T$. The time between M and N is PT.

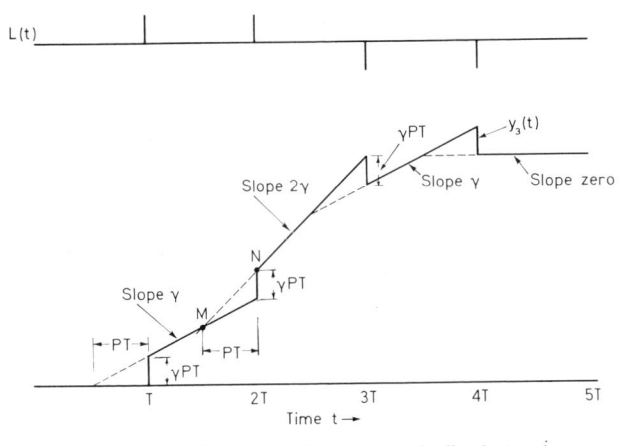

Fig. 2.9. The effect of prediction on the feedback signal

Decoder

It will be shown in the next section that prediction is employed in the encoder to damp high order idling patterns. The decoder is a stable open loop system and there is generally no necessity to use a similar predictor to that used in the encoder. Consequently the decoder consists of two integrators followed by a low pass filter, as shown in Figure 2.8.

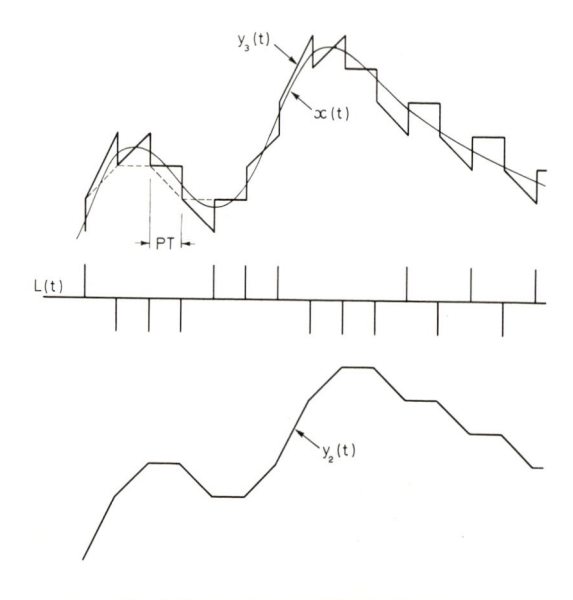

Fig. 2.10. Tracking with prediction, $P = 1$

Figure 2.10 shows $y_3(t)$ the feedback signal in the encoder and the $y_2(t)$ signal at the decoder or encoder, when an arbitrary signal $x(t)$ is applied to the encoder. Observe that the $y_2(t)$ signal does not contain the high frequency components which reside in $y_3(t)$ and that $y_2(t)$ is a close approximation to $x(t)$ delayed by PT seconds. The absence of the predictor in the decoder therefore has a minimal effect on the waveform at the output of the low pass filter, causing some slight phase distortion.

Damping the oscillatory idling response by prediction

Having briefly described how the predictor works consideration will now be given to the value of the prediction constant P which is necessary

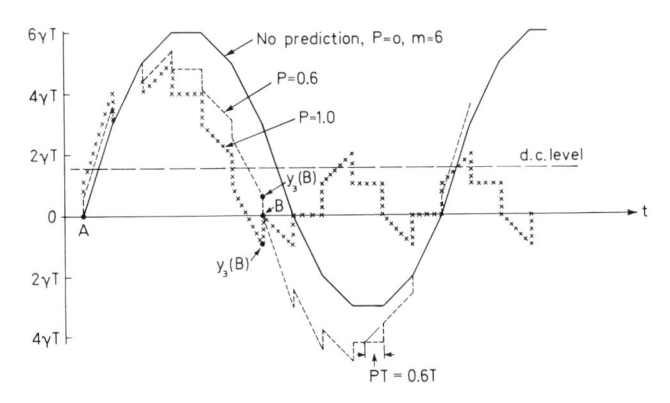

Double integration and exponential delta modulation 45

Fig. 2.11. *The effect of two values of prediction constant on an even order mode, $m = 6$, $\mu = 1$*

to ensure that the oscillatory idling waveforms are damped. Figure 2.11 shows the $y_2(t)$ waveform in the absence of prediction ($P = 0$ when $m = 6$, $\mu = 1$) and the effect of applying prediction to this waveform at time A. When the prediction constant $P = 0.6$ the feedback waveform $y_3(t)$ retains its oscillatory nature whereas when P is increased to unity a ...1 1 0 0 1 1 0 0... idling pattern is established instead of ...1 1 1 1 1 1 0 0 0 0 0 0 1 1 1 1 1 1.... Assuming that the prediction is applied at time A the maximum value of $y_3(t)$ with prediction is

$$y_p = \hat{y}_2 + PT\gamma - (mPT\gamma/2) \tag{2.11}$$

where $\hat{y}_2$ is the maximum value of the feedback voltage $y_2(t)$ without prediction. The $+PT\gamma$ term refers to the step at time A and $-mPT\gamma/2$ is the reduction in $y_2(t)$ due to the $m/2$ pulses from time A until the flat-top of the waveform occurs.

In order to damp the feedback waveform $y_2(t)$ the error must become negative at time B, Figure 2.11. Using similar arguments to those in Section 2.2.1 the value of $y_3(t)$ just prior to the next step change in this waveform due to the action of the predictor at time B is

$$y_3(B) = y_p - \left\{ \gamma T + 2\gamma T + \cdots + \left(\frac{m}{2} - 1\right)\gamma T \right\} - \left(\frac{m}{2} - 1\right)P\gamma T$$

i.e. at time B there have been $(m/2) - 1$ clock periods since $y_3(t) = y_p$ and the slope has increased progressively in the negative direction from $-\gamma, -2\gamma, \ldots$ to $-(m/2-1)\gamma$. The term $\{(m/2)-1\}P\gamma T$ represents the decrease in $y_3(t)$ due to the prediction process. $y_3(B)$ can be arranged as

$$y_3(B) = \left[\hat{y}_2 - \left\{\gamma T + 2\gamma T + \cdots + \left(\frac{m}{2} - 1\right)\gamma T\right\}\right] - \left\{\left(\frac{m}{2} - 1\right) + \left(\frac{m}{2} - 1\right)\right\}P\gamma T$$

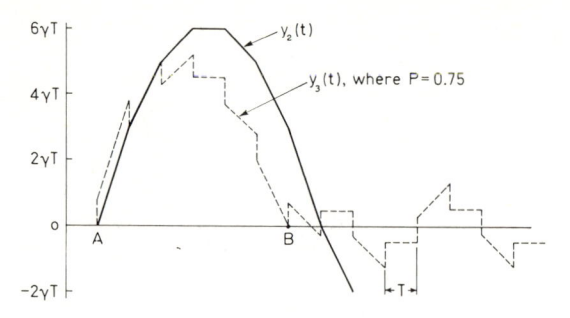

46   Double integration and exponential delta modulation

Fig. 2.12. Minimum value of prediction constant P(= 0·75) to damp $y_2(t)$ waveform where $m = 6, \mu = 1$

The square bracket term represents $y_2(t)$ at time B and is $m\gamma T/2$. Therefore
$$y_3(B) = (m\gamma T/2) - 2(m-2)(P\gamma T/2) \tag{2.12}$$
Damping of the oscillations will occur if $y_3(B) \leq 0$, giving
$$P \geq \frac{m}{2(m-2)}, \qquad m > 2 \tag{2.13}$$

For $m = 6$ the minimum value of P is 0·75 and it can be seen from Figure 2.12 that $y_3(B)$ is just zero at time B. The establishment of the above inequality assumes that the prediction is applied when the magnitude of the oscillation is zero but its slope is a maximum. Figure 2.13 shows that if the prediction process is considered to commence when $y_2(t)$ is at its maximum the oscillations are still damped using the minimum value $P = 0.75$.

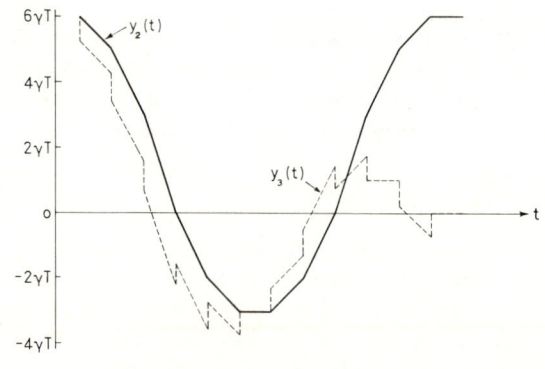

Fig. 2.13. Application of the minimum prediction constant $P = 0.75$ at the maximum amplitude of a 6th order mode, $y_2(t)$ waveform

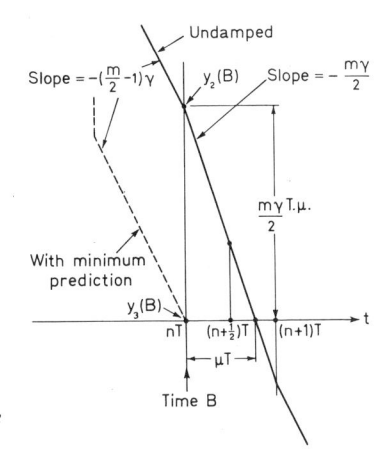

Double integration and exponential delta modulation 47

Fig. 2.14. Effect of minimum prediction on the undamped waveform $y_2(t)$

The general equation is established by observing that $Y_3(B)$ can be written as

$$Y_3(B) = y_2(B) - (m-2)P\gamma T \tag{2.14}$$

where $y_2(B)$ is the value of $y_2(t)$ at time B when no prediction is applied. From Figure 2.14, $y_2(B) = m\gamma T\mu/2$ and consequently when $Y_3(B) \leq 0$ to prohibit the mth consecutive 'one' occurring,

$$P \geq \frac{m\mu}{2(m-2)}, \quad m > 2 \tag{2.15}$$

This is the general result for damping $y_2(t)$ when it has a positive excursion.

Figure 2.15 shows $y_2(t)$ for $m = 5$, $\mu = 0.6$ and the response when the minimum value of prediction 0.5 is used.

If prediction is applied, not when $y_2(t)$ is embarking on a positive excursion as in Figure 2.11 for example, but when it is commencing a

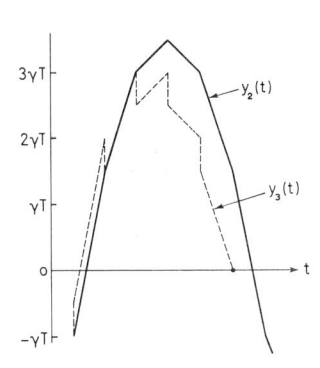

Fig. 2.15. The undamped $y_2(t)$ waveform where $m = 5$, $\mu = 0.6$ is damped using the minimum prediction constant $P = 0.5$ to produce the $y_3(t)$ waveform

negative excursion, then the value of P to prevent the mth consecutive zero occurring is

$$P > \frac{m(1-\mu)}{2(m-2)}, \quad m > 2, \tag{2.16}$$

This equation can be deduced by a similar argument to that used in establishing Equation 2.15.

Equations 2.15 and 2.16 are represented in Figure 2.16. The slopes of the lines in this figure are $\pm m/2(m-2)$. Observe that it is easier to damp $y_2(t)$ by applying prediction at the commencement of a negative excursion if it has a positive d.c. component and vice versa when prediction is applied at the beginning of a positive excursion. In order to ensure that $y_2(t)$ is damped to a lower order mode for a given μ, a suitable value of P can be found in the shaded part of Figure 2.16. The minimum value of P to ensure damping is $\mu = 0.5$.

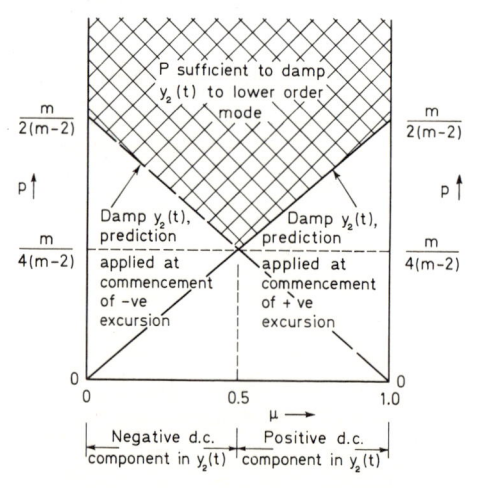

Fig. 2.16. *Rules for damping $y_2(t)$ to lower order modes by prediction in the idling condition*

Provided the above inequalities, 2.15 and 2.16 can be satisfied, an oscillatory pattern producing an output waveform $L(t)$ consisting of alternate sequences of m positive impulses and m negative impulses, $m > 2$, can be damped into a lower order mode. However, the idle channel mode is not necessarily the lowest mode as m is a function of both the prediction time and the initial conditions.

The preceding theory applies to the value of P required to damp the high order modes in $y_2(t)$. Prediction reduces even values of m by a

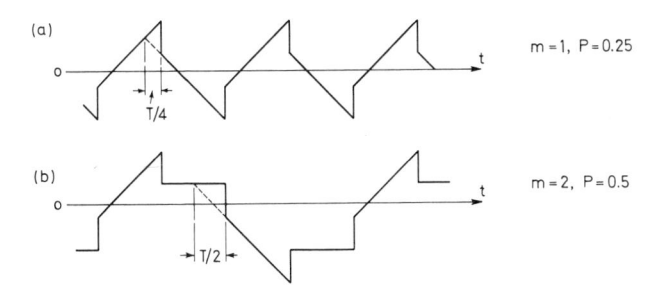

Double integration and exponential delta modulation

Fig. 2.17. $y_3(t)$ waveforms having low order modes in the presence of prediction: (a) $m = 1$, $P = 0.25$, (b) $m = 2$, $P = 0.5$

factor of 2, i.e. they remain even. Thus prediction cannot reduce an even mode below 2, i.e. ...1 1 0 0 1 1 0 0..., and similarly odd order modes are reduced by 2 and remain odd. The lowest order that they can be reduced to if $m = 1$ is the ...1 0 1 0 1 0... pattern associated with linear d.m. Figure 2.17 shows the low order modes $m = 1$ and $m = 2$ in the presence of prediction.

If P is selected to have a value greatly in excess of that required by the inequalities 2.15 and 2.16 then the high order modes will be damped into a $m = 2$ or $m = 1$ mode. Figure 2.18 shows the reduction of a fifth order mode into a first order mode. The $y_3(t)$ waveform in this figure has a large value of P, with high amplitude oscillations, but as the frequency is $f_p/2$

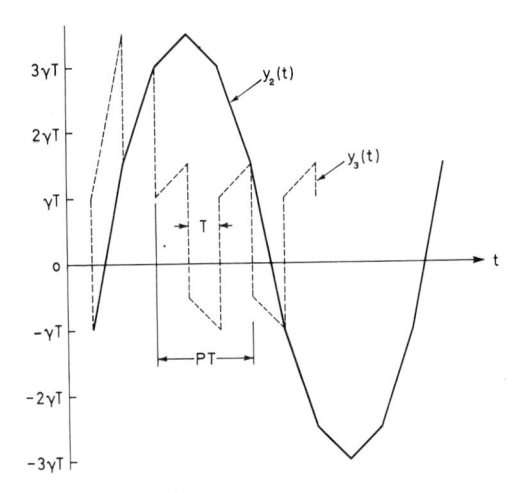

Fig. 2.18. The undamped $y_2(t)$ waveform where $m = 5$, $\mu = 0.6$ is damped using an excessive prediction constant $P = 3$ to give the $y_3(t)$ waveform

they will be rejected by the final filter in the decoder. When tracking an input signal the large steps in $y_3(t)$ due to the high value of P will cause a degradation of the s.q.n.r. at the output of the decoder. Consequently, P should be large enough to satisfy the inequality but not too large. A common choice[1] in practice is $P = 1$.

2.2.4. Shaping the overload characteristic

The overload characteristic of the double integration d.m. falls off at -12 dB/octave, as described by Equation 2.1. This implies that the encoder is only suitable for signals whose spectra are mainly composed of low frequency components. Ideally, the long term spectrum of the input signal should have the same shape as the overload characteristic of the encoder thereby minimising slope overload noise. The overload characteristic of the single integration d.m. falls off at 6 dB/octave and therefore is well suited to speech signals whose average spectrum decreases from about 800 Hz at approximately this rate.

In order to obtain an overload characteristic which falls off at 6 dB/octave over the message band it is arranged that the gain function of the second integrator with prediction has a Bode plot as shown in Figure 2.19. When the gain function of the first integrator is taken into consideration the resulting function of the complete network in the feedback loop, i.e. the local decoder, is as shown in Figure 2.20 and can be expressed as[3]

$$G(j2\pi f) = \frac{G(0)\{1+(jf/f_0)\}}{\{1+(jf/f_{c1})\}\{1+(jf/f_{c2})\}} \quad (2.17)$$

where $G(0)$ is the d.c. gain of the network. Replacing $j2\pi f$ by S in the above

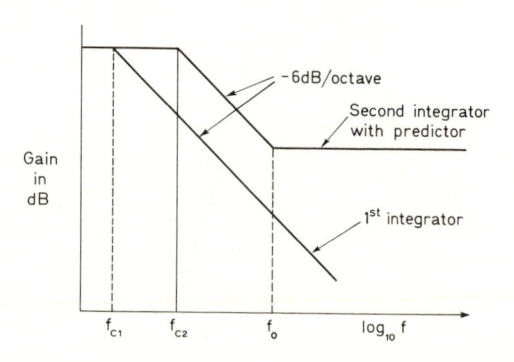

Fig. 2.19. Bode plots of first integrator and second integrator with prediction

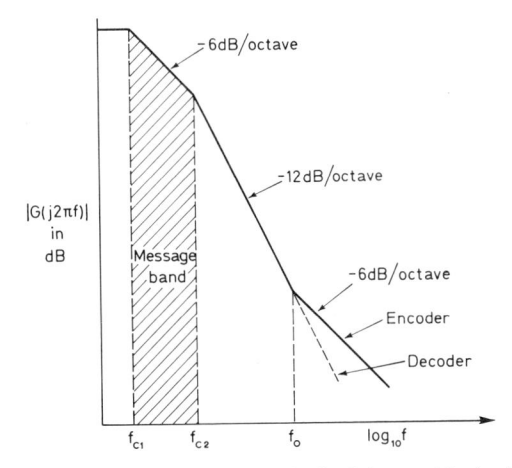

Double integration and exponential delta modulation 51

Fig. 2.20. *Bode plots of the network in the feedback loop and in the decoder*

equation the transfer function is

$$G(S) = \frac{G(0)2\pi f_{c1} f_{c2}}{f_0} \left\{ \frac{s + 2\pi f_0}{(s + 2\pi f_{c1})(s + 2\pi f_{c2})} \right\}$$

The partial fraction expansion of this equation is

$$G(S) = \frac{G(0)2\pi f_{c1} f_{c2}}{f_0} \left\{ \left(\frac{f_0 - f_{c1}}{f_{c2} - f_{c1}}\right) \frac{1}{(s + 2\pi f_{c1})} + \left(\frac{f_c - f_{c2}}{f_{c1} - f_{c2}}\right) \frac{1}{(s + 2\pi f_{c2})} \right\} \quad (2.18)$$

The impulse response $h(t)$ is the inverse Laplace transform of this equation

$$h(t) = \frac{G(0) f_{c1} f_{c2}}{f_0 (f_{c2} - f_{c1})} \left\{ 2\pi(f_0 - f_{c1}) \exp(-2\pi f_{c1} t) \right.$$
$$\left. - 2\pi(f_0 - f_{c2}) \exp(-2\pi f_{c2} t) \right\} \quad (2.19)$$

and $h(t) = 0$ for $t < 0$

$$t = 0, \quad h(0) = \frac{G(0)2\pi f_{c1} f_{c2}}{f_0} \quad (2.20)$$

Thus, when an impulse occurs at the output of the encoder $y_2(t)$ instantaneously increases by $V\tau h(0)$. In the encoder described in Section 2.2.3

$$y_3(0) = V\tau h(0) = \gamma PT$$

and

$$PT = 1/(2\pi f_0)$$

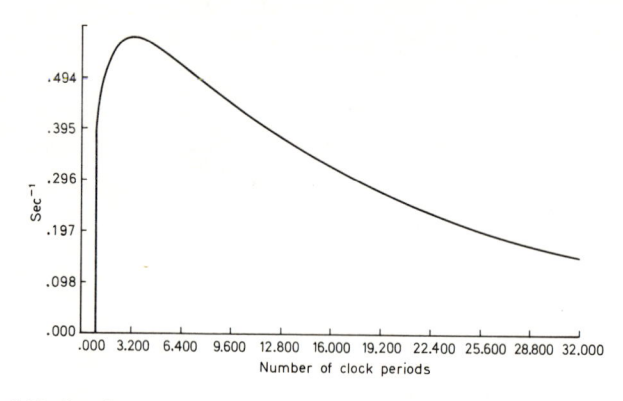

52 Double integration and exponential delta modulation

Fig. 2.21. Impulse response of the network in the feedback loop of the encoder

The local decoder has two RC integrators. The break frequencies of the integrators are f_{c1} and f_{c2}, i.e. there is no prediction introduced due to the Bode plot straightening out at the break frequency f_0 as in the case of the second integrator in the encoder. The gain function of the decoder is shown in Figure 2.20 and can be expressed as

$$D(j2\pi f) = \frac{G(0)}{\{1+(jf/f_{c1})\}\{1+(jf/f_{c2})\}} \quad (2.21)$$

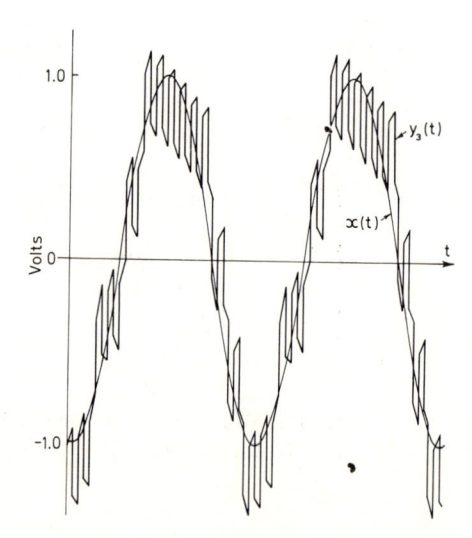

Fig. 2.22. Feedback waveform at encoder when tracking a sinusoidal input signal. Prediction time is $0 \cdot 643T$

Double integration and exponential delta modulation   53

Replacing $j2\pi f$ by S and resolving into partial fractions

$$D(S) = \frac{G(0)2\pi f_{c1} f_{c2}}{f_{c2} - f_{c1}} \left\{ \frac{1}{S + 2\pi f_{c1}} - \frac{1}{S + 2\pi f_{c2}} \right\} \qquad (2.22)$$

Taking the inverse Laplace transform of $D(S)$ gives the impulse response $d(t)$ of the decoder.

$$d(t) = \frac{G(0)2\pi f_{c1} f_{c2}}{f_{c2} - f_{c1}} \{\exp(-2\pi f_{c1} t) - \exp(-2\pi f_{c2} t)\} \qquad t \geq 0 \qquad (2.23)$$

When $t \leq 0$, $d(t) = 0$, i.e. when the decoder receives an impulse its output does not instantaneously increase.

The following waveforms are presented.[3] Figure 2.21 shows $h(t)$ for $P = 0.643$. The graph of $d(t)$ does not have an initial step like $h(t)$ but asymptotically approaches the latter with time. The $y_3(t)$ waveform at the

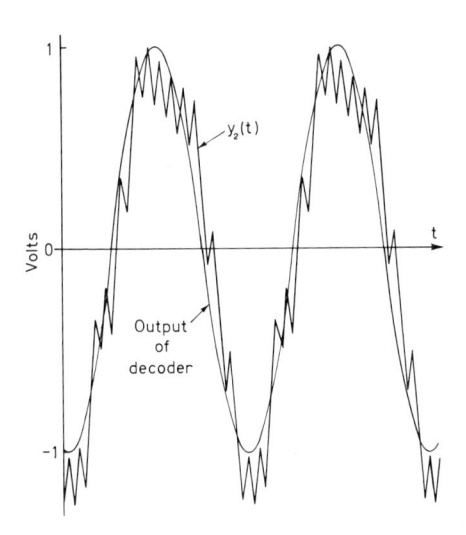

Fig. 2.23. Decoder waveforms $y_2(t)$ and $o(t)$ corresponding to Fig. 2.22

encoder for a prediction constant of 0·643 is shown in Figure 2.22 and the corresponding $y_2(t)$ waveform at the output of the second integrator in the decoder is shown in Figure 2.23 for a sinusoidal input signal. These $y_3(t)$ and $y_2(t)$ waveforms are produced for $f_{c1} = 200$ Hz, $f_{c2} = 3800$ Hz and a clock rate $f_p = 25$ kbits/s.

2.2.5. RC arrangement of local decoder

de Jager[1] arranged for the double integration and prediction system in the feedback loop to be implemented by an RC circuit as shown in Figure 2.24. From Equation 2.10

$$y_3(t) = y_2(t) + PT\frac{i}{C_2} \qquad (2.24)$$

where i is the current through R_2. From Figure 2.24

$$y_3(t) = y_2(t) + ir \qquad (2.25)$$

Equating Equations 2.24 and 2.25 gives the prediction time $PT = rC_2$.

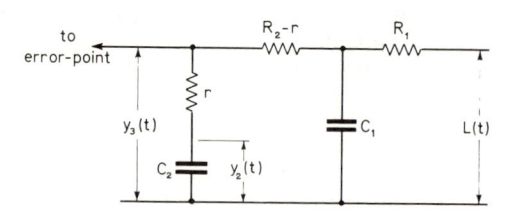

Fig. 2.24. *RC arrangement of double integration with prediction system used in the feedback loop*

In order to obtain the Bode plot shown in Figure 2.20 a series resonant circuit, whose resonant frequency is f_{c2}, is connected across C_2. The decoder can either have the circuit of Figure 2.24, thereby using a predictor at the decoder followed by a low-pass filter F_0 whose critical frequency is f_0, or a single integrator followed by a low-pass filter whose cut-off frequency is f_{c2}.

2.2.6. Signal-to-quantization noise ratio

The Bode plot, Figure 2.20, shows that the decoder behaves like a single integrator in the message band, i.e. the characteristic falls at 6 dB per octave. Consequently calculation of the quantization noise can be achieved by considering the decoder to be a single integrator followed by a low pass filter.[1]

Using similar reasoning to that in the establishment of Equation 1.7, i.e. that the distortion spectral density has a flat spectrum in the message band, the quantization noise power N_q^2 is consequently proportional to f_{c2}. If the clock rate is increased N_q^2 is reduced. Since N_q^2 is dependent on

Double integration and exponential delta modulation   55

the size of the steps $h(0)$ in the error waveform due to the predictor

$$N_q^2 \propto \frac{f_{c2}}{f_p} y_3^2(0)$$

$h(0)$ and f_0 are proportional to γ and j_p respectively, i.e., if the prediction time is altered the clock rate must be changed in order to maintain the same N_q^2. Since

$$y_3(0) \propto \frac{\gamma f_{c2}}{f_p}$$

$$N_q^2 = C_q \gamma^2 \left(\frac{f_{c2}}{f_p}\right)^3 \tag{2.26}$$

where C_q is the empirical constant of proportionality.

The maximum amplitude E_{sm} of a sinusoid of frequency f_s at the output of the solitary integrator in the decoder has a peak slope of $2\pi f_s E_{sm} = \gamma f_p$. The mean square value of this sinusoid is

$$S^2 = \frac{1}{8\pi^2} \cdot \frac{\gamma^2 f_p^2}{f_s^2} \tag{2.27}$$

From Equations 2.26 and 2.27 the maximum s.q.n.r. is

$$\widehat{\text{s.q.n.r.}} = \frac{1}{8\pi^2 C_q} \cdot \frac{f_p^5}{f_s^2 f_{c2}^3} \tag{2.28}$$

where C_q is a constant. The value of C_q is about 20 although this figure has a large tolerance, similar to the tolerance on the value of K_q used in Equation 1.7.

Although the constant C_q is approximately 60 times greater than K_q, the s.q.n.r. is higher for double integration d.m. than single integration d.m. because of the complex manner in which the s.q.n.r. and the frequency parameters are related. The double integration d.m. shows a significant improvement, 5 to 10 dB, over the single integration encoder (i.e. the linear d.m.) as the clock rate is increased above $12 f_{c2}$ when $f_s = 800$ Hz.

Figure 2.25 gives a visual appreciation of why double integration d.m. shown in Figure 2.8 has a superior s.q.n.r. to single integration d.m. The prediction time is one clock period and the total change in $y_3(t)$ is made identical to the change in the $y(t)$ feedback signal of the linear d.m. encoder shown in Figure 1.1. The binary patterns for linear and double integration d.m. systems are different when encoding the same $x(t)$. The decoded signal $y_2(t)$ in double integration d.m. has smaller low frequency components than $y(t)$ because the former deviates less from $x(t)$ than the

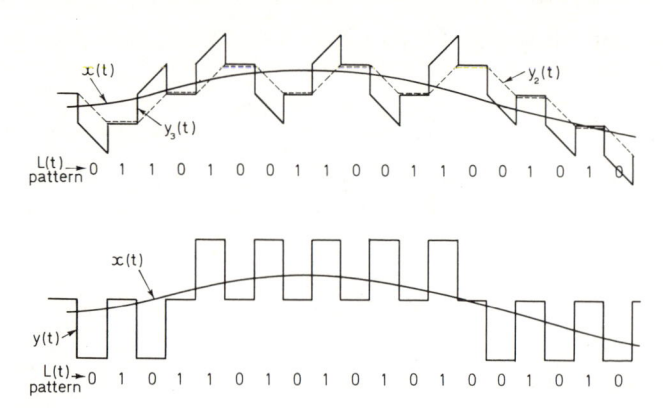

Fig. 2.25. Comparison between the encoding behaviour of linear d.m. and double integration d.m. systems for the same input signal

latter. In other words $y_2(t)$ is arranged about $x(t)$ in a more symmetrical way than $y(t)$.†

When speech is encoded it is necessary to modify[7] Equation 2.28. This is because the spectrum of speech falls off at a rate which is somewhat higher than 6 dB per octave, especially near f_{c2}. The Bode plot of Figure 2.20 needs to be modified such that the 12 dB per octave decrease in the plot starts at a frequency between f_{c1} and f_{c2}. In order to ensure stability f_0 is made equal to $f_p/(2\pi)$. The expression for peak s.q.n.r. then becomes

$$\widehat{\text{s.q.n.r.}} = \frac{1}{8\pi^2 C_q} \cdot \frac{f_p^5}{f_s^2 f_m^2 f_{c2}}; \qquad f < f_m \qquad (2.29)$$

Typical values are $f_m = 1800$ Hz and $f_{c2} = 3400$ Hz. This formula is satisfactory for $f_p/f_{c2} > 10$.

The expression for peak s.q.n.r. given by Equation 2.28 can be modified for the case of low f_p/f_{c2} ratios, i.e. < 10, by multiplying this equation by a correction factor $0.8 + 4(f_{c2}/f_p)$. This modification leads to a simple empirical formula for peak s.q.n.r. for the values of f_m and f_{c2} stated above and $f_s = 800$ Hz.

$$\widehat{\text{s.q.n.r.}} = f_p^{4.3}/6.30$$
$$= 43 \log_{10} f_p - 28, \text{ dB} \qquad (2.30)$$

where f_p is in kbits/s.

Experimental observations show that the d.m. encoder can transmit speech signals without noticeable overloading if the amplitude of these

† Repeating the comments made in the introduction, note that elsewhere in the Book the title 'linear d.m. codec' refers to encoders having the single integrator in the feedback loop. The double integration d.m. systems described in this section will be referred to as such throughout the Book.

Double integration and exponential delta modulation

signals does not exceed the maximum sinewave amplitude of a 800 Hz tone. It is for this reason that Equation 2.30 applies for an input having a frequency of 800 Hz.

2.3. EXPONENTIAL DELTA MODULATION SYSTEMS

In the exponential delta modulation system described by Johnson[4] perfect integrators (Figure 1.1) are replaced by an RC circuit. A sample and hold circuit in the form of a D-type flip-flop is introduced after the quantizer. This means that the output consists of binary levels rather than pulses. The encoder is shown in Figure 2.26. The input voltage to the RC circuit shown in this figure is the voltage at the output of the sample and hold circuit, i.e. $L(t)$, which has binary levels of $\pm V$. These levels may or may not change at sampling instants depending on whether there has been a change of the error signal at these instants. If a perfect integrator rather than an RC circuit is used in the encoder the $y(t)$ waveform for the perfect integrators shown in Figure 1.1 is composed of straight lines which have either positive or negative slopes of equal magnitude. The polarity of these slopes only change when the binary levels of $L(t)$ change. When an RC circuit is used the shape of $y(t)$ becomes exponential, hence the name for this type of encoder. When a perfect integrator is used in the feedback loop of the encoder a continuous sequence of ones at its output results in a constant rise in $y(t)$. Further, there is no limit to the magnitude of $y(t)$. However, the same continuous sequence of ones applied to the RC circuit causes $y(t)$ to rise exponentially

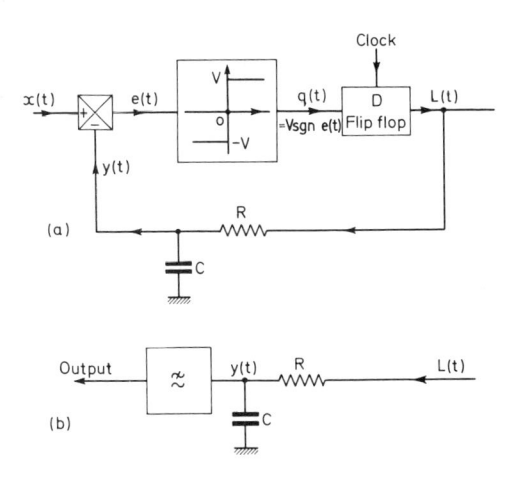

Fig. 2.26. Exponential delta modulation system: (a) encoder, (b) decoder

to the voltage representing logic one, i.e. to V. Thus when encoding, the increase or decrease in $y(t)$ following a change in the binary level of $L(t)$ depends on the actual value of $y(t)$.

Consider the value of $y(t)$ to be positive at a value close to $+V$ just prior to a sampling instant. If $L(t)$ becomes a logic one $y(t)$ will increase and if $L(t)$ becomes a logic zero it will decrease. If the changes in the values of $y(t)$ are compared one clock period later the negative change is larger than the positive change. This is because the capacitor C has a voltage difference which is small when $L(t)$ is a logic one but large when $L(t)$ is a logic zero; the change in $y(t)$ is a function of the difference between the voltage to which the capacitor is being charged and the actual voltage on the capacitor.

It is now apparent that by replacing a perfect integrator by an RC circuit in the feedback loop the values of $y(t)$ are more difficult to estimate.

Further the amplitude range and overload characteristics are modified. The RC circuit of Figure 2.26 has the transfer function

$$H(j2\pi f) = \frac{1}{1+j2\pi fCR}$$

Putting

$$f_1 = \frac{1}{2\pi RC} = \frac{1}{2\pi T_1} \qquad (2.31)$$

$$H(j2\pi f) = \frac{f_1}{f_1+jf}$$

and for input frequencies f, where $f \gg f_1$

$$H(j2\pi f) \doteqdot \frac{f_1}{jf} \qquad (2.32)$$

As the transfer function for a perfect integrator is $1/(j2\pi f)$ it follows that, except for a scaling factor, Equation 2.32 is the equation of a perfect integrator. Thus if f_1 is chosen to be much smaller than the lowest input frequency, f_{c1}, to be encoded then the RC circuit performs as an integrator and $y(t)$ is composed of straight line segments. However, it will be shown in Section 5.2 that exponential variations in the $y(t)$ waveform can be an advantage in the presence of transmission errors.

2.3.1. Idling characteristic of exponential d.m. codec

Consider the situation when the power is first switched on to the encoder and there is no input signal. The voltage at the output of the encoder will either go to $+V$ or $-V$. Suppose it initially goes to $-V$; the

Double integration and exponential delta modulation 59

capacitor C will start to charge from 0 to $-V$ and when subtracted from $x(t)$, which is zero, will produce a large positive error signal. This will cause $L(t)$ to switch to $+V$ and $y(t)$ to start to charge to this positive voltage. Provided $y(t)$ is positive by the next sampling time, thereby ensuring that the error is negative, $L(t)$ will switch to $-V$. However, if $y(t)$ is still negative at the next sampling instant $L(t)$ will maintain a voltage of $+V$. Eventually $y(t)$ will go positive at a sampling instant and $L(t)$ will have a negative level $-V$. After a settling down time $L(t)$ will acquire a ... 1 0 1 0 1 0 1 0... idling pattern. This will occur irrespective of the initial polarity of the $L(t)$ signal.

This idling pattern is the same as that described in Section 1.2.4. The RC circuit will have a time constant much greater than the sampling period thereby ensuring that it behaves similarly to a perfect integrator, i.e. maintaining $y(t)$ close to zero at all times when idling. Consequently there is negligible difference in the idling situation between the linear delta modulator of Section 1.2 and the exponential delta modulator.

However, the effect of the sample and hold circuit means that $L(t)$ is a square waveform and $y(t)$ is a small triangular waveform. These waveforms are shown in Figure 2.27.

The time constant T_1, is often tens or hundreds of sampling periods. T_1 has been shown for convenience in Figure 2.27 to be very short relative to the sampling period T.

T_1 is found by drawing a tangent at the origin of the $y(t)$ curve to intersect the aiming potential V; T_1 is the time between the origin and this intersection.

By applying the properties of similar triangles to Figure 2.27

$$\frac{D}{(T/2)} = \frac{V}{T_1}$$

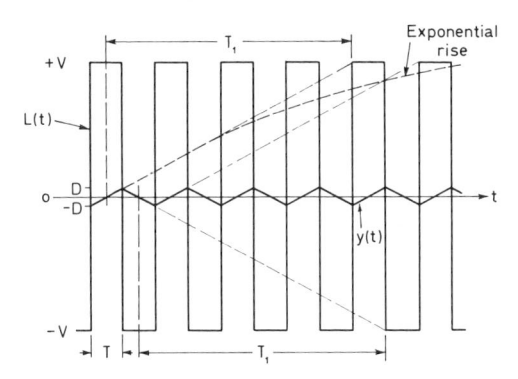

Fig. 2.27. Idling waveforms in exponential delta modulator

From Equation 2.31

$$D = \frac{\pi V f_1}{f_p} \qquad (2.33)$$

Effects of propagation

Practical quantizers require a minimum value of input amplitude in order to produce the voltages $\pm V$. They are also limited in frequency response, and conventional operational amplifiers cannot operate as quantizers at bit rates above say, 10 Mbit/s. At these high bit rates the comparator in a high speed D flip-flop can be used as the quantizer. This flip-flop must not have too high a propagation delay or the establishment of the required ... 1 0 1 0 1 0 ... idling pattern will be prohibited. This can be demonstrated as follows.

Suppose the minimum value of D which causes the flip-flop to function is $\pm d$, i.e. the error $> |d|$ at a clock instant for $L(t)$ to be $\pm V$. Let the propagation time of the flip-flop be τ_p, i.e. τ_p seconds after the clock has been applied the $L(t)$ level changes. The result is that the $y(t)$ waveform in Figure 2.27 has the same shape but is moved to the right τ_p seconds. The clock samples $q(t)$ when $e(t)$ is less than $|D|$, although $e(t)$ continues to increase in value. As the slope of $e(t)$ is $\pm 2Df_p$ it is easy to show by similar triangles that when $e(t)$ is sampled its magnitude is $D(1 - 2f_p\tau_p)$. This idling waveform $e(t)$ is maintained until the clock is increased above $(D-d)/(2\tau_p D)$ when the output binary pattern becomes ... 0 0 1 1 0 0 1 1 ... and $e(t)$ has peak-to-peak values of $\pm 2D$. In order to restore the ... 1 0 1 0 1 0 1 ... pattern a flip-flop having a low τ_p should be selected and f_1 made as high as the restrictions on amplitude range permit.

2.3.2. Overload characteristic

The overload condition is discussed for the case of the perfect integrator in Section 1.2.5. It is shown there that if the slope of the input signal becomes greater than the maximum rate of increase in the feedback signal $y(t)$ the encoder is overloaded. In the case of the exponential delta modulator the maximum rate of increase in $y(t)$ is exponential.

Suppose the input signal $x(t)$ is a sinewave $E_s \sin 2\pi f_s t$. The question to be answered is what value of E_s will cause the encoder to be overloaded? Now the slope of $x(t)$ with respect to time is

$$x'(t) = E_s 2\pi f_s \cos 2\pi f_s t = 2\pi f_s [E_s^2 - E_s^2 \sin^2 2\pi f_s t]^{1/2}$$
$$= 2\pi f_s [E_s^2 - x^2(t)]$$

Double integration and exponential delta modulation

Assuming that the encoder is tracking correctly the instantaneous value of $y(t)$, i.e., the voltage across the capacitor C, closely approximates to $x(t)$. When $x(t)$ is increasing monotonically with time $y(t)$ changes exponentially and its slope is the voltage difference between the aiming potential V and the voltage $x(t)$ on the capacitor divided by the RC time constant T_1, i.e.

$$\{V - x(t)\}/T_1$$

The difference in slopes between the signal $y(t)$ and $x(t)$ is

$$\Delta = \frac{V - x(t)}{T_1} - 2\pi f_s [E_s^2 - x^2(t)]^{1/2} \qquad (2.34)$$

When $x(t)$ is decreasing monotonically the slope of $y(t)$ is

$$-(V - x(t))/T_1$$

and the slope of $x(t)$ is negative, i.e. Equation 2.34 applies. Equation 2.34 is therefore valid for all $x(t)$.

Δ is seen to be a function of $x(t)$. Providing the slope of $y(t)$ remains greater than the slope of $x(t)$ the overload condition is avoided. To find the value of $x(t)$ which just causes slope overload we find the value of $x(t)$ for which Δ is a minimum, i.e. the value of the instantaneous input for which the two slopes are nearly equal, and equate Δ to zero.

Differentiating Equation 2.34 with respect to $x(t)$

$$\frac{d\Delta}{dx(t)} = -\frac{1}{T_1} + 2\pi f_s x(t) \{E_s^2 - x^2(t)\}^{-1/2} \qquad (2.35)$$

Putting Equation 2.35 equal to zero gives

$$x(t) = \frac{E_s}{\{1 + (2\pi f_s)^2 T_1^2\}^{1/2}} \qquad (2.36)$$

If $d^2\Delta/dx^2(t) > 0$ for $x(t)$ defined by Equation 2.36 then Δ is a minimum.

$$\frac{d^2\Delta}{dx^2(t)} = 2\pi f_s [E_s^2 - x^2(t)]^{-1/2} [1 + x^2(t)\{E_s^2 - x^2(t)\}^{-1}] \qquad (2.37)$$

Substituting the value of $x(t)$ from Equation 2.36 into Equation 2.37 gives

$$\frac{d^2\Delta}{dx^2(t)} = \frac{\{1 + (2\pi f_s T_1)^2\}^{3/2}}{E_s (2\pi f_s)^2 T_1^3} > 0 \qquad (2.38)$$

As Equation 2.38 is always positive, Equation 2.36 gives the value of $x(t)$ for Δ to be a minimum. Hence substituting $x(t)$ from Equation 2.36 in Equation 2.34 and rearranging gives

$$\Delta_{min} = \frac{V}{T_1} - \frac{E_s}{T_1} \{1 + (2\pi f_s T_1)^2\}^{1/2} \qquad (2.39)$$

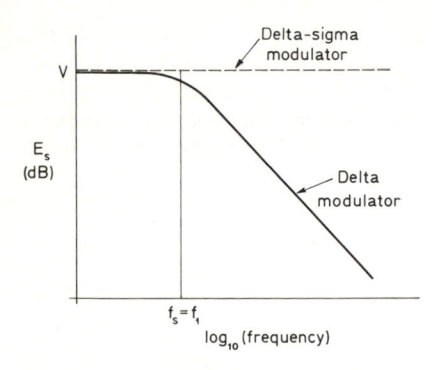

Double integration and exponential delta modulation

Fig. 2.28. Overload characteristics

If $\Delta_{min} = 0$ then for at least one value of $x(t)$ the slopes are equal and overload is just prohibited. Putting $\Delta_{min} = 0$, whence $E_s = E_{sm}$

$$E_{sm} = \frac{V}{\{1+(2\pi f_s T_1)^2\}^{1/2}} \quad (2.40)$$

From Equation 2.32

$$E_{sm} = V|H(j2\pi f)| \quad (2.41)$$

The overload characteristic is the graph of Equation 2.40 and is shown in Figure 2.28. From Equation 2.41 we see that this characteristic has the same shape as the transfer characteristic of the RC network and falls 6 dB per octave with frequency.

The overload characteristic gives the values of amplitude E_{sm} for each frequency to avoid the overload condition. Overload is avoided if values of E_s for a given frequency f_s are *below* the characteristic: it can be seen that for $f_s \ll f_1$, E_s can be as large as V before overload occurs. Also for these low frequencies overload is independent of frequency. At higher frequencies, $f \gg f_1$, the RC circuit behaves as a perfect integrator and the maximum value of E_s, i.e. E_{sm}, decreases with frequency. This also occurs with the encoder of Figure 1.1, as can be seen from Equation 1.5.

The value of E_{sm} has been found to be independent of clock rate (see Equation 2.40) but for the delta modulator of Figure 1.1 the value of E_{sm} can always be increased by increasing the clock rate f_p.

2.3.3. Amplitude range

Using the same definition as established in Section 1.2.6 for amplitude range AR

$$\text{AR} = \frac{E_{sm}}{D} \quad (2.42)$$

Double integration and exponential delta modulation 63

where E_{sm} and D are defined by Equations 2.40 and 2.33 respectively. Using Equation 2.31

$$AR = \frac{f_p}{\pi(f_1^2 + f_s^2)^{1/2}} \qquad (2.43)$$

In its slope-limiting mode, i.e. where $f_s \gg f_1$ the amplitude range of Equation 2.43 is the same as in Equation 1.6.

2.3.4. Quantization noise

In formulating Equation 1.7 it was stated that quantization noise is proportional to the square of the step height. With a perfect integrator this step height is a constant irrespective of the value of the analogue input. The changes in the voltage $y(t)$ at the output of the capacitor in the exponential delta modulator of Figure 2.26 depends on its magnitude. When idling $y(t)$ changes its magnitude by $2D$ every clock period, as can be seen from Figure 2.27. When $y(t)$ is close to its maximum value of $+V$ it increases by an amount $<2D$ when a logic one occurs at the encoder output whereas if a logic zero occurs the change is $>2D$ in the negative direction. This is because changes in $y(t)$ depend on $\pm V - y_0$, where $y_0 = y(t)$ when the logic change occurred. The converse argument applies if $y(t)$ is near to $-V$. However, if the input signal $x(t)$ has no d.c. component then over a long time interval $y(t)$ varies many times about zero and the average value of the changes in $y(t)$ during each sampling interval is approximately $2D$. Thus when calculating quantization noise an average value $2D$ should be used for the changes in $y(t)$ at sampling instants instead of a constant step height γ. This is reasonable as quantization noise only has meaning over a long time interval.

Applying the same assumptions used in establishing Equation 1.7

$$N_q^2 = K_q \cdot \frac{f_c(2D)^2}{f_p}$$

When D is inserted into the above equation

$$N_q^2 = K_q(f_c/f_p^3)V^2(2\pi f_1)^2$$
$$= \frac{K_q}{(RC)^2} \cdot \frac{f_c V^2}{f_p^3} \qquad (2.44)$$

K_q for exponential delta modulation is approximately the same as for the linear delta modulation described in Section 1.2.7. This similarity is expected from the discussion in Section 1.7 which showed that the zero-order hold circuit has negligible effect on the quantization noise.

2.3.5. Signal-to-quantization noise ratio

For a sinusoidal input which produces an approximately flat spectrum and has a mean square power of $E_s^2/2$ the signal-to-quantization noise ratio is found with the aid of Equation 2.44.

$$\text{s.q.n.r.} = \frac{E_s^2 \cdot (RC)^2}{2} \cdot \frac{f_p^3}{K_q} \cdot \frac{f_p^3}{f_c V^2} \tag{2.45}$$

This equation only applies when E_s is below E_{sm}.

The maximum s.q.n.r. occurs when $E_s \simeq E_{sm}$: the noise power is substantially uncorrelated with the input and does not increase with increasing E_s until E_s is large enough to overload the encoder.

For a sinusoid having an amplitude given by Equation 2.40

$$\widehat{\text{s.q.n.r.}} = \frac{f_p^3}{8\pi^2 K_q (f_1^2 + f_s^2) f_c}; \quad f_s \geq f_1 \tag{2.46}$$

When $f_s \gg f_1$

$$\widehat{\text{s.q.n.r.}} = \frac{1}{8\pi^2 K_q} \cdot \frac{f_p^3}{f_s^2 f_c} \tag{2.47}$$

For $K_q \doteq \frac{1}{3}$, $1/(8\pi^2 K_q) \doteq 0.04$.

This result was found originally by de Jager[1] and has been verified by Johnson[4] using computer simulation, and experimentally by the author[6] for sinusoidal inputs. It can be shown from Equation 2.47 that the maximum signal-to-quantization noise ratio improves at the rate of

(1) 9 dB/octave with increasing clock rate.
(2) 3 dB/octave with decreasing filter pass-band.
(3) 6 dB/octave with decreasing signal frequency.

2.4. EXPONENTIAL DELTA-SIGMA MODULATION

The exponential version of the delta-sigma modulation system shown in Figure 1.10 has an RC type integrator instead of a perfect integrator and a zero order hold circuit following the sampler.[4]

The overload characteristic is constant with frequency but the quantization noise is modified by the effect of the leaky integrator and the hold circuit. Using identical arguments to those given in Section 2.3.4 and 1.4, from Equation 2.44 the quantization noise is

$$N_q^2 = \frac{K_q V^2}{(RC)^2 f_p^3} \int_0^{f_{c2}} \frac{df}{|H(j2\pi f)|^2}$$

where $H(j2\pi f)$ is the transfer function of the RC circuit. It is defined by

Double integration and exponential delta modulation 65

Equation 2.32 for $f \gg f_1$

$$N_q^2 = \frac{4\pi^2}{3} K_q V^2 \left(\frac{f_{c2}}{f_p}\right)^3 \qquad (2.48)$$

From Equation 2.41 the maximum signal-to-quantization noise ratio is found by noting that the maximum input signal which does not overload the encoder

$$E_{sm} \cdot \frac{1}{|H(j2\pi f)|}$$

E_{sm} in Equation 2.41 applies to a delta modulator, and an exponential delta-sigma modulator is an exponential delta modulator preceded by an RC circuit. The result is that for an exponential delta-sigma modulator the overload characteristic

$$E_{sm} = V \qquad (2.49)$$

The characteristic is independent of frequency and is illustrated in Figure 2.28. The signal-to-quantization noise ratio is

$$\text{s.q.n.r.} = \frac{3}{8\pi^2 K_q} \left(\frac{E_s}{V}\right)^2 \left(\frac{f_p}{f_{c2}}\right)^3$$

with maximum value

$$\widehat{\text{s.q.n.r.}} = \frac{3}{8\pi^2 K_q} \left(\frac{f_p}{f_{c2}}\right)^3 \qquad (2.50)$$

Amplitude range of exponential delta-sigma modulator

The input signal amplitude E_{st} which just disturbs the idling pattern is the product of D, the value which disturbs the idling pattern for a delta modulator, (Equation 2.33) and the reciprocal of the modulus of the transfer function of the RC network given by Equation 2.32.

$$E_{st} = \pi V \cdot \frac{f_s}{f_p}$$

and from the above equation and Equation 2.49

$$\text{AR} = \frac{E_{sm}}{E_{st}} = \frac{f_p}{\pi f_s} \qquad (2.51)$$

The amplitude range AR is the same for delta modulators, delta-sigma modulators and their exponential versions provided $f_s \gg f_1$.

2.5. PULSE DELTA MODULATION

Pulse delta modulators have a slightly inferior performance to the exponential delta modulator but they can be simply constructed.[5] For example, two pulse delta modulators can be produced with one dual J.K. flip-flop integrated circuit package, six resistors and two capacitors.[6] However, such circuits made with a minimum of electronic components tend to have unstable idling characteristics but their value is that they can be built for little cost and can be conveniently used for laboratory experiments.

In pulse delta modulators the binary output comes from a monostable device, as illustrated in Figure 2.29. An RC circuit is shown rather than a perfect integrator. The subtractor is replaced by an adder–it should be emphasised that most practical encoders use adders rather than subtractors because they are easier to construct. The adder is directly equivalent to the subtractor of Figure 2.26, if instead of feeding the output $M(t)$ of the encoder into the RC circuit its inverse $\overline{M(t)}$ is used. This means that the voltage $y(t)$ across the capacitor is very nearly equal in magnitude and opposite in polarity to the input $x(t)$ and the error signal $e(t)$ is very small.

The monostable device has one stable state when $M(t)$ is a logic 1. This state is changed for a period τ after the device is triggered and then returns to its stable state. The period τ is a property of the monostable and is less than the sampling period T. The logic levels will be assumed to be positive, say $+E$, and zero volts as is the case with integrated circuit elements. The consequence of choosing these levels is that the voltage $y(t)$ across the capacitor is never negative and therefore the quantizer must be changed from a zero-crossing detector to the type shown in Figure 2.29

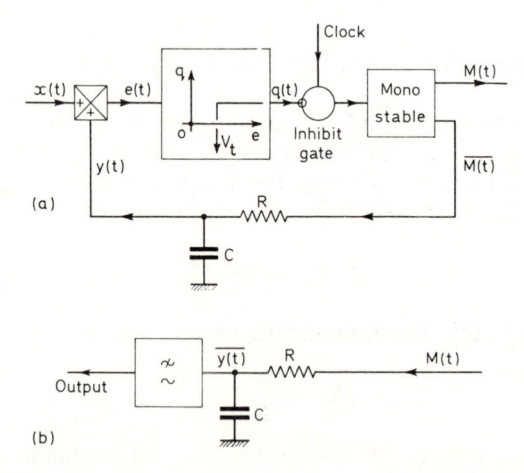

Fig. 2.29. Pulse delta modulation system; (a) encoder, (b) decoder

Double integration and exponential delta modulation 67

which has a positive threshold voltage V_t. When $e(t) \geq V_t$, $q(t)$ is a logic 1, and when $e(t) < V_t$, $q(t)$ is a logic 0. The value of V_t must satisfy the conditions of symmetrical encoding such that the encoder has an idling pattern $M(t)$ consisting of alternate logical ones and zeros. The average voltage across the capacitor in the idling state due to this $M(t)$ waveform is derived in the next section and is found to be $E\tau/2T$. As this voltage is the average value of $y(t)$, and $y(t) = e(t)$ in the idling condition, it follows that if

$$V_t = E\tau/2T$$

the instantaneous variations of $y(t)$ will alternate about V_t and ensure the correct idle channel pattern.

The encoder works as follows. Suppose there is a sudden increase in $x(t)$ which causes $e(t)$ to increase and the quantizer output to be in its logic 1 state for several clock periods. The clock pulse would then repeatedly be inhibited from triggering the monostable by the voltage at the output of the quantizer. During this time $M(t)$ remains at zero volts and capacitor C discharges via R and the output of the monostable. The decreasing $y(t)$ voltage will eventually cause $e(t)$ (equal to $x(t)+y(t)$) to fall below V_t and the output of the quantizer will go to its negative logic level. The next clock pulse will trigger the monostable producing a pulse of magnitude E and duration τ which charges C. While $e(t) < V_t$ the monostable will be triggered and C will charge. When $y(t)$ is sufficiently large, i.e. $e(t) \geq V_t$, the cycle is repeated.

Thus when tracking, pulses are either produced at the encoder output or inhibited such that $e(t)$ is always close to V_t and oscillates about this voltage.

The decoder is formed by passing $M(t)$ through an RC circuit, identical to that used in the encoder to give $y(t)$, and then filtering $y(t)$ to remove some of the effects of quantization to give $0(t)$, a close representation of $x(t)$.

2.5.1. Idling state

For the same reasons as given in Section 1.2.4, the pulse delta modulator idles with a waveform at the output of the encoder having ...1 0 1 0 1 0... pattern.

Figure 2.30 shows the waveforms $\overline{M(t)}$, $\overline{y(t)}$ and $y(t)$ in the idling state. Notice that the idling waveforms in the encoder and decoder are different.

Ignoring the output impedance of the monostable, the $y(t)$ waveform is calculated from the simple circuit shown in Figure 2.31 as follows. When the switch is in position A

$$E = i(t)R + (q(t)/C)$$

where $q(t)$ is the charge on the capacitor and $i(t)$ is the current flowing in

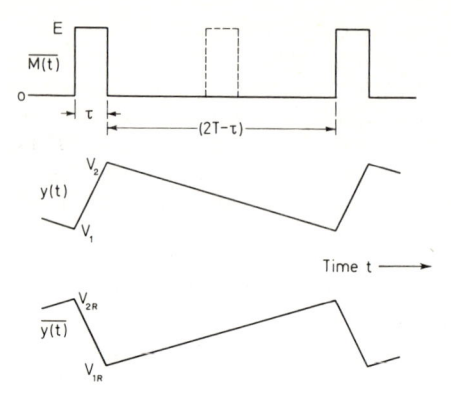

68   Double integration and exponential delta modulation

Fig. 2.30. Idling waveforms in pulse d.m. codec

the circuit. Since

$$i(t) = \frac{dq(t)}{dt}$$

$$\frac{dq(t)}{dt} + \frac{q(t)}{RC} = \frac{E}{R}$$

Taking the Laplace transform of the terms in this equation gives

$$sQ(s) - q(0) + \frac{Q(s)}{RC} = \frac{E}{Rs}$$

where s is the Laplacian variable, $Q(s)$ is the Laplace transform of $q(t)$ and $q(0)$ is the value of $q(t)$ at $t = 0$. Hence

$$Q(s) = \frac{E}{Rs[s + (1/RC)]} + \frac{q(0)}{s + (1/RC)}$$

Taking the inverse Laplace transform

$$q(t) = EC\left(1 - \exp\left(-\frac{t}{RC}\right)\right) + q(0)\exp\left(-\frac{t}{RC}\right)$$

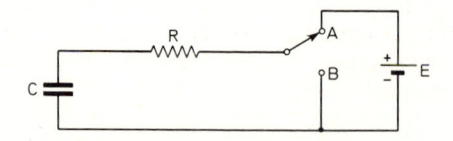

Fig. 2.31. Circuit used to calculate $y(t)$

Double integration and exponential delta modulation 69

Since
$$y(t) = q(t)/C$$

$$y(t) = E\left[1 - \exp\left(-\frac{t}{RC}\right)\right] + \frac{q(0)}{C}\exp\left(-\frac{t}{RC}\right) \quad (2.52)$$

In the steady state when the idling pattern is established and $y(t)$ has the periodic pattern shown in Figure 2.30

$$\frac{q(0)}{C} = V_1$$

If after τ seconds $y(t)$ reaches a value V_2 say then

$$V_2 = E\left[1 - \exp\left(-\frac{\tau}{RC}\right)\right] + V_1 \exp\left(-\frac{\tau}{RC}\right) \quad (2.53)$$

Directly $y(t)$ reaches V_2 the output of the monostable returns to its stable state and $\overline{M(t)}$ is at zero volts. In the model the switch changes to position B.

The capacitor C discharges according to the equation

$$i(t)R + \frac{q(t)}{C} = 0$$

The Laplace transform of this equation is

$$(sQ(s) - q(0)) + \frac{Q(s)}{RC} = 0$$

or

$$Q(s) = \frac{q(0)}{s + (1/RC)}$$

where $q(0)$ is the charge on C at the beginning of the discharge process. Taking the inverse Laplace transform of $Q(s)$ gives

$$q(t) = q(0) \exp\left(-\frac{t}{RC}\right)$$

where t is measured from the instant when the discharge process commenced.

$$y(t) = \frac{q(t)}{C} = \frac{q(0)}{C} \exp\left(-\frac{t}{RC}\right)$$

$$= V_2 \exp\left(-\frac{t}{RC}\right)$$

After time $(2T-\tau)$, $y(t) = V_1$ again, as can be seen from Figure 2.30. Hence

$$V_1 = V_2 \exp\{-(2T-\tau)/RC\} \qquad (2.54)$$

Substituting for V_1 from Equation 2.54 into Equation 2.53 gives

$$V_2 = \frac{E[1 - \exp(-\tau/RC)]}{1 - \exp(-2T/RC)} \qquad (2.55)$$

Substituting this value of V_2 into Equation 2.54

$$V_1 = \frac{E[1 - \exp(-\tau/RC)]}{1 - \exp(-2T/RC)} \exp\left\{-\frac{(2T-\tau)}{RC}\right\} \qquad (2.56)$$

Generally, $T_1 = RC \gg T$ or τ and hence the exponentials in Equations 2.55 and 2.56 can be replaced by the first two terms of their series expansions to give

$$V_2 \simeq \frac{E\tau}{2T} \qquad (2.57)$$

$$V_1 \simeq \frac{E\tau}{2T}\left(1 - \frac{2T-\tau}{RC}\right) \qquad (2.58)$$

When there is no input signal $x(t)$ the error voltage $e(t) = y(t)$ since $e(t) = x(t) + y(t)$. The threshold signal V_t for the quantizer is therefore selected to be the average value of $y(t)$. By doing this the waveforms of Figure 2.30 are established.

From Equations 2.57 and 2.58

$$V_t = \frac{V_1 + V_2}{2} = \frac{E\tau}{2T}\left(1 - \frac{2T-\tau}{2RC}\right) \qquad (2.59)$$

The peak-to-peak value of the idling error waveform is

$$V_2 - V_1 = \frac{E\tau(2T-\tau)}{2T\,RC} \qquad (2.60)$$

The idling waveform at the decoder

As the average value of $M(t)$ (the signal which is transmitted to the decoder) is greater than $M(t)$ the average value of $y(t)$ is greater than the average value of $y(t)$ and the voltages V_{1R} and V_{2R} in Figure 2.30 are greater than the voltages V_1 and V_2.

Double integration and exponential delta modulation

Proceeding as in Section 2.2.1, the peak values of the idling waveform across the capacitor in the decoder are

$$V_{2R} = \frac{E\{1 - \exp[-(2T-\tau)/RC]\}}{1 - \exp(-2T/RC)}$$

$$\simeq E\left(\frac{2T-\tau}{2T}\right) \quad (2.61)$$

and

$$V_{1R} = \frac{E\{1 - \exp[-(2T-\tau)/RC]\}}{1 - \exp(-2T/RC)} \exp\left(-\frac{\tau}{RC}\right)$$

$$\simeq E\left(\frac{2T-\tau}{2T}\right)\left(1 - \frac{\tau}{RC}\right) \quad (2.62)$$

Thus

$$V_{2R} - V_{1R} = \frac{E\tau}{2TRC}(2T-\tau) \quad (2.63)$$

and as expected the peak-to-peak values of the idling waveforms across the capacitors in the encoder and decoder are equal, as shown by Equations 2.60 and 2.63.

Case when $\tau = T$

In this case the pulse delta modulator becomes an exponential one whose binary signal has levels of $+E$ and O, rather than $\pm V$. This is an arrangement which is frequently used in practical exponential delta modulators which feedback $\overline{L(t)}$ rather than $L(t)$, and which have an adder rather than a subtractor to produce the error signal. The equations in Section 2.5.1 are therefore useful for many exponential d.m. systems.

A pulse d.m. system cannot of course be produced with $\tau = T$ because a monostable requires a recovery period, i.e. $\tau < T$.

2.5.2. Overload characteristic

When $x(t)$ has a negative slope which overloads the encoder the error signal remains negative and the monostable is repeatedly triggered. The result is that the capacitor C gets 'pumped-up' by the monostable and $y(t)$ rises exponentially approaching asymptotically to the voltage level $E\tau/T$ with a time constant RC. This condition is shown by Curve (a) in Figure 2.32.

Double integration and exponential delta modulation

Curve (c) in Figure 2.32 shows how the pulse delta modulator behaves when overloaded by a positive going input signal. The error $e(t)$ is positive and the monostable is inhibited for a number of clock periods by the logic 1 level at the output of the quantizer. As $M(t)$ is zero the capacitor C discharges towards earth via the monostable circuit with a time constant RC.

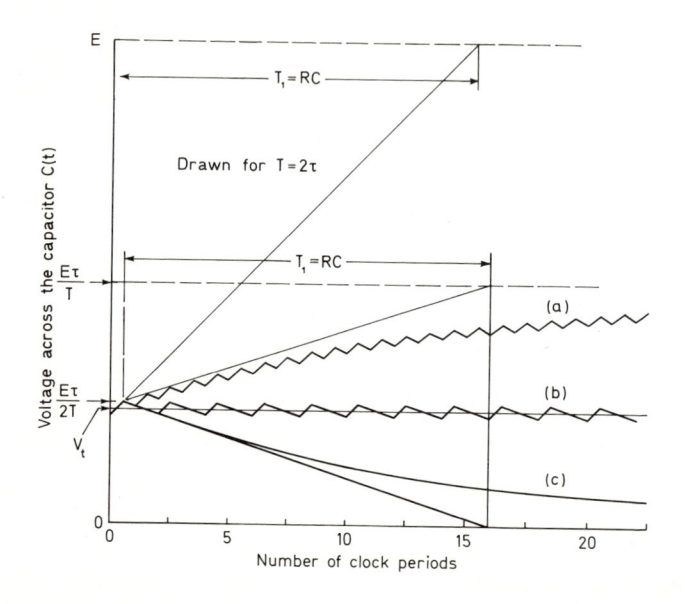

Fig. 2.32. Pulse d.m. waveforms, (a) and (c) encoder overloaded, (b) encoder idling

Curve (b) is the idling waveform (Fig. 2.32). It can be seen that the average of Curve (a) is symmetrical with Curve (c) about the idling waveform, i.e. the overload properties are symmetrical.

If $x(t)$ is a very low frequency sine wave it can be seen from Fig. 2.3.2 that the maximum amplitude which can be encoded is $E\tau/2T$. This is because $E\tau/2T$ is the difference between the idling value of $y(t)$ and its maximum value $E\tau/T$ or its minimum value of zero.

For this encoder, applying Equation 2.41 gives

$$E_{sm} = (E\tau/2T)H(j2\pi f)$$

$$= \frac{(E\tau/2T)}{\{1+(f_s/f_1)^2\}^{1/2}} \quad (2.64)$$

Double integration and exponential delta modulation

2.5.3. Signal-to-quantization noise ratio

Employing the arguments used in Section 2.3.4 and replacing $2D$ by $V_2 - V_1$ (Equation 2.60) the quantization noise is given by the approximate formula

$$N_q^2 = K_q \pi^2 E^2 k^2 (2-k)^2 \cdot (f_1^2 f_c / f_p^3) \qquad (2.65)$$

where $k = (\tau/T) < 1$.
By proceeding as in Section 2.3.5, the maximum s.q.n.r. is found with the aid of Equation 2.64

$$\widehat{s.q.n.r.} = f_p^3 / \{8\pi^2 K_q f_c f_s^2 (2-k)^2\}, \qquad f_s \gg f_1 \qquad (2.66)$$

The difference between $k = 1$ and $k = 0$ is 6 dB.

2.5.4. Practical pulse d.m. codec

These codecs are easily implemented, as shown in Figure 2.33. The only active element is seen to be a monostable which can be a single integrated circuit or simply formed using two transistors with resistors, etc.[5]

The interesting feature of the pulse d.m. system shown in Figure 2.33 is that the circuit appears to be simpler than the block schematic arrangement of Figure 2.29. This is because the capacitor C in Figure 2.29 is used for a number of operations.

The input signal $x(t)$ is passed through the d.c. blocking condenser C_1 which has negligible reactance for all frequencies in $x(t)$. The $M(t)$

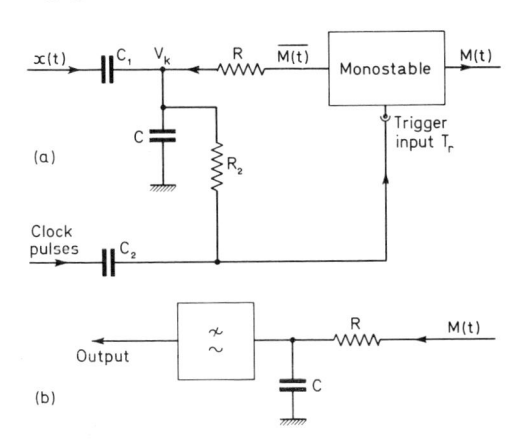

Fig. 2.33. A practical pulse d.m. codec: (a) encoder, (b) decoder
(Courtesy of Electronic Engineering[6])

waveform consisting of either a pulse or no pulse during each clock period is applied to the RC circuit, as described in Section 2.5. As both $x(t)$ and $M(t)$ are applied to capacitor C, it follows that capacitor C in Figure 2.33 is also part of the adder in Figure 2.29.

In order for the monostable to be driven from its stable state to its quasi-stable state and thereby produce a pulse in the $M(t)$ waveform it is necessary for a negative spike of sufficient amplitude to be applied to the trigger input of the monostable, assuming the monostable uses npn transistors. This negative trigger pulse is derived by passing a pulse through the differentiating circuit $C_2 R_2$ in Figure 2.33. As resistor R_2 does not have one end connected to earth but to positive voltage V_k the negative spike applied to T_r has its potential raised by V_k. The circuit is arranged such that if V_k is greater than a positive threshold V_t the negative spike at T_r will be insufficient to trigger the monostable. However, if $V_k < V_t$ the negative spike at T_r will be large enough to trigger the monostable and $\overline{M(t)}$ will produce a positive pulse.

The differentiating resistor R_2 is at a potential V_k which varies about V_t during the encoding process. If $V_k > V_t$ at the instant a clock pulse arrives the monostable will not be triggered. However, the clock pulse will effect V_k by reducing it when the clock pulse is decreasing and increasing it when the clock pulse is increasing. The effect of the clock pulse on the voltage V_k is ephemeral providing the time constant $C_2 R_2 \ll T$. Thus if the monostable is inhibited for several clock periods, when $V_k > V_t$, the value of V_k at each clock period is the value expected from the discharge of C via the monostable output and the charging of C due to the input signal. The effects of differentiating last for less than a clock period and consequently do not effect the decision process, i.e. whether or not V_k is greater or less than V_t.

The quantizer threshold level is now seen to be determined by the negative charge required to switch the monostable. The error signal is the difference between V_k and V_t, and V_t is the voltage which biases the negative differentiated spike until it will just trigger the monostable.

When $x(t)$ is zero the amplitude of the clock pulses is adjusted until the $M(t)$ pattern is 1 0 1 0 1 0 . . . . In this idling condition when the monostable is triggered C is charged up for τ seconds. It then starts to discharge, but by the time the next clock pulse arrives V_k is still greater than V_t and the monostable is not triggered. C continues to discharge and by the time the next clock pulse arrives $V_k < V_t$ and this time the monostable is triggered. The waveform in the idling state is shown in Figure 2.30.

The dependence on the amplitude of the clock pulses is not as critical as might first appear. If the amplitude of the clock pulses increases by a small amount the encoder will produce a short burst of 'ones' which raises V_k, enabling the desired idling pattern to be produced. Similarly, a small decrease in amplitude of the clock pulses results in a decrease in V_k, but

Double integration and exponential delta modulation 75

in spite of this effect the ... 1 0 1 0 1 0 ... idling pattern is maintained. However, the idling pattern will not be maintained if the change in the amplitude is too great. Similarly slight changes in the time constant and monostable pulse width are compensated for by the encoder.

If $x(t)$ is a positively increasing signal which overloads the encoder, $V_k > V_t$ and C discharges to earth for a number of clock periods until $V_k < V_t$. When overloaded by a negatively increasing input signal, $V_k < V_t$ and $M(t)$ consists of a train of pulses which raises V_k until $V_k > V_t$.

When the encoder is performing correctly the only distortion in the decoded output is that due to quantization noise, and the voltage V_k oscillates with small variations about the threshold signal V_t.

Setting-up procedure

This is achieved by reducing the input signal to zero volts and observing the binary pattern in the output signal $M(t)$. The pattern required is ... 1 0 1 0 1 0... and is attained by adjusting the amplitude of the clock pulse, the RC time constant, or the monostable period, whichever is the more convenient.

When the input signal is a sinusoid its amplitude is increased until the overload condition is reached. The distortion in the decoded signal should be symmetrical. It is not necessary to look at the decoded signal to observe overload effects. If V_k is observed it is found to consist of the initial portions of rising and falling exponentials which are close to V_t when the encoder is tracking. When overload begins to occur the positive and negative peaks of the sine wave occur relative to V_t in the V_k waveform, where the positive peak is smooth and the negative peak has many rising and falling exponentials superimposed on it due to the monostable being repeatedly triggered. Consequently to determine whether the encoder is behaving as required we observe the V_k waveform; there is no necessity to look at the decoded signal which in any case may be inaccessible.

Performance

The graph of decoded signal-to-noise against input signal power can be obtained by using the arrangement shown in Figure 2.34. White noise band limited between 450 and 550 Hz is encoded and decoded by the d.m. system under test.

With the switch in position 1 the integrating meter records the output signal power and distortion products, which for high signal-to-noise ratios approximates to the former output signal power. When the switch is changed to position 2 the integrating meter records the distortion

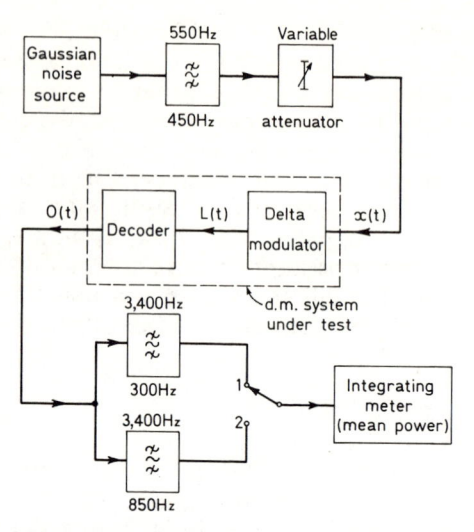

Fig. 2.34. Arrangement for measuring the signal-to-noise ratio

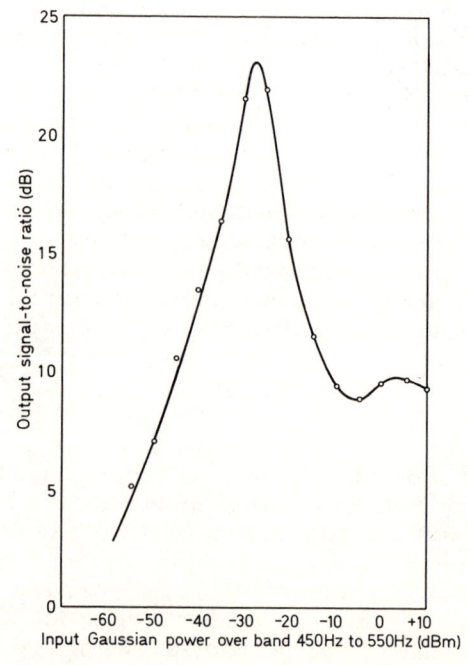

Fig. 2.35. Signal-to-noise ratio of pulse d.m. codec (Courtesy of Electronic Engineering[5])

products in the frequency band 850 to 3400 Hz. Assuming that the distortion spectral density function is flat a distortion power of from 300 to 850 Hz can be allowed for. The ratio of the two readings is a close approximation of the signal-to-noise ratio. By repeating the test for different settings of the variable attenuator the signal-to-noise ratio can be plotted as a function of the mean square value of $x(t)$. Using this measuring technique the curve shown in Figure 2.35 was obtained for the codec shown in Figure 2.33 for a clock rate of 60 kbit/s and $\tau = 2.4$ μs.

REFERENCES

1. de Jager, F., 'Delta modulation, a method of p.c.m. transmission using 1 unit code', *Philips Research Report*, **7**, 442–446, December (1952)
2. Flood J. E. and Hawksford, M. J., 'Exact model for delta modulation processes', *Proc. I.E.E.*, **118**, 1155–1161, (1971)
3. Boller D. J., Unpublished work, Electronic and Electrical Eng., Dept., Loughborough University of Technology, England
4. Johnson, F. B., 'Calculating delta modulator performance', *I.E.E.E. Trans.*, **AU-16**, 121–129, (1968)
5. Steele, R. and Thomas, M. W. S., 'Two-transistor delta modulator', *Electronic Engineering*, **40**, 513–516, (1968)
6. Steele, R., 'Pulse delta modulators-inferior performance but simpler circuitry', *Electronic Engineering*, **42**, 75–79, (1970)
7. Greefkes, J. A. and Riemens, K., 'Code modulation with digitally controlled companding for speech transmission,' *Philips Technical Review*, **31**, No. 11/12, 335–353, (1970)

Chapter 3

CALCULATION OF QUANTIZATION NOISE

3.1. INTRODUCTION

This chapter aims to establish the theoretical behaviour of the linear delta modulator when the input signal is band-limited white noise. This input signal is employed because it is widely used as a test signal for both p.c.m. and d.m. systems, and because its properties are well defined and understood. It must, however, be pointed out that this signal is difficult for a delta modulator to encode because its spectrum is not matched to that of the overload characteristic of the delta modulator.

The performance of the codec will be evaluated by using the Normal spectrum technique which is only applicable when the input signal has an amplitude probability density function which is Gaussian. This technique has been available for many years but it is surprisingly not widely known amongst engineers. Its virtue, in the case of the d.m. system, is that it offers a better insight into the behaviour of the system compared to other methods, although it is perhaps more complex. The Normal spectrum technique can also be used for calculating the distortion spectra in pulse code modulation systems and in general it is applicable to non-linear systems.

3.2. BRIEF HISTORY

The estimation of the quantization noise is essential in describing the performance of delta modulation. In Section 1.2.7 the quantization noise is briefly discussed and represented by the empirical Equation 1.7 which is repeated here for convenience.

$$N_q^2 = K_q \cdot \frac{f_c \gamma^2}{f_p}$$

where γ is the step height, f_c the bandwidth of the input signal, f_p the clock rate and K_q an empirical constant.

The establishment of this equation was first made by de Jager in 1952 in his classic paper[1] and was based on deduction and experiment. The value of K_q was found to be approximately $\frac{1}{3}$ for sinusoidal input signals.

Calculation of quantization noise   79

This equation has the supreme virtue of simplicity. It is adequate for a range of parameter variations provided accuracy is not essential. The equation has been used continuously in many publications since 1952, although K_q has been given differing values depending on various encoding conditions.

Soon after de Jager's paper, van de Wegg[2] theoretically established the quantization noise for band-limited white noise input signals.

Zetterberger's[3] comprehensive paper in 1955 compared delta modulation with pulse code modulation. His expression for quantization noise was derived from Bennett's[4] theory on the spectra of quantized signals. However, Zetterberg's results were less precise than those of van de Weg whose theory was satisfactory and experimentally verified for most practical purposes. Van de Weg's theory was refined and improved by Goodman[5] (1969) and also in the same year Iwersen[6] considered the effect of an asymmetrical encoder on the quantization noise.

M. Passot[7,8] and the Author,[7] supported by the Post Office, investigated the problem of quantization noise by using the Normal spectrum technique and it is this method which will be used here.

In this brief review only a few of the many contributors have been mentioned. The problem of calculating slope overload noise is difficult although many attempts have been made since the conception of d.m. The above remarks do not include slope overload noise which is considered in the next chapter.

Finally it is necessary to point out that in the literature *quantization noise* is sometimes referred to as *granular noise*.

3.3. REPRESENTATION OF THE DELTA MODULATOR

In order to calculate the spectral densities of the signals in the delta modulator the task is made very much easier by representing the delta modulator by an exact model which does not have a closed-loop. Such a model is the Infinite staircase model[(2)] described in Section 1.6.2 and depicted in Figure 1.14.

This model is invalidated if the encoder becomes slope overloaded. As the input signal $x(t)$ is band-limited white noise there is always a probability that the encoder will be slope overloaded (see Section 1.2.5). This probability will be insignificant if the r.m.s. value of the slope of the input signal is made sufficiently small in comparison with the maximum slope of the signal at the output of the integrator in the delta modulator.

From Equation 1.4 slope overload is avoided if

$$x'(t) = (\mathrm{d}x(t)/\mathrm{d}t) \leq \gamma f_p$$

As $x(t)$ is a Gaussian signal band-limited from 0 to f_{c2} Hz and σ^2 is the mean square value of $x(t)$, then

$$\sigma_d^2 = \langle\{x'(t)\}^2\rangle = \int_{-f_{c2}}^{f_{c2}} (2\pi f)^2 \left(\frac{\sigma^2}{2f_{c2}}\right) df$$

$$= \tfrac{4}{3}\pi^2 f_{c2}^2 \sigma^2$$

where $(\sigma^2/2f_{c2})$ is the spectral density function of band-limited white noise, and $(2\pi f)^2$ is introduced because of the act of differentiation, i.e. the power transfer function of a differentiator is $(2\pi f)^2$.

If $\sigma_d < (\gamma f_p/4)$, then the probability of the encoder experiencing slope overload is less than 6×10^{-5}. Hence

$$\frac{\sigma^2}{3}(2\pi f_{c2})^2 \leq (\gamma^2 f_p^2/16)$$

$$f_p \geq \frac{\sigma}{\gamma} \cdot \frac{8\pi}{\sqrt{3}} \cdot f_{c2} \qquad (3.1)$$

If the Gaussian input signal is band-limited white noise occupying the frequency band from f_{c1} to f_{c2} Hz then

$$\sigma_d^2 = 2\int_{f_{c1}}^{f_{c2}} \left\{\frac{\sigma^2}{2(f_{c2}-f_{c1})}\right\}(2\pi f)^2 \, df = \frac{4}{3}\pi^2\sigma^2\left\{\frac{f_{c2}^3 - f_{c1}^3}{f_{c2}-f_{c1}}\right\}$$

and consequently f_{c2} in Equation 3.1 is replaced by

$$f_{c2}\left\{\frac{1-\left(\frac{f_{c1}}{f_{c2}}\right)^3}{1-\left(\frac{f_{c1}}{f_{c2}}\right)}\right\}^{1/2}$$

By ensuring that these restrictions on σ for a given f_p, f_{c1}, f_{c2} and γ are obeyed the model of Figure 1.14 is valid.

Having prescribed certain limitations on the input signal in order to avoid overloading the linear delta modulator, it is now necessary to develop some results in order that the spectral density functions of the decoded signal and distortion noise can be found.

By calculating the spectral density functions at the output of the Infinite staircase model using the Normal spectrum technique it will be shown that the mathematical results, although somewhat complex, are relatively easy to interpret in terms of graphical presentation.

Difficulties in the calculation arise from the very nature of the Infinite staircase model. The autocorrelation functions have to be found at the output of the Infinite staircase characteristics by means of the Normal spectrum technique. The Fourier transform of these autocorrelation functions give the spectral density functions. Section 3.4 briefly reviews

Calculation of quantization noise 81

the Normal spectrum technique, and shows why it is necessary to take the nth order autoconvolution of the input spectral density function: the methods of calculation are dealt with in Sections 3.5 and 3.6. Because of the existence of samplers and a hold circuit in the model the effect on the autocorrelation function of a signal when the latter is sampled and when it is sampled and held are dealt with in Sections 3.7 and 3.8. These sections are mainly included for completeness as the effect of sampling or sampling and holding a signal are well known in terms of spectral density functions. Sections 3.4, 3.5 and 3.6 constitute the Normal spectrum technique and can be applied to other non-linear elements, in particular those encountered in pulse code modulation.

Sections 3.9 and 3.10 deal with the calculation of the spectral density functions of the signal at the output of the model and its distortion component, respectively. The final section presents the behaviour of the quantization noise and signal-to-noise ratio as a function of encoder parameters.

3.4. NORMAL SPECTRUM TECHNIQUE

The Infinite Staircase model of a delta modulator shown in Figure 1.14 has two non-linear elements. It is appropriate that before commencing the general analysis the autocorrelation function and power spectral density of the signal at the output of a non-linear element should be established when the input signal is Gaussian with zero mean, i.e. normal.

3.4.1. Properties of the input signal $x(t)$

The stationary statistical properties of the input signal are:
Gaussian amplitude probability density function, $f(x)$
Average value, 0
Variance, σ^2
Autocorrelation coefficient, $\rho_x(\tau)$
Normalised spectral density function, $\phi_x(f)$
Autocorrelation function, $R_x(\tau) = \sigma^2 \rho_x(\tau)$
Spectral density function, $S_x(f) = \sigma^2 \phi_x(f)$

3.4.2. Signal at the output of a non-linear element

Consider a signal $y(t)$ represented by a power series

$$y(t) = \sum_{n=0}^{N} b_n g_n[x(t)]$$

where $g_0, g_1, g_2, \ldots, g_N$ are called basis functions and $b_0, b_1, b_2, \ldots, b_N$ are their coefficients. By making these functions orthogonal for a given interval of time the property of 'finality of coefficients' can be evoked which allows the addition of extra terms in the above summation without the necessity of recalculating those coefficients which have previously been established. The exponential Fourier series representation of a signal is of this type when the $g_n[x(t)]$ basis functions are a set of mutually orthogonal exponential functions. There are many other useful basis functions, for example, Rademacher and Walsh functions, but for the case of Normal signals discussed here the basis functions are members of the Hermite polynomial.[9,10,11]

Consider an isolated non-linear element whose characteristic is memoryless. Figure 3.1 shows a particular non-linear function of this

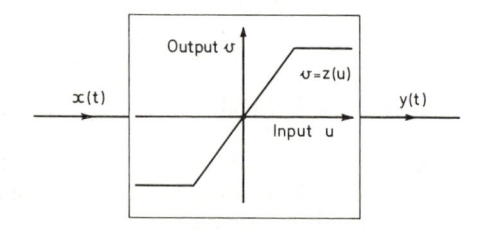

Fig. 3.1. *Non-linear element*

type where the input and output are designated u and v, respectively and v is a function of u, i.e. $v = z(u)$. When the input time signal to the non-linear element is $x(t)$ the output time signal is $y(t)$ say and can be expressed as $z[x(t)]$. Hence

$$y(t) = z[x(t)] = \sum_{n=0}^{\infty} b_n H_n[x(t)]$$

where $H_n[x(t)]$ are the Hermite polynomial basis functions. For convenience $x(t)$ is written as x, and the above equation becomes

$$y(t) = z(x) = \sum_{n=0}^{\infty} b_n H_n(x) \tag{3.2}$$

The Hermite polynomial is defined by

$$H_n(x) = (-1)^n \sigma^{2n} \exp\left(\frac{x^2}{2\sigma^2}\right) \cdot \frac{d^n}{dx^n}\left\{\exp\left(-\frac{x^2}{2\sigma^2}\right)\right\} \tag{3.3}$$

Calculation of quantization noise

The first 6 terms in this polynomial are

$$H_0(x) = 1$$
$$H_1(x) = x$$
$$H_2(x) = x^2 - \sigma^2$$
$$H_3(x) = x^3 - 3\sigma^2 x$$
$$H_4(x) = x^4 - 6\sigma^2 x^2 + 3\sigma^4$$
$$H_5(x) = x^5 - 10\sigma^2 x^3 + 15\sigma^4 x$$

The Hermite polynomial is used as the basis functions for $y(t)$ because it is orthogonal over the range of $x(t)$ from minus to plus infinity.

3.4.3. Two properties of the Hermite polynomial.

The orthogonality property of the Hermite polynomial is

$$\int_{-\infty}^{\infty} H_n(x) H_m(x) f(x) \, dx = 0, \quad \text{if } n \neq m$$
$$= N_n, \quad \text{if } n = m \quad (3.4)$$

where $f(x)$ is the Gaussian probability density function.

$$f(x) = \frac{1}{\sigma \sqrt{2\pi}} \exp\left(-\frac{x^2}{2\sigma^2}\right) \quad (3.5)$$

and

$$N_n = \int_{-\infty}^{\infty} H_n^2(x) f(x) \, dx = n! \sigma^{2n} \quad (3.6)$$

The odd-even property of the Hermite polynomial is that $H_n(x)$ is an even function of x if n is even, and an odd function of x if x is odd.

3.4.4. Autocorrelation function

At time t_1, $x(t) = x(t_1)$, and $z(x) = z(x_1)$. At a subsequent time $t_2 = t_1 + \tau$, $x(t) = x(t_2)$ and $z(x) = z(x_2)$. The autocorrelation function of $y(t)$ is

$$R_y(\tau) = \int_{-\infty}^{\infty} \int_{-\infty}^{\infty} z(x_1) z(x_2) f(x_1, x_2) \, dx_1 \, dx_2$$
$$= \langle z(x_1) z(x_2) \rangle \quad (3.7)$$

where $\langle \, \rangle$ means expected value.

$f(x_1, x_2)$ is the joint probability function. The expression for this function is relatively complex,[12] i.e.

$$f(x_1, x_2) = \frac{1}{2\pi\sigma^2(1-\rho_x^2(\tau))^{1/2}} \exp\left\{-\frac{1}{2} \cdot \frac{x_1^2 + x_2^2 - 2\rho_x(\tau)x_1 x_2}{\sigma^2(1-\rho_x^2(\tau))}\right\} \quad (3.8)$$

This expression can be simplified by using Mehler's formula

$$f(x_1, x_2) = f(x_1)f(x_2) \sum_{n=0}^{\infty} \frac{\rho_x^n(\tau)}{n!\sigma^{2n}} \cdot H_n(x_1)H_n(x_2) \quad (3.9)$$

This formula is derived from Equation 3.8 by taking the double Fourier transform of $f(x_1, x_2)$ thereby establishing a two-dimensional characteristic function. This characteristic function is then manipulated before taking the inverse double Fourier transform. The result enables Equation 3.3 to be applied leading to Equation 3.9.

Substituting Equation 3.9 into Equation 3.7 and using $f(x_1) = f(x_2) = f(x)$ and $H_n(x_1) = H_n(x_2) = H_n(x)$

$$R_y(\tau) = \sum_{n=0}^{\infty} \frac{\rho_x^n(\tau)}{n!\sigma^{2n}} \left(\int_{-\infty}^{\infty} f(x)z(x)H_n(x)\,dx\right)^2$$

Putting

$$b_n = \frac{1}{N_n} \int_{-\infty}^{\infty} H_n(x)z(x)f(x)\,dx \quad (3.10)$$

$$R_y(\tau) = \sum_{n=0}^{\infty} b_n^2 N_n \rho_x^n(\tau) \quad (3.11)$$

where N_n is defined by Equation 3.6.

The double integral and joint probability function shown in Equation 3.7 has been replaced in Equation 3.11 by a power series. The coefficients in this power series are given by Equation 3.10 and depend on the Hermite polynomial $H_n(x)$ the non-linearity $z(x)$ and the Gaussian probability density function $f(x)$; b_n depends only on the power of the input signal due to N_n, and not on the shape of its spectral density function $\phi_x(\omega)$. $R_y(\tau)$ is easier to calculate using Equation 3.11 rather than Equation 3.7. Note that although the Normal spectrum technique is confined to Gaussian signals the shape of their spectra is arbitrary.

3.4.5. Spectral density

The spectral density function of the signal $y(x)$ at the output of the non-linear element is found by taking the Fourier transform of the

Calculation of quantization noise 85

autocorrelation function of $R_y(\tau)$. From Equation 3.11

$$S_y(f) = \int_{-\infty}^{\infty} R_y(\tau) \exp(-j2\pi f\tau) d\tau$$

$$= \sum_{n=0}^{\infty} b_n^2 N_n \left\{ \int_{-\infty}^{\infty} \rho_x^n(\tau) \exp(-j2\pi f\tau) d\tau \right\}$$

$$= \sum_{n=0}^{\infty} b_n^2 N_n \phi_x^n(f) \quad (3.12)$$

where

$$\phi_x^n(f) = \int_{-\infty}^{\infty} \rho_x^n(\tau) \exp(-j2\pi f\tau) d\tau \quad (3.13)$$

Equation 3.13 shows that when $n = 1$, $\phi_x(f)$ is the Fourier transform of the normalised autocorrelation coefficient. When $n = 2$, $\phi_x^2(f)$ means the autoconvolution of $\phi_x(f)$. Similarly when $n = 3$

$$\phi_x^3(f) = \phi_x(f) * \phi_x(f) * \phi_x(f) \quad (3.14)$$

i.e. $\phi_x^3(f)$ is the second autoconvolution of $\phi_x(f)$. Thus $\phi_x^n(f)$ is really the $(n-1)$th autoconvolution of $\phi_x(f)$, although it is often referred to as the nth autoconvolution of $\phi_x(f)$.

3.5. nth ORDER AUTOCONVOLUTION OF LOW-PASS WHITE NOISE

Consider a signal $x(t)$ which represents low pass white noise. Let $\lambda_x(f)$ be the spectral density function of $x(t)$, where

$$\lambda_x(f) = 0.5 \text{ W/Hz}, \quad -1 \leq f \leq +1$$

$$= 0, \text{ elsewhere} \quad (3.15)$$

Fig. 3.2. Spectral density function of low-pass white noise

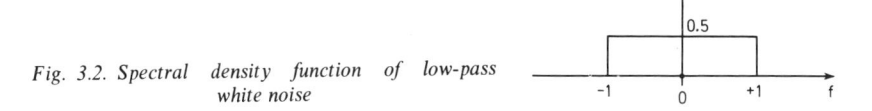

$\lambda_x(f)$ is sketched in Figure 3.2, and has been selected such that $x(t)$ develops one watt across a one ohm resistor, i.e. the area under $\lambda_x(f)$ is unity. $\lambda_x(f)$ has an autocorrelation function $\rho(\tau)$ and for $\rho(0)$ also to be one watt

$$\rho(\tau) = \int_{-\infty}^{\infty} \lambda_x(f) \exp(j2\pi f\tau) \, df$$

$$= \frac{\sin(2\pi\tau)}{2\pi\tau} \tag{3.16}$$

As convolution in the frequency domain corresponds to multiplication in the time domain, the nth convolution of $\lambda_x(f)$, denoted by $\lambda_x^n(f)$, is

$$\lambda_x^n(f) = \int_{-\infty}^{\infty} \left(\frac{\sin(2\pi\tau)}{2\pi\tau}\right)^n \exp(-j2\pi f\tau) \, d\tau \tag{3.17}$$

and

$$\left(\frac{\sin(2\pi\tau)}{2\pi\tau}\right)^n = \frac{\{\exp(j2\pi\tau) - \exp(-j2\pi\tau)\}^n}{(2\pi)^n (2j)^n \tau^n} \tag{3.18}$$

The numerator of Equation 3.18 can be expanded by means of the Binomial theorem, e.g.

$$(a+b)^n = \sum_{p=0}^{n} C_p^n \cdot a^{n-p} \cdot b^p \tag{3.19}$$

where

$$C_p^n = \frac{n!}{p!(n-p)!} \tag{3.20}$$

Consequently

$$\{\exp(j2\pi\tau) - \exp(-j2\pi\tau)\}^n$$

$$= \sum_{p=0}^{n} (-1)^p C_p^n \exp\{j2\pi\tau(n-p)\} \exp(-j2\pi\tau p)$$

$$= \sum_{p=0}^{n} (-1)^p C_p^n \exp\{j(n-2p)2\pi\tau\} \tag{3.21}$$

From Equations 3.18 and 3.21 substituting into Equation 3.17 gives

$$\lambda_x^n(f) = \frac{1}{(2\pi)^n} \cdot \frac{1}{(2j)^n} \int_{-\infty}^{\infty} \sum_{p=0}^{n} \frac{(-1)^p C_p^n \exp\{j(n-2p-f)2\pi\tau\}}{\tau^n} \, d\tau$$

$$= \frac{1}{(2\pi)^n} \cdot \frac{1}{(2j)^n} \sum_{p=0}^{n} (-1)^p C_n^p \int_{-\infty}^{\infty} \frac{\exp\{j(n-2p-f)2\pi\tau\}}{\tau^n} \, d\tau \tag{3.22}$$

Consider the integral term in Equation 3.22

$$I = \int_{-\infty}^{\infty} \frac{\exp(j\alpha\tau)}{\tau^n} \, d\tau \tag{3.23}$$

where

$$\alpha = (n-2p-f)2\pi \qquad (3.24)$$

Writing I in terms of the z-plane complex variable z, Equation 3.23 becomes

$$I = \int_{-\infty}^{\infty} \frac{\exp(j\alpha z)}{z^n} dz \qquad (3.25)$$

On solving for I by contour integration it is found that I has the same value for $\alpha < 0$ as $\alpha > 0$ except then it is negative. Thus the value of I for all α is

$$I = \frac{\pi j}{(n-1)!} \cdot \frac{d^{n-1}}{dz^{n-1}} \{\exp(j\alpha z)\}_{z=0} \cdot \text{sgn } \alpha \qquad (3.26)$$

This equation is substituted in Equation 3.22 to give

$$\lambda_x^n(f) = \frac{1}{(2\pi)^n} \cdot \frac{1}{(2j)^n} \cdot \frac{\pi j}{(n-1)!} \sum_{p=0}^{n} \left[(-1)^n C_p^n \frac{d^{n-1}}{dz^{n-1}} \right.$$

$$\left. \times [\exp\{j(n-2p-f)2\pi z\}]_{z=0} \text{ sgn } \{(n-2p-f)2\pi\} \right]$$

and

$$\frac{d^{n-1}}{dz^{n-1}} [\exp\{j(n-2p-f)2\pi z\}] = j^{n-1}(2\pi)^{n-1}(n-2p-f)^{n-1}$$

$$\times \exp\{j(n-2p-f)2\pi z\}$$

$$= j^{n-1}(2\pi)^{n-1}(n-2p-f)^{n-1}; \text{ when } z=0$$

whence[7]

$$\lambda_x^n(f) = \frac{1}{2^{n+1}(n-1)!} \sum_{p=0}^{n} (-1)^p C_p^n (n-2p-f)^{n-1} \text{ sgn } (n-2p-f) \qquad (3.27)$$

Example. To find the first autoconvolution of $\lambda_x(f)$. This problem is trivial but its solution is well known as a triangular curve, see Figure 3.3.

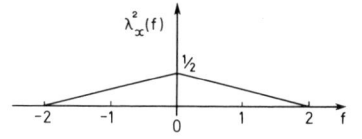

Fig. 3.3. *Autoconvolution of the spectral density function of low-pass white noise*

88   Calculation of quantization noise

Solving Equations 3.27 for $n = 2$ gives

$$\lambda_x^2(f) = \tfrac{1}{8} \sum_{p=0}^{2} (-1)^p C_p^2 (2-2p-f) \operatorname{sgn}(2-2p-f)$$
$$= \tfrac{1}{8}\{C_0^2(2-f) \operatorname{sgn}(2-f) - C_1^2(-f) \operatorname{sgn}(-f)$$
$$+ C_2^2(-2-f) \operatorname{sgn}(-2-f)\}$$

where

$$C_0^2 = 1, \quad C_1^2 = 2 \quad \text{and} \quad C_2^2 = 1.$$

By inserting different values of f into the above expression for $\lambda_x^2(f)$ the graph of Figure 3.3 is obtained.

Thus Equation 3.27 enables $\lambda_x^n(f)$ to be computed for all values of n. However, as n becomes very large the amount of computation becomes arduous; fortunately $\lambda_x^n(f)$ tends to a Gaussian curve and a more simple expression than Equation 3.27 can be used, i.e.

$$\lambda_x^n(f) = \frac{1}{\sigma\sqrt{(2\pi n)}} \exp\left\{-\frac{x^2}{2\sigma^2 n}\right\} \qquad (3.28)$$

This expression implies that the variance of the curve $\lambda_x^n(f)$ is $\sigma^2 n$.

3.6. *n*th ORDER AUTOCONVOLUTION OF BAND-PASS WHITE NOISE

Consider the signal $x(t)$ to be band-limited white noise and to develop a power of 1 W in a 1 Ω resistor. The power spectral density function associated with $x(t)$ is $\phi_x(f)$ and is shown in Figure 3.4.

The area under each rectangle in Figure 3.4 is 0·5 W. Consider the function $\phi_x(f)$ for positive frequencies

$$\phi_x(f) = \tfrac{1}{2}\lambda_x(f - f_{ab}) \qquad (3.29)$$

where $\lambda_x(f)$ is a power spectral density function centred about the origin, as in Figure 3.2. The $\tfrac{1}{2}$ is introduced in Equation 3.29 because the power associated with $\lambda_x(f)$ is 1 W, whereas the power for $\phi_x(f)$, $f > 0$ is $\tfrac{1}{2}$ W.

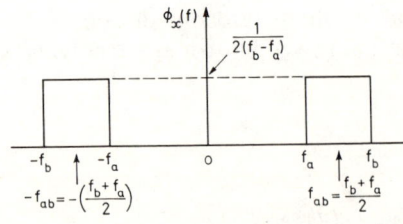

Fig. 3.4. *Spectral density function of band-pass white noise*

Calculation of quantization noise

For positive and negative frequencies

$$\phi_x(f) = \tfrac{1}{2}\lambda_x(f - f_{ab}) + \tfrac{1}{2}\lambda_x(f + f_{ab}) \qquad (3.30)$$

Consider the autoconvolution of $\phi_x(f)$

$$\phi_x^2(f) = \phi_x(f) * \phi_x(f)$$

$$\phi_x^2(f) = \tfrac{1}{2}\{\lambda_x(f - f_{ab}) + \lambda_x(f + f_{ab})\} * \tfrac{1}{2}\{\lambda_x(f - f_{ab}) + \lambda_x(f + f_{ab})\}$$

$$= \tfrac{1}{4}\{\lambda_x(f - f_{ab}) * \lambda_x(f - f_{ab}) + \lambda_x(f + f_{ab}) * \lambda_x(f - f_{ab})$$
$$+ \lambda_x(f - f_{ab}) * \lambda_x(f + f_{ab}) + \lambda_x(f + f_{ab}) * \lambda_x(f + f_{ab})\}$$

$$= \tfrac{1}{4}\{\lambda_x^2(f) + \lambda_x^2(f + 2f_{ab}) + \lambda_x^2(f - 2f_{ab}) + \lambda_x^2(f)\}$$

$$= \tfrac{1}{4}\{\lambda_x^2(f + 2f_{ab}) + 2\lambda_x^2(f) + \lambda_x^2(f - 2f_{ab})\} \qquad (3.31)$$

where $\lambda_x^2(f)$ is the autoconvolution of $\lambda_x(f)$ and can be found from Equation 3.27 by putting $n = 2$. The terms $\lambda_x^2(f + 2f_{ab})$ and $\lambda_x^2(f - 2f_{ab})$ are simply $\lambda_x^2(f)$ shifted in frequency by $\pm 2f_{ab}$, respectively. Hence

$$\phi_x^2(f) = \tfrac{1}{4}\lambda_x^2(f + 2f_{ab}) + \tfrac{1}{2}\lambda_x^2(f) + \tfrac{1}{4}\lambda_x^2(f - 2f_{ab}) \qquad (3.32)$$

Thus the computation of the autoconvolution of band-pass white noise has been expressed in terms of the autoconvolution of low-pass white noise. It can be observed that the coefficients of the $\lambda_x^2(\ )$ terms are satisfied by the binomial coefficients defined by C_p^2, see Equation (3.20). If $\phi_x^2(f)$ is convolved with $\phi_x(f)$ to give the third order autoconvolution, i.e.

$$\phi_x^3(f) = \phi_x(f) * \phi_x(f) * \phi_x(f) = \phi_x^2(f) * \phi_x(f)$$

then

$$\phi_x^3(f) = \tfrac{1}{8}\{\lambda_x^3(f - 3f_{ab}) + 3\lambda_x^3(f - f_{ab})$$
$$+ 3\lambda_x^3(f + f_{ab}) + \lambda_x^3(f + 3f_{ab})\}$$

and the coefficients are again binomial.

Extending the order of autoconvolution results in a general expression[7] for the nth autoconvolution

$$\phi_x^n(f) = \frac{1}{2^n} \sum_{p=0}^{n} C_p^n \lambda_x^n \{f - (n - 2p)f_{ab}\} \qquad (3.33)$$

Observe $\phi_x^n(f)$ contains $(n+1)$ terms each corresponding to the nth convolution of low-pass white noise which has been weighted by binomial coefficients and shifted in frequency. Thus the nth order autoconvolution of band-pass white noise is greatly simplified by computing it in terms of low-pass white noise. Figure 3.5(a) shows some autoconvolutions of low-pass white noise, while Figures 3.5(b) and (c) show the autoconvolutions of band-pass white noise obtained by using Equation 3.33 for odd and even values of n, respectively.

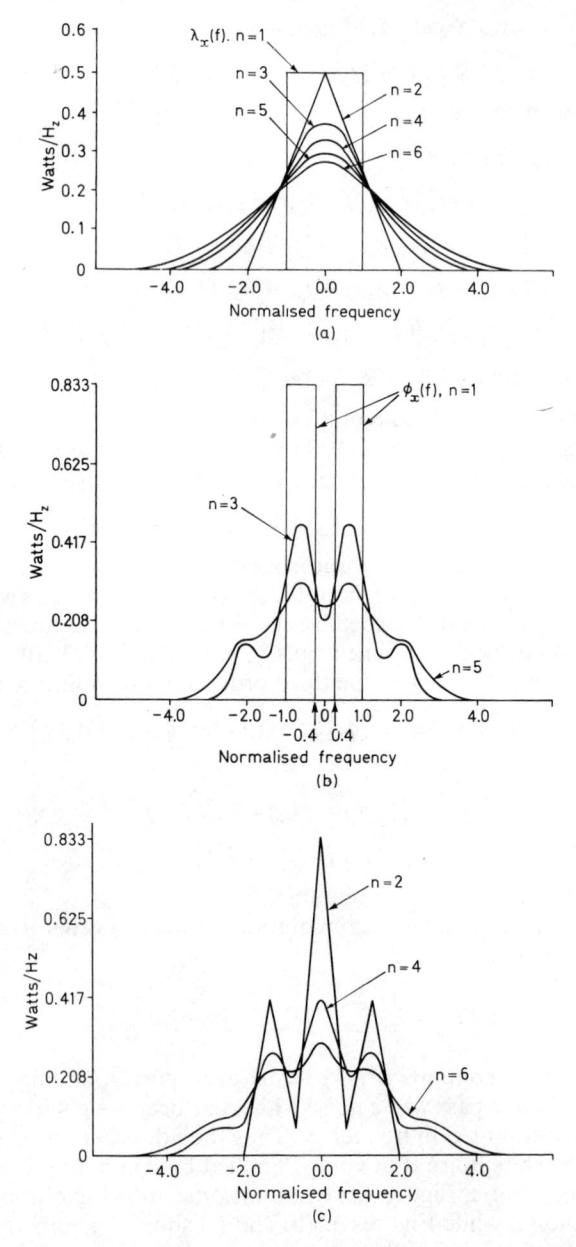

Fig. 3.5. nth *autoconvolution of* (a) *low-pass white noise,* $n = 1$ *to* 6; (b) *band-pass white noise,* $n = 1, 3, 5$; (c) *band-pass noise,* $n = 2, 4, 6$

Calculation of quantization noise 91

3.7. AUTOCORRELATION FUNCTION OF SAMPLED WAVEFORMS

In the staircase model of the delta modulator (see Figure 1.14) the autocorrelation functions and their corresponding spectral density functions at the output of the staircase non-linearities can be expressed in terms of Normal spectra with the aid of Equations 3.11, 3.12 and 3.33. However, as the waveforms of these staircase functions are sampled at the outputs it is necessary to determine the effect of sampling on the autocorrelation functions.

This section establishes the simple result that the sampled autocorrelation function is equal at sampling instants to the autocorrelation function multiplied by the clock rate f_p. There are no restrictions placed upon f_p, i.e. the sampling can be above or below the Nyquist rate. Expressing this result as an equation

$$R^*(nT) = f_p R(nT) \qquad (3.34)$$

The raised * indicates that the function has been sampled at the clock rate $f_p = 1/T$; n is an integer.

Observe that the above equation is general and not specifically related to the Normal spectrum technique. Equation 3.34 may be proved as follows. Consider a signal $z(t)$ which is sampled by an impulse sampling train $\delta_T(t)$ to give a set of impulses $z^*(t)$.

$$z^*(t) = z(t) \times \delta_T(t) \qquad (3.35)$$

$$\delta_T(t) = \sum_{m=-\infty}^{\infty} \delta(t+mT) \qquad (3.36)$$

Taking the Fourier transform of $\delta_T(t)$ an infinite series of frequency impulses is obtained.

$$F\{\delta_T(t)\} = f_p \sum_{m=-\infty}^{\infty} \delta(f+mf_p) \qquad (3.37)$$

These frequency impulses are spaced f_p Hz apart and are of strength f_p.

The spectrum of the sampled waveform $z^*(t)$ is achieved by convolving the spectrum of $z(t)$, i.e. $z(j2\pi f)$ with the spectrum of the impulse train

$$F\{z^*(t)\} = z^*(j2\pi f) = z(j2\pi f) * f_p \sum_{m=-\infty}^{\infty} \delta(f+mf_p)$$

Hence,

$$z^*(j2\pi f) = f_p \sum_{m=-\infty}^{\infty} z\{j2\pi(f+mf_p)\} \qquad (3.38)$$

The spectral density function of $z^*(t)$ is

$$S_z^*(f) = z^*(j2\pi f)\{z^*(j2\pi f)\}^C$$

Calculation of quantization noise

where the raised C indicates the complete conjugate.

$$S_z^*(f) = f_p^2 \sum_{m=-\infty}^{\infty} S(f+mf_p) \qquad (3.39)$$

The $S_z^*(f)$ function is composed of $S(f)$ spectra spaced f_p Hz apart and scaled in magnitude by f_p^2. Figure 3.6 shows the $S_z^*(f)$ function in terms of the original and displaced $S(f)$ spectra. The diagram has been drawn on the assumption that some aliasing, i.e. overlapping has occurred.

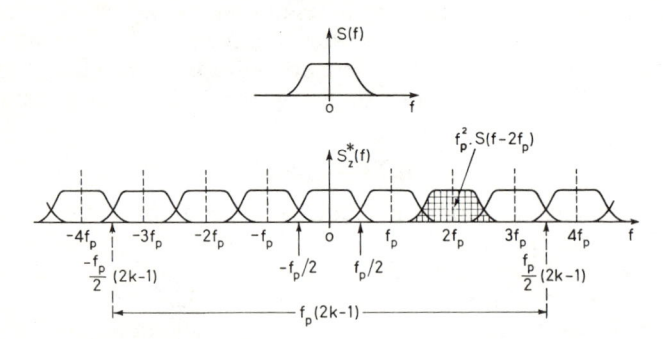

Fig. 3.6. Spectra $S_z^*(f)$ of the sampled signal $Z(t)$. $S(f)$ is the spectral density function of $Z(t)$

Due to the periodicity of $S_z^*(f)$, the sampled autocorrelation function over the frequency range $-f_p/2$ to $+f_p/2$, is

$$R_z^*(nT) = \frac{1}{f_p} \int_{-f_p/2}^{f_p/2} S_z^*(f) \exp(j2\pi fnT) \, df \qquad (3.40)$$

Eventually it will be necessary to make the limits in the above integral approach infinity so that all of the spectra will be considered thereby establishing the correct value of $R^*(nT)$. An intermediate step is to extend the integration interval to $(2k-1)f_p$ as shown in Figure 3.6. Substituting for $S_z^*(f)$ from Equation 3.39 in Equation 3.40

$$R_z^*(nT) = \frac{1}{(2k-1)f_p} \int_{-(2k-1)f_p/2}^{(2k-1)f_p/2} \{f_p^2 \sum_{m=-\infty}^{\infty} S(f+mf_p)\} \exp(j2\pi fnT) \, df \qquad (3.41)$$

which can be written as

$$R_z^*(nT) = \operatorname*{Lt}_{k \to \infty} \frac{1}{(2k-1)f_p} \int_{-(2k-1)f_p/2}^{(2k-1)f_p/2} f_p^2 \sum_{m=-k}^{k} S(f+mf_p) \exp(j2\pi fnT) \, df \qquad (3.42)$$

Calculation of quantization noise

Put

$$A_m = \underset{k \to \infty}{\text{Lt}} \int_{-(2k-1)f_p/2}^{(2k-1)f_p/2} S(f + mf_p)\exp(j2\pi fnT)\,df$$

$$= \int_{-\infty}^{\infty} S(f + mf_p)\exp(j2\pi fnT)\,df$$

where $S(f + mf_p)$ represents a spectral density function $S(f)$ located at $-mf_p$. Rearranging the above equation

$$A_m = \int_{-\infty}^{\infty} S(f)\exp\{j2\pi nT(f + mf_p)\}\,df$$

Since $f_p = 1/T$

$$\exp(j2\pi nTmf_p) = \cos 2\pi nm + j\sin 2\pi nm$$
$$= 1$$

as m and n are integer valued. Consequently

$$A_m = \int_{-\infty}^{\infty} S(f)\exp(j2\pi fnT)\,df \qquad (3.43)$$

Substituting for A_m in Equation 3.42

$$R_z^*(nT) = \frac{f_p}{2k-1}\sum_{m=-k}^{k} A_m = \frac{f_p}{2k-1}(2k-1)A_m$$
$$= f_p A_m \qquad (3.44)$$

The autocorrelation function of the signal $z(t)$ prior to sampling is $R_z(\theta)$

$$S(f) = \int_{-\infty}^{\infty} R_z(\theta)\exp(-j2\pi f\theta)\,d\theta \qquad (3.45)$$

where θ is a variable. Substituting Equation 3.45 in 3.43 and the resulting equation in Equation 3.44

$$R_z^*(nT) = f_p \int_{-\infty}^{\infty}\int_{-\infty}^{\infty} R_z(\theta)\exp\{-j2\pi f(\theta - nT)\}\,df\,d\theta$$

and

$$\int_{-\infty}^{\infty} \exp\{-j2\pi f(\theta - nT)\}\,df = \delta(\theta - nT)$$

Hence

$$R_z^*(nT) = f_p \int_{-\infty}^{\infty} R_z(\theta)\,\delta(\theta - nT)\,d\theta$$

$$= f_p R_z(nT)$$

Thus Equation 3.34 is established.

3.8. AUTOCORRELATION FUNCTION OF THE SIGNAL AT THE OUTPUT OF A SAMPLE AND HOLD CIRCUIT

Consideration will now be given to the calculation of the autocorrelation function at the output of a sample and hold circuit in terms of the input autocorrelation function.

Consider a signal $z(t)$ which is sampled every T seconds and these samples held in a zero-order hold circuit. This arrangement is shown in Figure 3.7 where the symbols for the time signals and their spectral

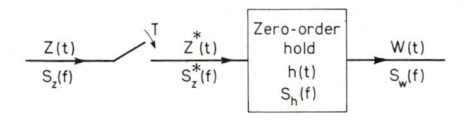

Fig. 3.7. *Sample and hold circuit*

densities are presented. The impulse response $h(t)$ of the hold circuit must be of unit magnitude over the time range 0 to T seconds and zero outside, i.e. an impulse of magnitude $z(nt)$ must be converted by the circuit to a pulse of amplitude $z(nt)$ and duration T. $h(t)$ is shown in Figure 3.8.

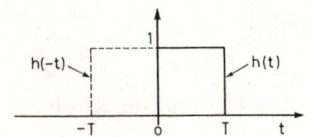

Fig. 3.8. *Impulse response of zero-order hold circuit*

The Fourier transform of $h(t)$ is $H(j2\pi f)$ and the power transfer function of the hold circuit is $S_h(f)$. The output spectral density $S_w(f)$ is

$$S_w(f) = S_z^*(f) \cdot S_h(f) \tag{3.46}$$

As the spectral density functions and their corresponding autocorrelation functions are a Fourier transform pair the autocorrelation functions

Calculation of quantization noise

relating to the above equations are

$$R_w(\theta) = R_z^*(\theta) * F^{-1}\{S_h(f)\} \quad (3.47)$$

where F^{-1} is the inverse Fourier transform.

$$F^{-1}\{S_h(f)\} = F^{-1}\{H(j2\pi f) \cdot H(j2\pi f)^c\}$$
$$= h(t) * h(-t)$$

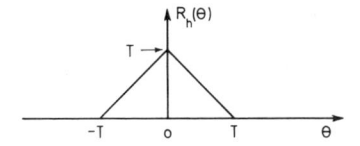

Fig. 3.9. *Autocorrelation function of hold circuit*

where $H(j2\pi f)^c$ is the Fourier transform of $h(-t)$. The convolution of $h(t)$ with $h(-t)$ yields $R_h(\theta)$, as shown in Figure 3.9.

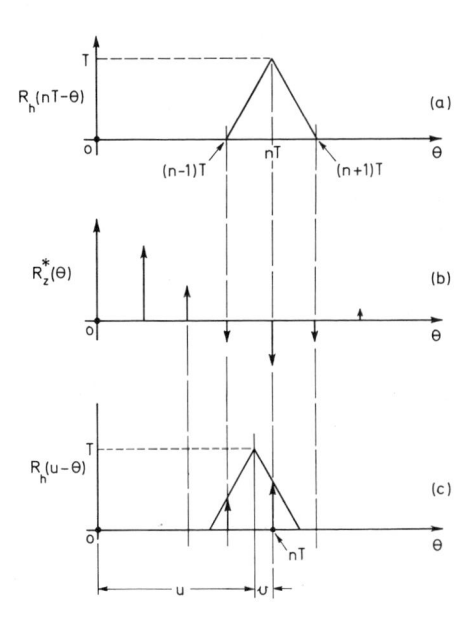

Fig. 3.10. (a) *Autocorrelation function of hold circuit shift by nT* ; (b) *Sampled autocorrelation function at the input to the sample and hold circuit* ; (c) *Autocorrelation function of hold circuit shifted by u*

$R_h(\theta)$ can also be found by taking the autocorrelation function (symbol ⊛) of $h(t)$

$$F^{-1}\{S_h(f)\} = R_h(\theta) = h(t) * h(-t) = h(t) \circledast h(t) \tag{3.48}$$

From Equations 3.47 and 3.48

$$R_w(\theta) = R_z^*(\theta) * R_h(\theta) \tag{3.49}$$

When $\theta = nT$, i.e. the autocorrelation function of the signal at the output of the sample and hold circuit at sampling times, is found by solving Equation 3.49. As $R_h(\theta)$ is a symmetrical function, $R_h(\theta) = R_h(-\theta)$. Figure 3.10 (a) and (b) shows $R_h(-\theta)$ shifted by nT, and a typical set of $R_z^*(\theta)$ samples. The value of $R_w(\theta)$ is then found by writing Equation 3.49 as

$$R_w(nT) = \int_{-\infty}^{\infty} R_z^*(\theta) \cdot R_h(nT-\theta) \, d\theta \tag{3.50}$$

and noting that for the particular case $R_h(nT-\theta))$ is by definition zero over the ranges $-\infty < \theta \le (n-1)T$ and $(n+1)T \le \theta < \infty$. From Figure 3.10 it can be seen that the maximum width of $R_h(nT-\theta)$ is two clock periods. Consequently the value of $R_w(nT)$ when the variable $\theta = nT$ is the multiple of the value of $R_h(nT-\theta)$, namely T, and $R_z^*(nT)$. Therefore

$$R_w(nT) = T R_z^*(nT)$$

Using the result of Equation 3.34

$$R_w(nT) = R_z(nT) \tag{3.51}$$

Result

The autocorrelation function of the waveform at the output of a sample and hold circuit is equal to the autocorrelation function of the input waveform at multiples of the sampling period.

3.8.1. Autocorrelation function of the waveform between sampling intervals

Consider $R_h(u-\theta)$ shown in Figure 3.10(c). The solution of Equation 3.49 requires that the magnitude of $R_h(u-\theta)$ be known as $\theta = (n-1)T$ and nT, as $R_z^*(\theta)$ exists only at sampling times. Let

$$R_h(u-\theta) = T\left(1 - \frac{u-\theta}{T}\right), \quad \theta > u$$

$$R_h(u-\theta) = T\left(1 + \frac{u-\theta}{T}\right), \quad \theta < u$$

Calculation of quantization noise

and
$$(n-1)T < u < nT$$
$$u + v = nT$$

When $\theta = u+v = nT$, i.e. $\theta - u = v$, $\theta > u$

$$R_h(v) = T\left(1 - \frac{v}{T}\right) \quad (3.52)$$

When $\theta = u+v-T = (n-1)T$, i.e. $\theta - u = v - T$

$$R_h(v-T) = T\left(1 + \frac{v-T}{T}\right) \quad (3.53)$$

From Equations 3.49, 3.52 and 3.53 the autocorrelation function of the waveform at the output of the sample and hold circuit between sampling periods is

$$R_w(u-\theta) = R_z^*\{(n-1)T\}R_h(v-T) + R_z^*(nT)R_h(v)$$

$$= R_z^*\{(n-1)T\} \cdot T\left(1 + \frac{v-T}{T}\right) + R_z^*(nT) \cdot T\left(1 - \frac{v}{T}\right)$$

$$= T\left[R_z^*(nT) + \frac{v}{T}\{R_z^*((n-1)T) - R_z^*(nT)\}\right] \quad (3.54)$$

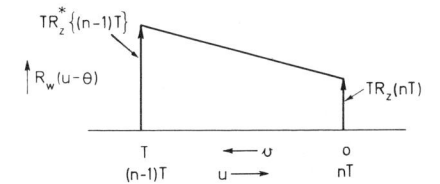

Fig. 3.11. *Autocorrelation function at the output of the sample and hold circuit during a sampling period*

Figure 3.11 shows that when $R_w(u - \theta)$ is plotted as a function of v, i.e. the variation of $R_w(u - \theta)$ over a sampling period, the graph is a straight line. The autocorrelation function $R_z(\theta)$ prior to sampling is a smooth curve.

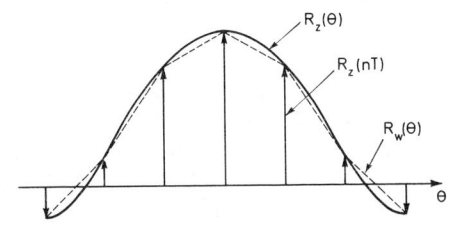

Fig. 3.12. *Comparison between the autocorrelation functions $R_z(\theta)$ and $R_w(\theta)$ of the signals at the input and output of a sample and hold circuit respectively*

Calculation of quantization noise

However, although the autocorrelation function $R_w(\theta)$ of the signal at the output of the hold circuit is equal to the samples of $R_z(\theta)$ scaled by T at a sampling instant, the autocorrelation function is not a smooth curve but is composed of straight lines which exist for one sampling period. $R_w(\theta)$ is a good approximation to $R_z(\theta)$ as shown in Figure 3.12.

3.9. THE SIGNAL FROM THE LOCAL DECODER

Having derived some useful results in the previous five sections we now proceed to calculate the spectral density function of the waveform $y(t)$ at the output of the integrator in the delta modulator shown in Figure 1.1.

3.9.1. Autocorrelation function

Using the Infinite staircase model of Figure 1.14, the sampled autocorrelation function of the signal $y^*(t)$ will now be found as the initial step in determining the spectral density function of $y(t)$.

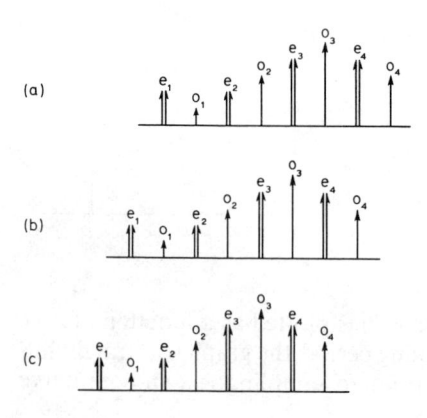

Fig. 3.13. Impulse train $y^*(t) = y_e^*(t) + y_o^*(t)$

In Figure 3.13 the thick arrows represent the impulse samples $y_e^*(t)$ obtained by sampling the signal $y_e(t)$ at the output of the even staircase function. Similarly the thin arrows correspond to the impulse train $y_o^*(t)$ derived by sampling the output signal $y_o(t)$ from the odd staircase function. The impulses $y^*(t)$ in Figure 3.13 are obtained by adding $y_e^*(t)$ to $y_o^*(t)$. In order to find the autocorrelation function of $y^*(t)$ the following procedure is adopted.

Calculation of quantization noise

The impulse trains in Figure 3.13 have a short sequence for the purpose of illustration. The autocorrelation function $R^*_{yy}(0)$ is found by multiplying the impulse train (a) of Figure 3.13 by itself and finding the sum of resulting impulses, i.e.

$$R^*_{yy}(0) = e_1e_1 + o_1o_1 + e_2e_2 + o_2o_2 + e_3e_3 + o_3o_3 + e_4e_4 + o_4o_4$$

To find $R^*_{yy}(T)$ the impulse sequence (a) is shifted by one clock period to the left to produce sequence (b) in Figure 3.13. Sequences (a) and (b) are then multiplied together and the result summed

$$R^*_{yy}(T) = e_1o_1 + o_1e_2 + e_2o_2 + o_2e_3 + e_3o_3 + o_3e_4 + e_4o_4$$

Continuing to shift the impulse train by one clock period to the left gives impulse sequence (c). From sequence (c) and (a)

$$R_{yy}(2T) = e_1e_2 + o_1o_2 + e_2e_3 + o_2o_3 + e_3e_4 + o_3o_4$$

Similarly

$$R_{yy}(3T) = e_1o_2 + o_1e_3 + e_2o_3 + o_2e_4 + e_3o_4$$

etc.

Some general conclusions become apparent. At even clock times the autocorrelation function is

$$R^*_{yy}(2mT) = R^*_{y_ey_e}(2mT) + R^*_{y_oy_o}(2mT) \tag{3.55}$$

This is in accord with $R^*_{yy}(0)$ and $R^*_{yy}(2T)$ derived in the example. At odd clock times

$$R^*_{yy}[(2m-1)T] = R^*_{y_ey_o}[(2m-1)T] + R^*_{y_oy_e}[(2m-1)T] \tag{3.56}$$

For the actual $y^*(t)$, Equation 3.56 can be expressed as

$$R^*_{yy}[(2m-1)T] = 2R^*_{y_ey_o}[(2m-1)T] \tag{3.57}$$

Equations 3.55 and 3.57 represent the autocorrelation function of $y^*(t)$ at even and odd sampling times respectively. Thus the autocorrelation of $y^*(t)$ at every sampling time is simply the sum of Equations 3.55 and 3.57, i.e.

$$R^*_{yy}(mT) = R^*_{yy}(2mT) + R^*_{yy}[(2m-1)T] \tag{3.58}$$

The Fourier transform of $R^*_{yy}(mT)$ yields the spectral density $S^*_{yy}(f)$. Figure 3.14(a) shows an impulse train $R_e(2mT)$ where the magnitude of each impulse is unity. The Fourier series of this impulse train is a frequency impulse train as shown in Figure 3.14(b). The magnitude of an impulse is $\frac{1}{2}T$ or $f_p/2$, and the spacing between impulses is $f_p/2$. The impulse train $R_0[(2m-1)T]$ shown in Figure 3.14(c) does not have an impulse at $\theta = 0$. The Fourier series representation is again a frequency impulse train where the frequency impulses are spaced $f_p/2$ Hz apart and of magnitude $f_p/2$,

Calculation of quantization noise

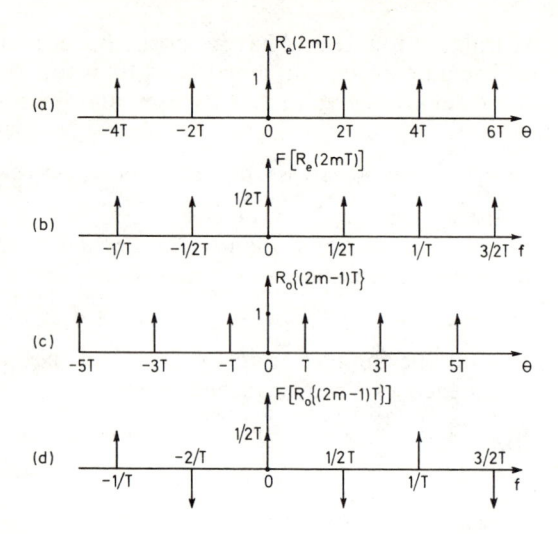

Fig. 3.14. Sampled autocorrelation functions and their Fourier transforms: (a) and (b) at even sampling times, (c) and (d) at odd sampling times

but the polarity of the impulse train alternates as shown in Figure 3.14(d). Hence, if

$$F[R_{yy}^*(2mT)] = S_e^*(f)$$

where F is the Fourier transform, then

$$F[R_{yy}^*((2m-1)T)] = (-1)^m S_o^*(f)$$

The $(-1)^m$ term takes into account the alternating polarity of the frequency impulse train, Figure 3.14(d).

The Fourier transform of Equation 3.58 is therefore

$$S_{yy}^*(f) = S_e^*(f) + (-1)^m S_o^*(f) \qquad (3.59)$$

Using Equations 3.39 and 3.59

$$S_{yy}^*(f) = \left(\frac{f_p}{2}\right)^2 \sum_{m=-\infty}^{\infty} \left\{ S_e\left(f + m\frac{f_p}{2}\right) + (-1)^m S_o\left(f + m\frac{f_p}{2}\right) \right\} \qquad (3.60)$$

3.9.2. Sampled spectral density function in terms of Normal spectra

Using the results so far obtained a general result for sampled Normal spectra can be determined. From Equations 3.12 and 3.39 the spectral

Calculation of quantization noise

density function of a sampled signal $y(t)$ is

$$S_y^*(f) = f_p^2 \sum_{m=-\infty}^{\infty} \sum_{n=0}^{\infty} b_n^2 N_n \phi_x^n(f+mf_p)$$

Let

$$\{\phi_x^n(f)\}^* = f_p^2 \sum_{m=-\infty}^{\infty} \phi_x^n(f+mf_p) \quad (3.61)$$

Hence

$$S_y^*(f) = \sum_{n=0}^{\infty} b_n^2 N_n \{\phi_x^n(f)\}^* \quad (3.62)$$

Using Equation 3.62 the sampled spectral density function $S_{yy}^*(f)$ of the signal $y^*(t)$ will now be expressed in terms of Normal spectra. $S_e(f)$, the first term in Equation 3.60, forms a Fourier transform pair with $R_{yy}^*(2mT)$, see Equation 3.55. Thus at even sampling times

$$S_e^*(f) = \left(\frac{f_p}{2}\right)^2 \sum_{n=0}^{\infty} S_e\left(f+m\frac{f_p}{2}\right) = F\{R_{y_e y_e}^*(2mT) + R_{y_o y_o}^*(2mT)\} \quad (3.63)$$

and from Equation 3.62

$$S_e^*(f) = \sum_{n=0}^{\infty} b_{1,n}^2 N_n \{\phi_x^n(f)\}^* + \sum_{n=0}^{\infty} b_{2,n} N_n \{\phi_x^n(f)\}^* \quad (3.64)$$

where $b_{1,n}$ and $b_{2,n}$ are related to the non-linear even and odd staircase functions, respectively and are defined by Equation 3.10 for $z(x)$ equal to $z_e(x)$ and $z_o(x)$, respectively.

Applying Equation 3.61 the sampled nth autoconvolution of the input spectral density function $\phi_x(f)$ is

$$\{\phi_x^n(f)\}^* = \left(\frac{f_p}{2}\right)^2 \sum_{m=-\infty}^{\infty} \phi_x^n\left(f+m\frac{f_p}{2}\right) \quad (3.65)$$

At odd sampling instants, from Equations 3.57 and 3.60

$$S_o^*(f) = \left(\frac{f_p}{2}\right)^2 \sum_{m=-\infty}^{\infty} (-1)^m S_o\left(f+m\frac{f_p}{2}\right) = F\{2R_{y_e y_o}[(2m-1)T]\} \quad (3.66)$$

3.9.3. Cross-spectral density function in terms of Normal spectra

In order to evaluate Equation 3.66 a general result describing cross-spectral density functions in terms of Normal spectra will now be established.

Calculation of quantization noise

Let $A(t)$ and $B(t)$ be two non-linear functions of $C(t)$. Representing $A(t)$ and $B(t)$ in Hermite polynomial basis functions as described by Equation 3.2

$$A(t) = \sum_{n=0}^{\infty} b_{A,n} \cdot H_n[x(t)] \tag{3.67}$$

$$B(t) = \sum_{m=0}^{\infty} b_{B,m} \cdot H_m[x(t)] \tag{3.68}$$

The cross-correlation function $R_{A,B}(\theta)$ is

$$R_{A,B}(\theta) = \langle A(t)B(t) \rangle$$

where $\langle \, \rangle$ symbolises time averaging.

$$R_{A,B}(\theta) = \left\langle \sum_{n=0}^{\infty} b_{A,n} \cdot H_n[x(t)] \sum_{m=0}^{\infty} b_{B,m} \cdot H_m[x(t)] \right\rangle$$

$$= \sum_{n=0}^{\infty} \sum_{m=0}^{\infty} b_{A,n} \cdot b_{B,m} \langle H_n[x(t)] H_m[x(t)] \rangle$$

Now the Hermite polynomial of order n is orthogonal with the probability density function

$$\langle H_n[x(t)] \cdot H_m[x(t)] \rangle = 0, \qquad m \neq n$$
$$= N_n \rho_x^n(\theta), \qquad m = n$$

where N_n and $l_x^n(\theta)$ are defined in Section 3.6. Hence

$$R_{A,B}(\theta) = \sum_{n=0}^{\infty} b_{A,n} b_{B,n} N_n \rho_x^n(\theta) \tag{3.69}$$

The cross-spectral density function is found by taking the Fourier transform of Equation 3.69

$$S_{A,B}(f) = \sum_{n=0}^{\infty} b_{A,n} b_{B,n} N_n \phi_x^n(f) \tag{3.70}$$

where $\phi_x^n(f) = F[\rho^n(\theta)]$. Applying this result to Equation 3.66 yields

$$S_o^*(f) = \sum_{n=0}^{\infty} 2 b_{1,n} b_{2,n} N_n \{\phi_x^n(f)\}^* (-1)^m \tag{3.71}$$

where $\{\phi_x^n(f)\}^*$ is given by Equation 3.65.

Hence, the spectral density function $S_{yy}^*(f)$ of the sampled $y(t)$ signal defined by Equation 3.59 is formed by the addition of Equations 3.64

and 3.71 and replacing $\{\phi_x^n(f)\}^*$ by Equation 3.61.

$$S_{yy}^*(f) = \left(\frac{f_p}{2}\right)^2 \sum_{n=0}^{\infty} \sum_{m=-\infty}^{\infty} \left\{ b_{1,n}^2 N_n \phi_x^n \left(f + m\frac{f_p}{2}\right) \right.$$
$$\left. + b_{2,n}^2 N_n \phi_x^n \left(f + m\frac{f_p}{2}\right) + 2b_{1,n} b_{2,n} N_n \phi_x^n \left(f + m\frac{f_p}{2}\right)(-1)^m \right\} \quad (3.72)$$

3.9.4. Alternative representation of sampled spectral density function

Equations 3.72 can be simplified by considering the odd and even integer values of m over the range $-\infty$ to $+\infty$.

When m is odd, let $m = 2k - 1$, and when m is even, let $m = 2k$. Equation 3.72 becomes

$$S_{yy}^*(f) = \left(\frac{f_p}{2}\right)^2 \left[\sum_{n=0}^{\infty} \sum_{k=-\infty}^{\infty} \left\{ b_{1,n}^2 N_n \phi_x^n \left(f + 2k\frac{f_p}{2}\right) + b_{2,n}^2 N_n \phi_x^n \left(f + 2k\frac{f_p}{2}\right) \right.\right.$$
$$\left. + 2b_{1,n} b_{2,n} N_n \phi_x^n \left(f + 2k\frac{f_p}{2}\right)(-1)^{2k} \right\} + \sum_{n=0}^{\infty} \sum_{k=-\infty}^{\infty} \left\{ b_{1,n}^2 N_n \phi_x^n \right.$$
$$\times \left(f + (2k-1)\frac{f_p}{2}\right) + b_{2,n}^2 N_n \phi_x^n \left(f + (2k-1)\frac{f_p}{2}\right)$$
$$\left.\left. + 2b_{1,n} b_{2,n} N_n \phi_x^n \left(f + (2k-1)\frac{f_p}{2}\right)(-1)^{2k-1} \right\} \right]$$

Since $(-1)^{2k} = 1, (-1)^{2k-1} = -1$

$$S_{yy}^*(f) = \left(\frac{f_p}{2}\right)^2 \left\{ \sum_{n=0}^{\infty} \sum_{k=-\infty}^{\infty} (b_{1,n} + b_{2,n})^2 N_n \phi_x^n(f + kf_p) \right.$$
$$\left. + \sum_{n=0}^{\infty} \sum_{k=-\infty}^{\infty} (b_{1,n} - b_{2,n})^2 N_n \phi_x^n \left(f + (2k-1)\frac{f_p}{2}\right) \right\}$$

which can be written in the shortened form of

$$S_{yy}^*(f) = \tfrac{1}{4} \sum_{n=0}^{\infty} (b_{1,n} + b_{2,n})^2 N_n \{\phi_x^n(f)\}^*$$
$$+ \tfrac{1}{4} \sum_{n=\infty}^{\infty} (b_{1,n} - b_{2,n})^2 N_n \left\{\phi_x^n\left(f - \frac{f_p}{2}\right)\right\}^* \quad (3.73)$$

where in Equation 3.73

$$\{\phi_x^n(f)\}^* = f_p^2 \sum_{k=-\infty}^{\infty} \phi_x^n(f + kf_p)$$

and

$$\left\{\phi_x^n\left(f-\frac{f_p}{2}\right)\right\}^* = f_p^2 \sum_{k=-\infty}^{\infty} \phi_x^n\left(f+(2k-1)\frac{f_p}{2}\right)$$

The feedback signal $y(t)$ in the delta modulator at sampling instants has a spectral density function $S_{yy}^*(f)$ which is represented either by Equation 3.72 or Equation 3.73. In Equation 3.72 $S_{yy}^*(f)$ has been established by sampling at half the clock rate at odd and even clock times and results in spectra located at multiples of half the clock frequency f_p. However, Equation 3.73 shows that the spectra are located at multiples of f_p because $\{\phi_x^n(f)\}^*$ is identical to $\{\phi_x^n[f-(f_p/2)]\}^*$, except for a frequency shift $f_p/2$. The terms $(b_{1,n} \pm b_{2,n})^2$ are relatively easy to evaluate and consequently the computation of $S_{yy}^*(f)$ is more easily accomplished using Equation 3.73 than Equation 3.72.

3.9.5. Spectral density function of the feedback signal

The signal $y(t)$ at the output of the integrator in the delta modulator decoder (Figure 1.1) is identical with the signal at the output of the hold circuit in the staircase model of Figure 1.14.

The spectral density function of $y(t)$ is

$$S_{yy}^*(f) \cdot S_h(f)$$

where

$$S_h(f) = |H(j2\pi f)|^2$$

and $H(j2\pi f)$ is the transfer function of the zero-order hold circuit.

3.9.6. Power in the decoded signal

The power at the output of the integrator in the message band is

$$P^2 = 2 \int_{f_{c1}}^{f_{c2}} S_{yy}^*(f) |H(j2\pi f)|^2 \, df \tag{3.74}$$

where $f_{c1} = f_a$, $f_{c2} = f_b$ in Figure 3.4, and

$$|H(j2\pi f)| = \frac{1}{f_p} \cdot \frac{\sin(\pi f/f_p)}{(\pi f/f_p)} \tag{3.75}$$

Note that P^2 includes noise power as well as signal power, i.e. it is the input signal power plus the error power, suitably modified by the encoder's characteristics.

3.9.7. Summary

The sampled autocorrelation function at the output of the adder in the Staircase model of the delta modulator (Figure 1.14) has been established in terms of the sampled autocorrelation and sampled cross correlation functions of the signals at the output of the even and odd staircase functions.

Taking the Fourier transform of the sampled autocorrelation function at the output of the adder gave the corresponding sampled spectral density function $S_{yy}^*(f)$ in terms of spectra located at harmonics of half the clock rate (Equation 3.60).

The next step was to express $S_{yy}^*(f)$ in terms of Normal spectra. This resulted in Equation 3.72 and then its simplified version, Equation 3.73. This latter equation expresses $S_{yy}^*(f)$ in terms of two sets of sampled spectral densities functions, where the sampling rate is equal to the delta modulator clock rate, and the sets are displaced in frequency by half the clock rate.

Having determined $S_{yy}^*(f)$, the power in the message band of the signal at the output of the integrator in the delta modulator was established. This power has to be split into a term linearly related to the input power plus a term dependent on quantization effects which is the object of the next section.

3.10. NOISE POWER AT THE DECODER OUTPUT

The distortion in the signal at the output of the delta modulator decoder is due to amplitude and time quantization. The time quantization is the result of sampling and the amplitude quantization is due to the input signal being passed through the two staircase functions. Accordingly when dealing with amplitude distortion it is convenient to consider the difference between the input signal and the signal at the output of the staircase functions. Let

$$\Delta y_e(t) = y_e(t) - x(t) \qquad (3.76)$$

$$\Delta y_o(t) = y_o(t) - x(t) \qquad (3.77)$$

The signals $\Delta y_e(t)$ and $\Delta y_o(t)$ can be considered to occur at the output of infinite sawtooth functions.

Thus the quantized signal at the output of an infinite staircase function can be considered as the sum of the input signal $x(t)$ and the signal at the output of the infinite sawtooth function. These infinite sawtooth functions are formed by subtracting the values of the abscissa from the ordinates in the infinite staircase functions (see Figure 1.13(a)). A simple relationship will now be established between the Normal spectrum coefficients b_n associated with the infinite staircase functions and those related to the corresponding infinite sawtooth functions. This relationship between the b_n coefficients used in the Normal spectrum technique will enable the spectral density function $S_{yy}^*(f)$ given by Equation 3.73 to be expressed as two separate terms, one dependent on the input signal and the other representing the effects of quantization.

3.10.1. Normal spectrum coefficients for infinite sawtooth and staircase functions

Let the Normal spectrum coefficients relating to the signals $\Delta y_e(t)$ and $\Delta y_o(t)$ be $b_{e,n}$ and $b_{o,n}$ respectively. The corresponding coefficients related to the $y_e(t)$ and $y_o(t)$ signals are $b_{1,n}$ and $b_{2,n}$.

The relationship between the Normal spectrum coefficients of the even infinite staircase and sawtooth functions will now be established. The result will also apply to the odd version of these functions.

From Equation 3.10, the $b_{n,e}$ coefficients for the even infinite sawtooth function are obtained by replacing $z(x)$ by $z_e(x) - x$, where $z_e(x)$ applies to the even infinite staircase function. Thus

$$b_{n,e} = \frac{1}{N_n} \int_{-\infty}^{\infty} H_n(x)\{z_e(x) - x\} f(x) \, dx \tag{3.78}$$

$$b_{n,e} = b_{1,n} - \frac{1}{N_n} \int_{-\infty}^{\infty} H_n(x) x f(x) \, dx$$

$$b_{n,e} = b_{1,n} - b_{x,n} \tag{3.79}$$

where $b_{1,n}$ are the coefficients applying to the even infinite staircase function.

Examining $b_{x,n}$ for different values of n using Equations 3.3 and 3.6

$$b_{x,1} = \frac{1}{\sigma^2} \int_{-\infty}^{\infty} x^2 f(x) \, dx$$

$$b_{x,2} = \frac{1}{2\sigma^4} \int_{-\infty}^{\infty} x^3 f(x) \, dx - \frac{1}{2\sigma^2} \int_{-\infty}^{\infty} x f(x) \, dx$$

$$b_{x,3} = \frac{1}{6\sigma^6} \int_{-\infty}^{\infty} x^4 f(x) \, dx - \frac{1}{2\sigma^4} \int_{-\infty}^{\infty} x^2 f(x) \, dx$$

Calculation of quantization noise

and so on. Now[12]

$$\int_{-\infty}^{\infty} x^n f(x) \, dx = \begin{cases} 1, 3, \ldots, (n-1)\sigma^n, & \text{for } n \text{ even} \\ 0, & \text{for } n \text{ odd} \end{cases}$$

and consequently the $b_{x,n}$ term simplifies to

$$b_{x,1} = 1, \quad b_{x,n} = 0 \quad \text{for } n \geq 2 \qquad (3.80)$$

The relationship between the Normal spectrum coefficients of the infinite staircase and sawtooth functions is now established by applying the result of Equation 3.80 to Equation 3.79, i.e.

$$\begin{aligned} b_{1,e} &= b_{1,1} - 1 \\ b_{n,e} &= b_{1,n}, \quad n \geq 2 \end{aligned} \qquad (3.81)$$

The result for the odd infinite staircase and sawtooth functions is

$$\begin{aligned} b_{1,o} &= b_{2,1} - 1 \\ b_{n,o} &= b_{2,n}, \quad n \geq 2 \end{aligned}$$

Equation 3.81 states that for $n \geq 2$ the Normal spectrum coefficients of both functions are identical–the first order coefficient for the infinite sawtooth function is the first order coefficient for the infinite staircase function minus one.

In the next section we show how only the odd order Normal spectrum coefficients need be calculated, and then proceed to utilise the simple relationship between the Normal spectrum coefficients for the sawtooth and infinite staircase functions to develop the expression for $S_{yy}^*(f)$ given by Equation 3.73. This re-arranged expression for $S_{yy}^*(f)$ isolates the sampled distortion spectral density function $S_{nn}^*(f)$ which results from time and amplitude quantization effects in the encoding process.

3.10.2. Output spectral density function as the sum of two components

It will be recalled that the expression for $S_{yy}^*(f)$ given by Equation 3.73 was for the sampled spectral density of the signal $y(t)$ in the delta modulator, see Figures 1.1 and 1.14. The disadvantage of this expression is that it does not distinguish between the wanted signal and the distortion as does Equation 3.81.

We shall now recall an important property of odd non-linear functions. With the non-linearity $z(x) = -z(-x)$ which is of the type associated with delta modulators and pulse code modulators, the coefficients defined by Equation 3.10 are zero when n is *even*. This can be shown simply from Equation 3.10 by dividing the integration interval into two parts

$$b_n = -\frac{1}{N_n}\int_{-\infty}^{0} z(x)H_n(x)f(x)\,dx$$
$$+\frac{1}{N_n}\int_{0}^{\infty} z(x)H_n(x)f(x)\,dx = 0$$

Thus, when $z(x) = -z(-x)$ only the odd order b_n coefficients need be calculated.

Substituting the b_n coefficients from Equation 3.81 into Equation 3.73 and remembering that even order coefficients can be ignored, then

$$S_{yy}^*(f) = \frac{\sigma^2}{4}(b_{e,1}+b_{o,1}+2)^2\{\phi_x(f)\}^* + \frac{1}{4}\sum_{n=2}^{\infty} N_n(b_{e,n}+b_{o,n})^2\{\phi_x^n(f)\}^*$$
$$+\frac{1}{4}\sum_{n=0}^{\infty} N_n(b_{e,n}-b_{o,n})^2\left\{\phi_x^n\left(f-\frac{f_p}{2}\right)\right\}^*$$

and $(b_{e,1}+b_{o,1}+2)^2 = 4(b_{e,1}+b_{o,1}+1)+(b_{e,1}+b_{o,1})^2$ hence,

$$S_{yy}^*(f) = \sigma^2(b_{e,1}+b_{o,1}+1)\{\phi_x(f)\}^* + \frac{1}{4}\sum_{n=0}^{\infty} N_n(b_{e,n}+b_{o,n})^2\{\phi_x^n(f)\}^*$$
$$+\frac{1}{4}\sum_{n=0}^{\infty} N_n(b_{e,n}-b_{o,n})^2\left\{\phi_x^n\left(f-\frac{f_p}{2}\right)\right\}^* \quad (3.82)$$

This representation of $S_{yy}^*(f)$ has an advantage over Equation 3.73 in that the first term is proportional to the input spectral density function, while the second and third terms correspond to the effects of quantization.

Writing Equation 3.82 as

$$S_{yy}^*(f) = K\{\phi_x(f)\}^* + S_{nn}^*(f) \quad (3.83)$$

where

$$K = \sigma^2(b_{e,1}+b_{o,1}+1) \quad (3.84)$$

K is the gain factor for the encoder and is generally close in value to σ^2. $S_{nn}^*(f)$ is the sampled distortion spectral density function and is conveniently represented by

$$S_{nn}^*(f) = \sum_{n=0}^{\infty} C_{1,n}\{\Phi_x^n(f)\}^* + \sum_{n=0}^{\infty} C_{2,n}\left\{\phi_x^n\left(f-\frac{f_p}{2}\right)\right\}^* \quad (3.85)$$

or

$$S_{nn}^*(f) = S_{CEN}^*(f) + S_{OFF}^*(f) \quad (3.86)$$

where $S_{CEN}^*(f)$ and $S_{OFF}^*(f)$ are the first and second terms respectively in

Calculation of quantization noise 109

Equation 3.85, and from Equations 3.82 and 3.85

$$C_{1,n} = N_n \left(\frac{b_{e,n}+b_{o,n}}{2}\right)^2 \quad (3.87)$$

$$C_{2,n} = N_n \left(\frac{b_{e,n}-b_{o,n}}{2}\right)^2 \quad (3.88)$$

$S^*_{CEN}(f)$ and $S^*_{OFF}(f)$ are referred to as the centred and off-centred sampled spectral density functions respectively. $S^*_{CEN}(f)$ spectra occur at zero and harmonics of the clock rate f_p, whereas $S^*_{OFF}(f)$ spectra occur at harmonics of $f_p/2$. It is the representation of the distortion spectral density function into two components which enables the experimental results to be easily explained.

3.10.3. Output power in the message band

With the establishment of Equation (3.83) the calculation of the performance of the encoder is virtually completed.

Substituting $S^*_{yy}(f)$ from Equation 3.83 and $S^*_{nn}(f)$ from Equation 3.86 in Equation 3.74 gives

$$P^2 = 2\int_{f_{c1}}^{f_{c2}} K\{\phi_x(f)\}^*|H(j2\pi f)|^2 \, df + 2\int_{f_{c1}}^{f_{c2}} [\{S^*_{CEN}(f)+S^*_{OFF}(f)\} \\ \times |H(j2\pi f)|^2] \, df$$

$$= P^2_{so} \text{ (signal power)} + N^2_q \text{ (distortion power)} \quad (3.89)$$

Observe from Equation 3.61 that $\{\phi_x(f)\}^*$ in P_{so} can be represented as

$$\{\phi_x(f)\}^* = f_p^2 \sum_{m=-\infty}^{\infty} \phi_x(f+mf_p)$$

and the shifted spectra, for $m \neq 0$, are outside the passband f_{c1} to f_{c2}. Hence

$$\{\phi_x(f)\}^* = f_p^2 \cdot \phi_x(f), \qquad f_{c1} \leq f \leq f_{c2}$$

The situation with the noise spectra is more complex as the nth order convolution of the input spectral density function results in the functions having 'skirts', see Figure 3.5. This effect together with that of sampling produces power in the frequency band f_{c1} to f_{c2}.

3.10.4. Signal-to-quantization noise ratio

The signal-to-quantization noise ratio, s.q.n.r. can be defined, as in Section 1.2.8, as the ratio of the signal power at the input to noise power at the output in the presence of the signal, i.e.

$$\text{s.q.n.r.} = \frac{\sigma^2}{N_q^2} \qquad (3.90)$$

Alternatively, s.q.n.r. can be defined as the ratio of the signal power in the presence of noise at the decoder output to the noise power at the output of the decoder in the presence of the signal, i.e.

$$\text{s.q.n.r.} = \frac{P_{so}^2}{N_q^2} \qquad (3.91)$$

There are other definitions of signal-to-noise ratio but the two definitions given above are most suitable for a d.m. system which is symmetrical and is operating without any channel errors.

It should be noted that the above two s.q.n.r.'s are generally equivalent, as K in Equation (3.84) approximates to σ^2.

3.11. PERFORMANCE OF THE DELTA MODULATOR

3.11.1. Presentation of results

The distortion sampled density function $S_{nn}^*(f)$ in Equation 3.85 can be written with the aid of Equation 3.61 as

$$\frac{S_{nn}^*(f)}{f_p^2} = \sum_{n=0}^{\infty} C_{1,n} \sum_{k=-\infty}^{\infty} \phi_x^n(f + kf_p)$$
$$+ \sum_{n=0}^{\infty} C_{2,n} \sum_{k=-\infty}^{\infty} \phi_x^n\left(f + (2k-1)\frac{f_p}{2}\right) \qquad (3.92)$$

The value of the distortion spectral density $S_{nn}(f)$ at the output of the delta modulator is found by multiplying $S_{nn}^*(f)$ by the power transfer function of the hold circuit

$$S_{nn}(f) = S_{nn}^*(f) \cdot |H(j2\pi f)|^2$$

$|H(j2\pi f)|^2$ is given by Equation 3.75, thus

$$S_{nn}(f) = \frac{S_{nn}^*(f)}{f_p^2} \cdot \frac{\sin(\pi f/f_p)}{(\pi f/f_p)} \qquad (3.93)$$

The graphs of distortion spectral density functions used in this section have ordinates $S_{nn}^*(f)/f_p^2$, i.e. Equation 3.92 is plotted as a function of frequency.

The highest normalised frequency f_{c2} in the input signal is unity. Consequently the abscissa of the $S_{nn}^*(f)/f_p^2$ graph is f/f_{c2}.

Calculation of quantization noise 111

The lowest frequency in the input signal has different values in different graphs. The values vary from zero to $0.8 f_{c2}$. The input power σ^2 is maintained at unity unless otherwise stated.

3.11.2. Distortion coefficients

The C_n coefficients described by Equations 3.87 and 3.88 are plotted in Figure 3.15 against the order of convolution n for two values of step-

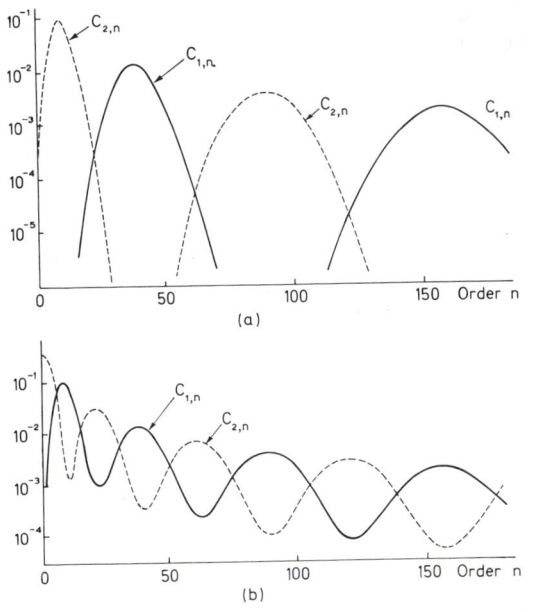

Fig. 3.15. *The C_n coefficients as a function of n for $\sigma = 1$ and* (a) $\gamma = 1$, (b) $\gamma = 2$

size γ. From Equations 3.87 and 3.88 it can be observed that the C_n coefficients depend on the Normal spectrum coefficients $b_{e,n}$ and $b_{o,n}$ which are related to the infinite sawtooth functions whose periodicity and peak-to-peak amplitude is 2γ. Hence the C_n coefficients are dependent on the delta modulator step-size γ as demonstrated in Figure 3.15.

Observations from Figure 3.15

(1) The C_n coefficients have maxima and minima as n is increased. The difference between adjacent maxima and minima is greatest and

the spacing between adjacent peaks increases when the step-size is reduced.

(2) The average magnitude of the C_n coefficients decrease with increasing n, and decreasing step-size.

(3) As n increases from zero $C_{2,n}$ has a maximum before $C_{1,n}$, irrespective of the step size.

Small step-size γ corresponds to conditions of good encoding. From Figure 3.15(a) it is apparent that for n up to about 25 only the $C_{2,n}$ coefficients are significant. As the magnitude of the nth autoconvolutions of the input spectral density decreases with increasing n (see Figure 3.5) then for good encoding the distortion spectral density function depends on the off-centred spectral density function $S_{OFF}(f)$, and is virtually independent of the centred spectral density function $S_{CEN}(f)$. As $S_{OFF}(f)$ decreases with frequency from its location at half the clock rate the value of $S'_{OFF}(f)$ in the message band is small. This characteristic will be emphasised in the following sub-sections.

3.11.3. Variations of step size

When the input signal is band pass white noise the nth autoconvolution of the spectral density function of this signal is shown in Figure 3.5 for various values of n. If $x(t)$ is narrow-band white noise, the nth order autoconvolution of $\phi_x(f)$ is composed of peaks and valleys as shown by $\phi_x^3(f)$ and $\phi_x^5(f)$ in Figure 3.5(b). When the lowest frequency f_{c1} in $x(t)$ is made to approach the highest frequency in this signal, these peaks and valleys in $\phi_x^n(f)$ become more exaggerated. However, if f_{c1} is zero $\phi_x(f) = \lambda_x(f)$ is a low-pass white noise spectral density function and $\phi_x^n(f)$ is given by the smooth curves shown in Figure 3.5(a).

Figure 3.16 shows the variation of $S_{nn}^*(f)/f_p^2$ as a function of normalised frequency f/f_{c2}. The curves are drawn with the aid of Equation 3.92 and are therefore dependent on $\phi_x^n(f)$. When $x(t)$ is wide-band white noise the curves in Figure 3.16(a) apply and are smooth because $\phi_x^n(f)$ is smooth for this type of input signal. However, when $x(t)$ is narrow-band white noise, $\phi_x^n(f)$ is composed of maxima and minima, as previously described, and this results in Curve (b) in Figure 3.16 having peaks and valleys.

Consider the effect of γ on the $S_{nn}^*(f)/f_p^2$ curves. Because the C_n coefficients are given by Equations 3.87 and 3.88, they are a function of γ since they depend on the Normal spectrum coefficients $b_{e,n}$, $b_{o,n}$ of the infinite sawtooth characteristics $\Delta y_e(t)$ and $\Delta y_o(t)$ which are a function of γ. The $S_{nn}^*(f)/f_p^2$ curves shown in Figure 3.16 for two values of γ are calculated by using Equation 3.92. However, as an aid to understanding why $S_{nn}^*(f)/f_p^2$ has a high value in the message band when γ is high, but not when it is low, and why it has a high value for frequencies in the vicinity

Calculation of quantization noise

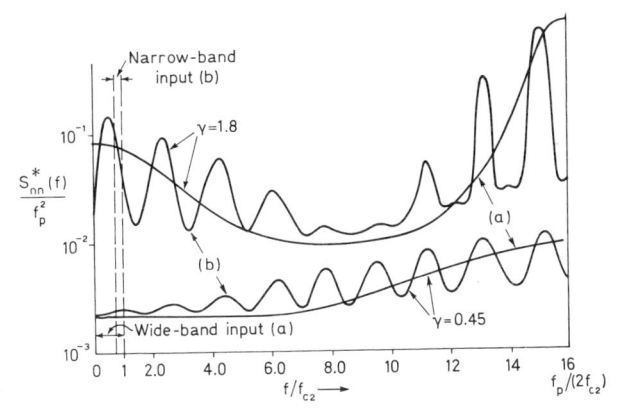

Fig. 3.16. Distortion spectral density functions for $\sigma = 1$, $\gamma = 1·8$ and $0·45$: (a) low-pass and (b) narrow-band white noise input signals

of half the clock rate, some approximations will be made to Equation 3.92.
When γ is small $C_{1,n}$ is small compared to $C_{2,n}$ for values of n up to approximately 25 (Section 3.11.2). This means that the significant spectra are off-centred and located at $\pm f_p/2$, $3f_p/2$, $\pm 5f_p/2$, etc. Assuming that f_p is sufficiently large so that over the frequency band $\pm f_p/2$ only the off-centred spectra at $\pm f_p/2$ need be considered. Then Equation 3.92 becomes

$$\frac{S^*_{nn}(f)}{f_p^2} \doteq \sum_{n=0}^{\infty} C_{2,n}\phi_x^n\left(f-\frac{f_p}{2}\right) + \sum_{n=0}^{\infty} C_{2,n}\phi_x^n\left(f+\frac{f_p}{2}\right) \quad (3.94)$$

The curve drawn in Figure 3.16 for a wide-band input signal and $\gamma = 0·45$ indicates the presence of spectra at $\pm f_p/2$ as the main influence on the shape of this curve. When γ is made large both the $C_{1,n}$ and $C_{2,n}$ coefficients need to be considered, and assuming that centred spectra at $\pm f_p$, $\pm 2f_p$, $\pm 3f_p$, etc. and offcentred spectra at $\pm 3f_p/2$, $\pm 5f_p/2$, $\pm 7f_p/2$, etc. can be ignored over the frequency range $\pm f_p/2$

$$\frac{S^*_{nn}(f)}{f_p^2} \doteq \sum_{n=0}^{\infty} C_{1,n}\phi_x^n(f) + \sum_{n=0}^{\infty} C_{2,n}\phi_x^n\left(f-\frac{f_p}{2}\right)$$
$$+ \sum_{n=0}^{\infty} C_{2,n}\phi_x^n\left(f+\frac{f_p}{2}\right) \quad (3.95)$$

Hence for large γ and f_p there is a spectrum located at d.c. and others located at $\pm f_p/2$. Consequently the $S^*_{nn}(f)/f_p^2$ have maxima at d.c. and $\pm f_p/2$ with the result that a minimum occurs close to $\pm f_p/4$.

When a narrow band white noise signal is applied to the encoder the $S^*_{nn}(f)/f_p^2$ curve oscillates as shown in Figure 3.16 due to the peaky nature of $\phi_x^n(f)$. These oscillatory curves have mean values similar to the corresponding smooth curves resulting from wide band input signals.

The distortion noise power N_q^2 at the output of the final filter in the decoder is to a good approximation

$$N_q^2 = 2 \int_{f_{c1}}^{f_{c2}} \frac{S_{nn}^*(f)}{f_p^2} df$$

because over the frequency band f_{c1} to f_{c2} Equation 3.93 is

$$S_{nn}(f) \doteq \frac{S_{nn}^*(f)}{f_p^2}$$

Hence the performance of the d.m. codec is determined by the magnitude and shape of the curves displayed in Figure 3.16 over the message band. A low quantization noise is therefore achieved by minimising the magnitude of the spectrum located at d.c. by using a small value of γ, and also by minimising the contribution of the spectra sited at $\pm f_p/2$ to the value of $S_{nn}^*(f)/f_p^2$ in the message band by increasing f_p. This point is dealt with in the next section.

3.11.4. Effect of clock rate on the distortion spectral density function

The effect of clock rate f_p on the distortion spectral density function is illustrated in Figure 3.17 for a low-pass white noise input signal. At low clock rates a substantial improvement in the reduction of the inband distortion noise can be achieved by increasing the clock rate. This fact is used in establishing the simplified equation of s.q.n.r. (Equation 1.7). The clock rate does not effect the shape of centred and off-centred spectra but it does control their location as can be seen in Figure 3.17. By increasing

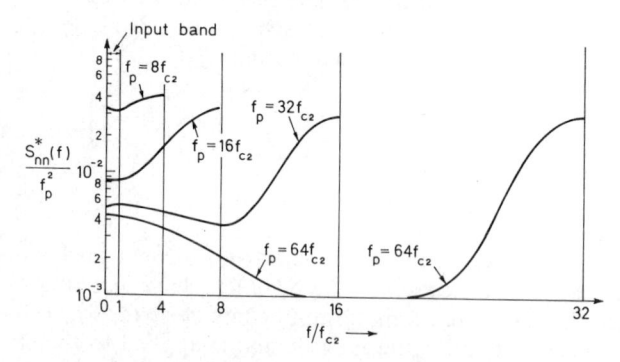

Fig. 3.17. Effect of f_p/f_{c2} ratios on distortion spectral density functions for $\sigma = 1, \gamma = 0.65$

Calculation of quantization noise

the clock rate a point is reached where the $S_{OFF}(f)$ spectra located at $\pm f_p/2$ contribute a negligible amount to the distortion spectral density in the message band.

Because $S_{OFF}(f)$ and $S_{CEN}(f)$ are dependent on the step-size and bandwidth of the input signal it follows that the value of f_p, above which no decrease of in-band distortion can be achieved, also depends on these parameters.

3.11.5. Signal-to-noise ratio

Using Equation 3.91 the signal-to-quantization noise ratio, s.q.n.r. as a function of input signal power P_i in dBm, i.e. dB referred to 1 mW, is represented by the solid part of the curves shown in Figure 3.18 Each curve is drawn for a constant clock rate f_p when the input signal is low-pass white noise. The encoder step-size is 50 mV.

Each curve increases in an approximately linear way until it reaches a maximum. This is because the noise power is substantially independent of the input power level, i.e. it remains virtually constant with increasing input signal power P_i until the peak s.q.n.r. is reached. This means that there is one input power level which gives the maximum signal-to-noise ratio. The dotted part of the curves show the signal-to-noise ratio decreasing due to the delta modulator experiencing slope overload. This condition is discussed in depth in the next chapter.

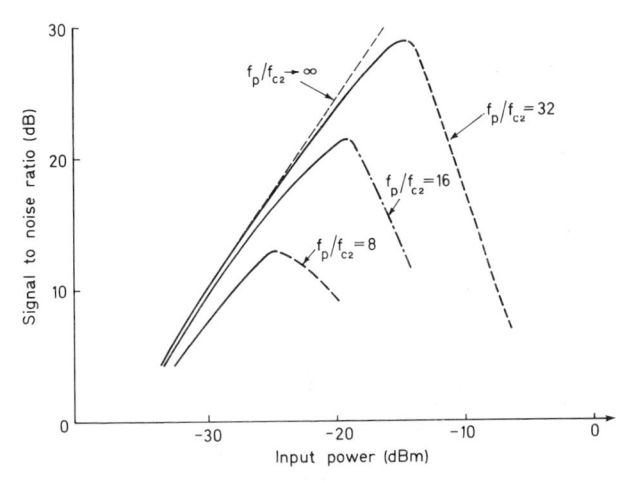

Fig. 3.18. Signal-to-noise ratio against input signal power for various values of f_p/f_{c2}. Low-pass white noise input signal, $\gamma = 50\ mV$

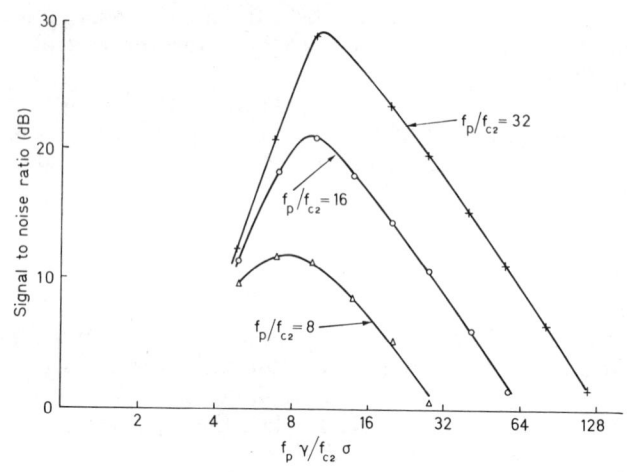

Fig. 3.19. Signal-to-noise ratio against $f_p\gamma/f_{c2}\sigma$ for various values of f_p/f_{c2}. Low-pass white noise input signal, $\sigma = 1$

Those curves with higher values of f_p/f_{c2} have larger s.q.n.r.'s. However, there is not much improvement to be gained in the s.q.n.r. by increasing f_p above $32f_{c2}$ as the $S_{OFF}(f)$ and $S_{CEN}(f)$ spectra have effectively separated for $f_p > 32f_{c2}$.

Figure 3.19 shows the signal-to-noise ratio as a function of the normalised parameter $(f_p\gamma/f_{c2}\sigma)$ for different f_p/f_{c2} ratios when $x(t)$ is a low-pass white noise signal. The maximum signal-to-noise ratio occurs for each of these curves when the encoder is tracking $x(t)$ for the majority of the time, but on occasions the encoder is slope overloaded. The parts of the curves having negative and positive slopes correspond to the system mainly experiencing quantization and slope overload effects, respectively.

If the input signal is RC shaped Gaussian noise the curves in Figure 3.19 are displaced to the left and elevated.[13] This means that higher s.q.n.r.'s are obtained for a given f_p/f_{c2} ratio when compared to the low-pass white noise signal, although γ/σ is reduced.

REFERENCES

1. de Jager, F., 'Delta modulation—a new method of p.c.m. transmission using the 1 unit code', *Philips Research Report*, **7**, 442–446, December (1952)
2. van de Weg, H., 'Quantizing noise of a single integration delta modulation system with an N-digit code'. *Philips Research Report*, No. 8, 367–385, (1953)
3. Zetterberg, L. H., 'A comparison between delta and pulse code modulation', *Ericsson Technics*, **11**, No. 1, 95–154, January (1955)
4. Bennett, W. R., 'Spectra of quantized signals', *Bell Systems Tech. J.*, **27**, No. 3, 446–472, July (1948)

5. Goodman, D. J., 'Delta modulation granular quantization noise', *Bell Systems Tech. J.*, 1197–1218, May–June (1969)
6. Iwerson, J. E., 'Calculated quantizing noise of single integration delta-modulation coders', *Bell Systems Tech. J.*, **48**, No. 7, 2359–2389, September (1969)
7. Passot, M. and Steele, R., 'Application of the Normal Spectrum technique to the calculation of distortion noise in delta modulators'. *Memorandum* No. 76, Electronic and Electrical Eng. Dept., University of Technology, Loughborough, England, May (1973)
8. Passot, M., 'Normal spectra application to the study of delta modulation', *Masters Thesis*, Electronic and Electrical Eng. Dept., University of Technology, Loughborough, England, March (1973)
9. Thomson, W. E., 'The response of a linear system to random noise', *Proc. I.E.E. Monograph*, **106**, Pt. C, September (1954)
10. Barrett, J. F. and Lampard, D. G., 'An expansion for some second-order probability distributions and its application to noise problems', *I.R.E. Trans. Information Theory*, **IT-1**, No. 1, 10–15, March (1955)
11. Kettel, E., 'The non-linear distortion of a noise signal with normal distribution', *Archiv der Elektrischen Ubertragung*, **22**, No. 4, 165–170, April (1968)
12. Papoulis, A., *Probability, Random Variables, and Stochastic Process*, McGraw-Hill (1965)
13. O'Neal, J. B. Jr., 'Delta modulation quantizing noise analytical and computer simulation results for Gaussian and television signals,' *Bell Systems Tech. J.*, **45**, No. 1, 117–141, January (1966)

Chapter 4

SLOPE OVERLOAD NOISE IN LINEAR d.m. SYSTEMS

4.1. INTRODUCTION

When the slope of the input signal exceeds the maximum slope of the feedback signal in the delta modulator the latter is said to be in slope overload. This condition has been briefly discussed in Section 1.2.5. In this chapter we examine the distortion in the decoded signal when the linear d.m. is slope overloaded by either sinusoidal or Gaussian signals. The former signal is periodic and the distortion of slope overload noise is relatively easy to estimate. The performance of the linear d.m. system is more difficult to analyse when the input Gaussian signal frequently causes it to be overloaded. Research workers, notably Protonotarios[1], O'Neal Jr,[2] Slepian[8] and Greenstein[9] have made a direct analytical approach to the problem but the theoretical results are complex. Others[3,4] have represented the delta modulator by models, and it is this technique that is used here.

4.2. SLOPE OVERLOAD WITH SINUSOIDAL INPUT SIGNALS

The general availability of signal generators in laboratories make them convenient for setting-up delta modulators using a sinusoidal input. However, for a more accurate estimate of performance it is necessary to use a signal more representative of the type which the decoder will encounter in service. This section establishes how linear and exponential d.m. behave when overloaded by a sinusoidal input signal. It will be found that when severe overloading occurs the output from the local decoder tends to be a triangular waveform.

4.2.1. Linear delta modulator

The $y(t)$ waveform at the output of the integrator in the linear delta modulator will now be calculated for the situation when a sinusoidal input causes overload for part of each cycle of the input,[5] as shown in Figure 1.4.

The quantization effects which result in the staircase nature of the $y(t)$ waveform will be ignored. This is justified on the basis that the significant

Slope overload noise in linear d.m. systems   119

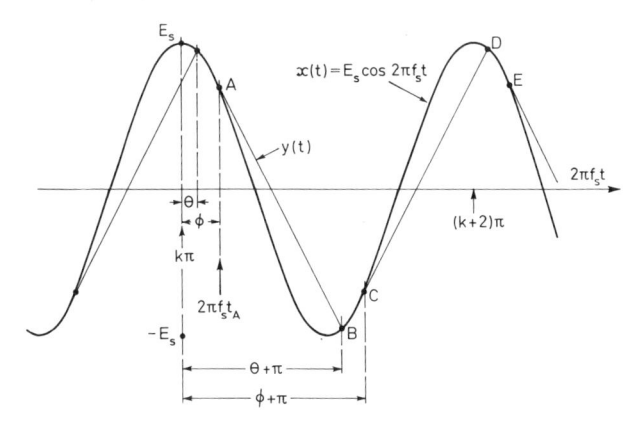

Fig. 4.1. *Slope overload of linear modulator: x(t) is the input signal and y(t) is the signal at the output of the local decoder*

distortion in the $y(t)$ waveform is due to the slope overload condition. Accordingly the staircase waveform shown in Figure 1.4 is reproduced in Figure 4.1 where the $y(t)$ curve is shown as a smooth line. Because the effects of quantization are ignored the waveform $y(t)$ is equal to the input waveform $x(t)$ at all times provided that the delta modulator is not overloaded, i.e.

$$y(t) = x(t) = E_s \cos 2\pi f_s(t) \qquad (4.1)$$

where $2\pi f_s E_s \leq \gamma f_p$.

When the delta modulator becomes slope overloaded on the downward part of the sinusoid, point A in Figure 4.1, the output $y(t)$ of the local decoder consists of a sequence of negative pulses which when perfectly integrated result in a $y(t)$ waveform of constant slope $-\gamma f_p$.

Point A occurs when the phase angle of $x(t)$ is ϕ radians measured from the peak immediately preceding where the phase is $k\pi$ radians. Now at point A

$$dx(t)/dt = x'(t) = -E_s 2\pi f_s \sin(\phi + k\pi) = -E_s 2\pi f_s \sin \phi$$

where k is assumed to be even, and as this is the point where slope overload occurs

$$\zeta = \gamma f_p = E_s 2\pi f_s \sin \phi \qquad (4.2)$$

The equation representing $y(t)$ between points A and B is the straight line

$$y(t) = -\zeta(t - t_A) + E_s \cos \phi$$

$$= -\frac{\zeta}{2\pi f_s}(2\pi f_s t - \phi - k\pi) + E_s \cos \phi \qquad (4.3)$$

where
$$2\pi f_s t_A = \phi + k\pi \tag{4.4}$$
and $t \geq t_A$.

The condition of slope overload ceases at point B where $t = t_B$ and
$$y(t) = E_s \cos 2\pi f_s t_B$$

Consequently
$$-\zeta(t_B - t_A) + E_s \cos \phi = E_s \cos 2\pi f_s t_B \tag{4.5}$$
and
$$2\pi f_s t_B = \theta + (k+1)\pi \tag{4.6}$$

Substituting the values of t_A and t_B in Equation 4.5
$$E_s \cos \phi - \frac{\zeta}{2\pi f_s}(\theta - \phi + \pi) = E_s \cos\{\theta + (k+1)\pi\}$$
$$= E_s \cos \theta \cdot (-1)^{k+1}$$

k must be even to agree with Figure 4.1, which is a condition already assumed, giving the obvious result that ζ is the voltage difference between A and B divided by the time between A and B.

$$E_s(\cos \phi + \cos \theta)\frac{2\pi f_s}{(\theta + \pi) - \phi} = \zeta \text{ V/s} \tag{4.7}$$

The delta modulator now tracks the input waveform for $(\phi - \theta)/2\pi f_s$ seconds until point C is reached when the derivative of $x(t)$ exceeds ζ and slope overload recommences.

By applying similar arguments to those employed in the derivation of Equation 4.3, the equation of $y(t)$ over the interval CD is

$$y(t) = -E_s \cos \phi + \frac{\zeta}{2\pi f_s}(2\pi f_s t - \phi - (k+1)\pi) \tag{4.8}$$

Thus the waveform $y(t)$ can be represented when tracking, as
$$y(t) = x(t) = E_s \cos 2\pi f_s t \tag{4.9}$$
this condition applies over the range
$$\theta + k\pi \leq 2\pi f_s t \leq \phi + k\pi \tag{4.10}$$
and when in slope overload, as
$$y(t) = (-1)^k E_s \cos \phi + (-1)^{k+1}\frac{\zeta}{2\pi f_s}\{2\pi f_s t - (\phi + k\pi)\} \tag{4.11}$$

Slope overload noise in linear d.m. systems 121

this condition applies over the range

$$\phi + k\pi \le 2\pi f_s t \le \theta + (k+1)\pi \tag{4.12}$$

where $k = 0, 1, 2, \ldots$

Condition when $\theta = \phi$

When this condition occurs the delta modulator cannot track the input sinewave over any part of its curve.
Substituting $\theta = \phi = \lambda$ in Equation 4.7 gives

$$\cos \lambda = \zeta/(4f_s E_s) \tag{4.13}$$

$$\zeta/(4f_s E_s) = (1 - \sin^2 \lambda)^{1/2} \tag{4.14}$$

From Equation 4.2

$$\sin \lambda = \frac{\zeta}{E_s 2\pi f_s} \tag{4.15}$$

Substituting from Equation 4.15 in 4.14

$$\frac{2\pi f_s E_s}{\zeta} = \left(1 + \frac{\pi^2}{4}\right)^{1/2}$$

or

$$2\pi f_s E_s = 1{\cdot}86\zeta \tag{4.16}$$

The equality sign in Equation 1.4 represents the condition where slope overload just begins to occur, while Equation 4.16 denotes the condition when the delta modulator is completely overloaded and the decoded output waveform $y(t)$ becomes a triangular waveform having a periodicity equal to that of the cosine wave input and a magnitude of

$$E_\Delta = E_s \cos \lambda$$

From Equations 4.13 and 4.16

$$\cos \lambda = 0{\cdot}844 \tag{4.17}$$

hence

$$E_\Delta = 0{\cdot}844 E_s \tag{4.18}$$

The phase angle of the triangular $y(t)$ waveform is from Equation 4.17, 32° 30′ or 0·567 rad.

Observations

(1) The waveform $y(t)$ at the output of the integrator is a close replica of the input signal $x(t)$ provided the slope of $x(t)$ is less than the maximum slope ζ of $y(t)$.

(2) If the slope of the input waveform exceeds the value ζ over part of its cycle thereby causing slope overload then $y(t)$ consists of straight lines of constant slope in juxtaposition with sinusoidal segments. The error in the decoded waveform $y(t)$ is the difference between $y(t)$ and the cosine wave input $x(t)$. The error signal $e(t) = x(t) - y(t)$ is shown in Figure 4.2 and clearly cross-correlates with the input signal. This is in contrast to the error signal when the encoder is tracking a cosine wave

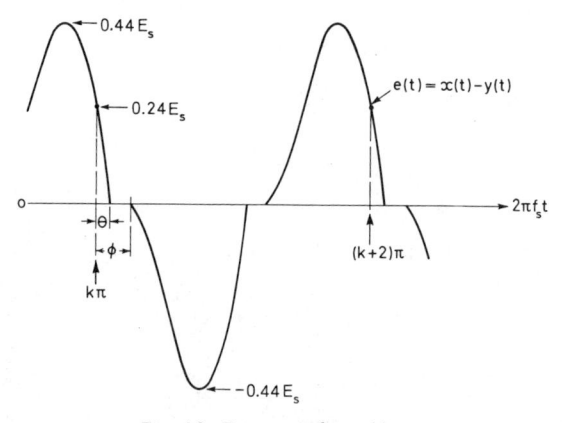

Fig. 4.2. Error waveform $e(t)$

input, when the cross-correlation function associated with the input and error signal is negligible, as discussed in Chapter 3. The fact that slope overload noise cross-correlates with the input, whereas quantization noise does not, is of importance when estimating the total noise, i.e. the error signal $e(t)$ in a delta modulator encoding a Gaussian signal which frequently causes slope overload. This will be demonstrated in Section 4.3.

(3) The periodicity of the error signal as shown in Figure 4.2 means that a sinewave overloading a delta modulator produces a line spectrum where each line is integer related to every other line, and the lowest line frequency is the same as that of the sine-wave being encoded. Thus, if the sharp cut-off filter, which is normally inserted after the integrator in order to remove the sharp angular changes in the $y(t)$ waveform, has a pass-band such that its upper critical frequency is greater than kf_s, then clearly k harmonics of f_s will pass to the output.

Slope overload noise in linear d.m. systems

When the slope of the input sinusoid reaches the value given by Equation 4.16, the delta modulator remains in a condition of slope overload, and the integrator has a triangular output waveform.

4.2.2. Exponential delta modulator

An exponential delta modulator has a local decoder which consists of a resistor–capactor circuit, as shown in Figure 4.3, in place of the perfect

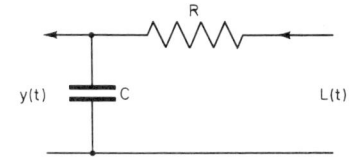

Fig. 4.3. *Local decoder in exponential delta modulator*

integrator in the delta modulator of Figure 1.1. The output pulses of amplitude V or $-V$ volts are applied via the resistor R to the capacitor C such that the voltage on the latter, namely $y(t)$, varies in an attempt to make it a facsimile of $x(t)$.

When the encoder becomes overloaded by a series of positive output pulses, instead of $y(t)$ increasing at a constant rate, which is what happens when a perfect integrator is inserted in the feedback loop, it proceeds as an exponential function of time.

Figure 4.4 shows the voltage waveform at the capacitor when the encoder is overloaded for part of the input cycle by a cosine wave. Consider the commencement of overload at point A when the sinusoid has a negative slope. The capacitor C discharges with each negative pulse, and

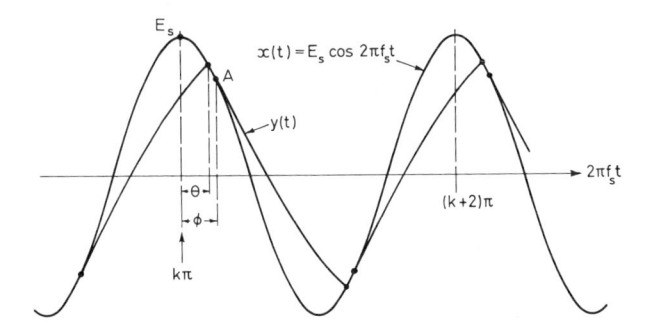

Fig. 4.4. *Slope overload of exponential delta modulator: $x(t)$ is the input signal and $y(t)$ is the signal at the output of the local decoder*

although the time constant RC is very much larger than the clock period T, some of the charge on C leaks away between pulses. In other words the RC circuit behaves as a leaky integrator, as discussed in Chapter 2. It is this leaky charge between successive negative pulses which causes the mean fall of the $y(t)$ waveform in Figure 4.4 to be exponential rather than linear.

An exponential decrease of $y(t)$ at the point A in Figure 4.4 has an initial slope equal to that of the input waveform $x(t)$ at the point A, as prior to overload $x(t)$ is equal to $y(t)$.

It has been shown in Section 2.5.1 that applying a sequence of identical pulses to an RC network results in step-like waveform whose mean value is that of an exponential. If the pulse amplitude is V, of width τ seconds and duration between pulses T seconds, then the capacitor charges to a voltage V_1, where,

$$V_1 = \frac{V\tau}{T}$$

with time constant $T_1 = RC$. Thus the mean voltage across the capacitor is the same as if a voltage step V_1 had been applied to the RC circuit rather than a sequence of pulses.

Now from Equation 2.52 the value of $y(t)$ after applying a voltage step $+E$ to the RC circuit is

$$y(t) = V_c \exp\left(-\frac{t}{RC}\right) + E\left\{1 - \exp\left(-\frac{t}{RC}\right)\right\} \tag{4.19}$$

where V_c is the initial voltage on the capacitor C.

The initial slope of $y(t)$ at $t = 0$ is

$$y'(0) = -\frac{V_c}{RC} + \frac{E}{RC} = \frac{E - V_c}{RC} = r \tag{4.20}$$

From Equation 4.20

$$E = V_c + rRC \tag{4.21}$$

Substituting for E in Equation 4.19

$$y(t) = V_c \exp\left(-\frac{t}{RC}\right) + (V_c + rRC)\left\{1 - \exp\left(-\frac{t}{RC}\right)\right\}$$

$$= V_c + rRC\left\{1 - \exp\left(-\frac{t}{RC}\right)\right\} \tag{4.22}$$

Equation 4.22 describes the variation of voltage across the capacitor as a function of time in terms of its initial voltage V_c and initial slope r. Since it has been argued that Equation 4.22 expresses how the mean value

Slope overload noise in linear d.m. systems   125

of $y(t)$ varies during slope overload, it only remains to insert the values of V_c and r into this equation. Using the same nomenclative for angles as in Figure 4.1 and 4.4

$$V_c = E_s \cos \phi \tag{4.23}$$

The value of r is simply the derivative of the input cosine wave at the angle ϕ

$$r = -E_s 2\pi f_s \sin \phi = -\zeta \tag{4.24}$$

By modifying Equation 4.22 to start at time t_a, i.e. at the point A in Figure 4.4, and proceeding in a similar manner to that used to establish Equations 4.3 and 4.4 the equation for $y(t)$ is

$$y(t) = E_s \cos \phi - \zeta RC \left\{ 1 - \exp\left(-\frac{2\pi f_s t - (\phi + k\pi)}{2\pi f_s RC}\right) \right\} \tag{4.25}$$

By following a similar argument to that used in the derivation of Equations 4.3 to 4.12 the equation for $y(t)$ when the system is overloaded is

$$y(t) = (-1)^k E_s \cos \phi + (-1)^{k+1} RC\zeta \left\{ 1 - \exp\left(-\frac{2\pi f_s t - (\phi + k\pi)}{2\pi f_s RC}\right) \right\} \quad 4.26$$

over the range

$$\phi + k\pi \le 2\pi f_s t \le \theta + (k+1)\pi$$

When the encoder is always in overload

Here $\phi = \theta = \lambda$, $2\pi f_s t = \pi + \theta + k\pi$ and

$$y(t) = E_s \cos(\theta + \pi) = -E_s \cos \lambda$$

Substituting in Equation 4.25

$$-E_s \cos \lambda = E_s \cos \lambda - \zeta RC \left\{ 1 - \exp\left(-\frac{\pi}{2\pi f_s RC}\right) \right\}$$

giving

$$\cos \lambda = \frac{\zeta RC}{2E_s} \left\{ 1 - \exp\left(-\frac{\pi}{2\pi f_s RC}\right) \right\} = \{1 - \sin^2 \lambda\}^{1/2} \tag{4.27}$$

From Equation 4.2

$$\sin^2 \lambda = \left(\frac{\zeta}{E_s 2\pi f_s}\right)^2$$

Hence

$$\left(\frac{\zeta RC}{2E_s}\right)^2 \left\{1 - \exp\left(-\frac{1}{2f_s RC}\right)\right\}^2 = 1 - \left(\frac{\zeta}{E_s 2\pi f_s}\right)^2$$

$$\frac{E_s 2\pi f_s}{\zeta} = \left[1 + \frac{\pi f_s RC}{2}\left\{1 - \exp\left(-\frac{1}{2f_s RC}\right)\right\}^2\right]^{1/2} \quad (4.28)$$

If the time constant RC is large such that

$$(1/2f_s RC) \ll 1$$

then

$$\exp\left(-\frac{1}{2f_s RC}\right) \simeq 1 - \frac{1}{2f_s RC}$$

and Equation 4.28 becomes

$$\frac{E_s 2\pi f_s}{\zeta} \doteq \left(1 + \frac{\pi^2}{4}\right)^{1/2} \quad (4.29)$$

When Equation 4.29 is compared with Equation 4.16 it is apparent that the effect of making the time constant of the RC circuit large compared to the period of the input sinusoid is to approximate this circuit to a close representation of a perfect integrator.

Comparison between Figures 4.1 and 4.4 show that when the local decoder is changed from a perfect integrator to a resistor capacitor circuit the power in the error signal is increased provided that the condition of slope overload commences at the same part of the cycle, i.e. when $x(t) = E_s \cos \phi$.

4.3. SLOPE OVERLOAD WITH GAUSSIAN SIGNALS

When the input signal is stationary band-limited white noise the delta modulator will sometimes track this signal and at other times it will experience slope overload. The ratio of the time the encoder spends tracking the signal to the periods when it is overloaded depends on the r.m.s. value of the input signal. Because overloading is occurring the staircase quantization functions for a delta modulator with single integration, shown in Figure 1.13, and discussed in Section 1.6.2, are not applicable. This means that no simple analytical technique can be employed in estimating the $e(t)$ signal. However, a theoretical investigation will be made to discover if a partial analytical solution exists to the determination of $e(t)$.

Slope overload noise in linear d.m. systems 127

4.3.1. Linear filter model

A useful approach[3] to the solution of this problem is to represent the delta modulator by some simple model which preferably does not have any closed loops associated with it in order that the required calculations of $y(t)$ and $e(t)$ can be more easily accomplished. On deciding what form this model shall take cognisance will be given to computer simulation results which indicate that if the input signal is band-limited white noise having a stationary Gaussian amplitude density function then the instants when this signal causes the encoder to be overloaded will result in an error signal which will correlate with the input signal. Thus the part of the model which represents the effects of overloading in the encoder can be a simple filter having a transfer function $G(j2\pi f)$. Suppose that the effects of quantization are temporarily disregarded and the input to the filter is $x(t)$ and the output from the filter is $y(t)$. If the encoder is not overloaded then $y(t) = x(t)$, which means that the filter must have zero insertion loss when tracking occurs. At the instants when overloading occurs $y(t)$ differs from $x(t)$ and intuitively it might be expected that the transfer function of the equivalent filter is closely associated with the overload characteristic of the encoder. The actual association becomes evident in Section 4.3.2.

In Chapter 3 the theory for estimating the error waveform in a d.m. system is prescribed for the situation when a Gaussian input signal is employed which has the same properties as that of the input signal described above. However, in Chapter 3, the r.m.s. value of the input signal is allowed to have only low values which means that the probability of the encoder becoming overloaded is negligible. By restricting the r.m.s. value of the input signal to these low values errors are due to quantization effects only.

Hence in order to allow for the effects of quantization, the model is amended to that of a low pass filter at whose output a signal $n(t)$ is added. This is the noise component of the error signal which can be considered as independent of the overload component, i.e. it does not correlate with the input signal $x(t)$. The complete model is shown in Figure 4.5. An alternative representation of this model is that given in Figure 4.6. In this

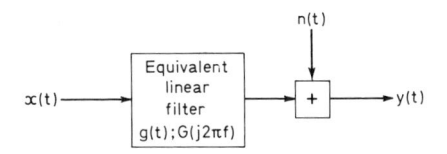

Fig. 4.5. Linear filter model

Slope overload noise in linear d.m. systems

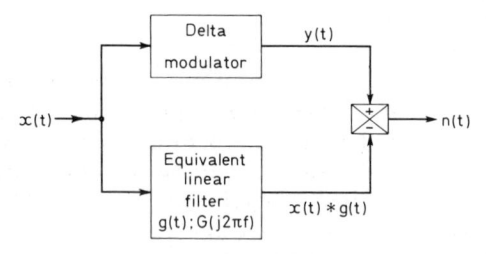

Fig. 4.6. *Alternative representation of linear filter model*

model the output of the filter is the convolution of $x(t)$ with the impulse response of the filter $g(t)$. The * symbol shown in Figure 4.6 indicates convolution. The output of this filter when subtracted from the output $y(t)$ of the local decoder gives the quantization noise signal $n(t)$. This is because $y(t)$ is composed of two components, one which correlates with the input signal and represents the effects of slope overload and is allowed for by the equivalent filter, and a second component $n(t)$ which does not correlate with the input signal and is due to the effects of quantization.

The output of the equivalent linear filter is found by convolving the input signal $x(t)$ with the impulse response $g(t)$ of the filter.

$$g(t) * x(t) = \int_{-\infty}^{\infty} g(\lambda)x(t-\lambda)\,d\lambda \tag{4.30}$$

where λ is the variable. The output $y(t)$ from the integrator is found from Figure 4.5.

$$y(t) = x(t) * g(t) + n(t) \tag{4.31}$$

The cross-correlation function relating with the component of $x(t)$ associated with quantization noise $n(t)$ is $R_{xn}(\tau)$

$$R_{xn}(\tau) = \langle x(t+\tau)n(t) \rangle \tag{4.32}$$

where the symbol $\langle \rangle$ denotes time averaging. Substituting $n(t)$ from Equation 4.31 in Equation 4.32

$$R_{xn}(\tau) = \langle x(t+\tau)[y(t) - g(t) * x(t)] \rangle \tag{4.33}$$

Substituting the convolution integral from Equation 4.30 into Equation 4.33

$$R_{xn}(\tau) = \langle x(t+\tau)y(t) \rangle - \langle x(t+\tau) \int_{-\infty}^{\infty} g(\lambda)x(t-\lambda)\,d\lambda \rangle$$

and observing that by definition

$$\langle x(t+\tau)y(t) \rangle = R_{xy}(\tau) \tag{4.34}$$

Slope overload noise in linear d.m. systems

where $R_{xy}(\tau)$ is the cross-correlation function between the input signal and the output signal of the delta modulator. Hence

$$R_{xn}(\tau) = R_{xy}(\tau) - \int_{-\infty}^{\infty} \langle x(t+\tau) \cdot x(t-\lambda) \rangle g(\lambda) \, d\lambda \qquad (4.35)$$

where the term $\langle x(t+\tau) \cdot x(t-\lambda) \rangle$ can be evaluated by putting $\mu = t - \lambda$ to give

$$\langle x\{\mu + (\lambda + \tau)\} x(\mu) \rangle = R_{xx}(\lambda + \tau) \qquad (4.36)$$

Thus Equation 4.35 becomes

$$R_{xn}(\tau) = R_{xy}(\tau) - \int_{-\infty}^{\infty} R_{xx}(\lambda + \tau) g(\lambda) \, d\lambda \qquad (4.37)$$

The integral in Equation 4.37 is of the convolution type described by Equation 4.30, and in order to express it in convolution notation the substitution $\lambda = -u$ is made.

$$\int_{-\infty}^{\infty} R_{xx}(\tau - u) g(-u) \, d(-u) = g(-\tau) * R_{xx}(\tau) \qquad (4.38)$$

Hence

$$R_{xn}(\tau) = R_{xy}(\tau) - g(-\tau) * R_{xx}(\tau) \qquad (4.39)$$

Assuming there is no cross correlation between $x(t)$ and $n(t)$, $R_{xn}(\tau)$ is zero and Equation (4.39) can be written as

$$R_{xy}(\tau) = g(-\tau) * R_{xx}(\tau) \qquad (4.40)$$

The cross-correlation function between $x(t)$ and $y(t)$ is therefore the convolution of the autocorrelation function of the input with the impulse function of the linear filter when the variable τ is made negative. Equations 4.34 and 4.40 are of course equivalent.

The objective however is to isolate the transfer function $G(j2\pi f)$, and to do this the next stage is to take the Fourier transform of both sides of Equation 4.40

$$\int_{-\infty}^{\infty} R_{xy}(\tau) e^{-j2\pi f \tau} \, d\tau = \int_{-\infty}^{\infty} \{g(-\tau) * R_{xx}(\tau)\} e^{-j2\pi f \tau} \, d\tau \qquad (4.41)$$

The Fourier transforms of correlation functions result in power spectral functions. Accordingly the left hand side of Equation 4.41 becomes the cross-energy density function $S_{xy}(f)$. Substituting the integral representation of the convolution of $g(-\tau)$ with $R_{xx}(\tau)$ described by Equation 4.38.

in Equation 4.41

$$S_{xy}(f) = \int_{-\infty}^{\infty} \left\{ \int_{-\infty}^{\infty} R_{xx}(\tau - u) g(-u) \, du \right\} e^{-j2\pi f\tau} \, d\tau$$

$$= \int_{-\infty}^{\infty} g(-u) \left\{ \int_{-\infty}^{\infty} R_{xx}(\tau - u) e^{-j2\pi f(\tau - u)} \, d\tau \right\} e^{-j2\pi fu} \, du \quad (4.42)$$

By substituting $\psi = (\tau - u)$ into the integral enclosed by the curly brackets in Equation 4.42, $S_{xy}(f)$ simplifies to

$$S_{xy}(f) = \int_{-\infty}^{\infty} g(-u) S_{xx}(f) e^{-j2\pi fu} \, du$$

where

$$S_{xx}(f) = \int_{-\infty}^{\infty} R_{xx}(\psi) e^{-j2\pi f\psi} \, d\psi$$

$S_{xx}(f)$ is the spectral energy density of the input signal and is a real function. Consequently

$$S_{xy}(f) = S_{xx}(f) \int_{-\infty}^{\infty} g(-u) e^{-j2\pi fu} \, du$$

$$= S_{xx}(f) G(-j2\pi f) \quad (4.43)$$

where $G(-j2\pi f)$ is the Fourier transform of $g(-u)$. Taking the complex conjugate of Equation 4.41 in order to find $G(+j2\pi f)$

$$S_{xy}^c(f) = \{S_{xx}(f) G(-j2\pi f)\}^c$$

where the raised c sign above a symbol means the complex conjugate of that symbol.

Now $S_{xx}(f)$ is a real function, i.e. $S_{xx}(f) = S_{xx}^c(f)$, and consequently

$$S_{xy}^c(f) = S_{xx}(f) G^c(-j2\pi f) = S_{xx}(f) G(j2\pi f)$$

Hence the transfer function of the equivalent linear filter is

$$G(j2\pi f) = \frac{S_{xy}^c(f)}{S_{xx}(f)} \quad (4.44)$$

An alternative expression for $G(j2\pi f)$ is now derived. Because $S_{xy}^c(f) = S_{yx}(f)$

$$G(j2\pi f) = \frac{S_{yx}(f)}{S_{xx}(f)} \quad (4.45)$$

$$S_{yx}(f) = X(j2\pi f) \cdot Y^c(j2\pi f) \quad (4.46)$$

where $X(j2\pi f)$ and $Y(j2\pi f)$ are the Fourier transforms of $x(t)$ and $y(t)$ respectively.

Slope overload noise in linear d.m. systems 131

The Fourier transform of the error signal $e(t)$ is $E(j2\pi f)$ and as

$$e(t) = x(t) - y(t)$$

then

$$E(j2\pi f) = X(j2\pi f) - Y(j2\pi f) \qquad (4.47)$$

Taking the complex conjugate of Equation 4.47 and substituting $Y^c(j2\pi f)$ in Equation 4.46

$$S_{yx}(f) = X(j2\pi f)X^c(j2\pi f) - X(j2\pi f)E^c(j2\pi f)$$
$$= S_{xx}(f) - S^c_{xe}(f)$$

Substituting for $S_{yx}(f)$ in Equation 4.45

$$G(j2\pi f) = 1 - \frac{S^c_{xe}(f)}{S_{xx}(f)} \qquad (4.48)$$

The transfer function $G(j2\pi f)$ has been shown in Equation 4.45 to be related to the input spectral density function $S_{xx}(f)$ and the cross-spectral density function between the signals at the output and input of the delta modulator, i.e. $S_{yx}(f)$. Alternatively it can be described in terms of $S_{xx}(f)$, and the complex conjugate of the cross-spectral density function between the input signal and the error signal is shown by Equation 4.48. Unfortunately the cross-spectral density functions in Equations 4.45 and 4.48 are not analytically realisable. They can however be determined by means of computer simulation and hence $G(j2\pi f)$ can be found. If $G(j2\pi f)$ is calculated by making acceptable approximations then the above theory enables $S_{yx}(f)$ and $S^c_{xe}(f)$ to be found.

For example, an appropriate expression for $G(j2\pi f)$ is found in the next section, Equation 4.79. From Equation 4.45 an expression for the cross-spectral density function between $y(t)$ and $x(t)$ when the delta modulator is in slope overload is

$$S_{xx}(f)/[1 + j(f/\mu)]$$

where μ is defined by Equation 4.80.

4.3.2. Determination of linear filter using quasi-linearisation

Having experienced the difficulty of determining the transfer function of the equivalent linear filter $G(j2\pi f)$ by analytical means, an alternative approach to this problem is now made based on a model of the delta modulator in a condition of slope overload.

Consider the block diagram in Figure 4.7. This diagram is the linear d.m. system displayed in Figure 1.1 with two important exceptions,

132   Slope overload noise in linear d.m. systems

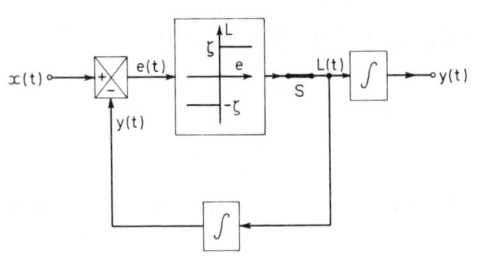

Fig. 4.7. *Representation of delta modulator in slope overload. Sampler is permanently closed*

namely that the sampling switch is permanently closed and the limits of the quantizer have been altered from V to the maximum rate of increase that the encoder can achieve, i.e. ζ V/s. The absence of the sampler in Figure 4.7 is consistent with the analysis given in Section 4.2 where the output of the local decoder was calculated for a sinusoidal input. It was argued that when the encoder is in a frequent state of slope overload the effects of quantization are trivial in comparison. This conclusion is also valid when Gaussian inputs are applied to the encoder.

The value of the quantization levels in Figure 4.7 have been selected as ζ. This means that t seconds after a polarity change in the error signal $e(t)$ the magnitude agrees with the situation in the true encoder of Figure 1.1, which when in slope overload causes the $y(t)$ waveform to increase at a rate of ζ volts per second.

Consider the case when no input signal is applied to the asynchronous encoder shown in Figure 4.7. The system oscillates at a high frequency limited by the propagation delays of the closed loop. The high frequency variations at the output of the local decoder are prohibited from attaining large magnitudes by the effect of the closed loop system. For example, as soon as the error changes sign the output of the quantizer has a value of $+\zeta$ say, which when integrated causes $y(t)$ to increase. However, $y(t)$ only needs to become large enough for the error to change polarity, i.e. $L(t) = -\zeta$ causing $y(t)$ to change sign. Thus the $y(t)$ oscillations are of low amplitude and high frequency and can therefore be neglected.

If a sinusoid input is applied to the asynchronous encoder, then when the slope of the input is in excess of ζ the value of $y(t)$ increases at a rate of ζ V/s. However, when $y(t)$ catches $x(t)$ it consists of the low amplitude high frequency oscillations described above, and its mean value is equal to the input $x(t)$. This waveform $y(t)$ has smaller amplitude oscillations than the quantization variations associated with the waveform $y(t)$ of Figure 1.1, due to the absence of the sampler.

Thus for the sinusoidal input signal, the encoder of Figure 4.7 would produce a $y(t)$ waveform which is a very close approximation to the smooth $y(t)$ waveform shown in Figure 4.1.

Having argued the viability of Figure 4.7 as a representation of the delta modulator for all values of the slope of the input signal, provided

Slope overload noise in linear d.m. systems

the effects of quantization are ignored, the next step is to calculate the transfer function $G(j2\pi f)$ of this asynchronous coder. Now

$$G(j2\pi f) = \frac{Y(j2\pi f)}{X(j2\pi f)} \quad (4.49)$$

where $X(j2\pi f)$ and $Y(j2\pi f)$ are the Fourier transforms of the signals $x(t)$ and $y(t)$ shown in Figure 4.7, respectively. In order to determine $G(j2\pi f)$ the first step is to replace the quantizer by an equivalent linear network. Consider the case of an isolated, amplitude sensitive, memoryless non-linear network having an input signal $e(t)$ whose amplitude distribution is Gaussian. The output signal $L(t)$ from the non-linear element will clearly be distorted, i.e.

$$L(t) = K_{eq}e(t) + e_H(t) \quad (4.50)$$

where K_{eq} is the equivalent gain of the non-linear element, and $e_H(t)$ represents the distortion terms.

If $e_H(t)$ is neglected then $L(t)$ is proportional to $e(t)$ and the non-linear device has been replaced by a linear one having a 'gain' of K_{eq}. The representation of $L(t)$ as the equivalent of $K_{eq}e(t)$ is attributed to Booton[6] and because K_{eq} is not truly independent of the input signal $e(t)$ the term *quasi*-linearisation is used to describe this method.

The value of K_{eq} is chosen so as to optimise the mean square error between the actual output $L(t)$ and the wanted term in the output $K_{eq}e(t)$

$$[L(t) - K_{eq}e(t)]^2 = [f\{e(t)\} - K_{eq}e(t)]^2 \quad (4.51)$$

where $L(t) = f\{e(t)\}$, i.e. the output signal is a function of the input signal. The mean square error M is

$$M = \int_{-\infty}^{\infty} [f(e(t)) - K_{eq}e(t)]^2 P_1\{e(t)\} \, de(t)$$

$$= \int_{-\infty}^{\infty} f^2\{e(t)\} P_1\{e(t)\} \, de(t) - 2K_{eq} \int_{-\infty}^{\infty} f\{e(t)\}e(t)P_1\{e(t)\} \, de(t)$$

$$+ K_{eq}^2 \int_{-\infty}^{\infty} e^2(t) P_1\{e(t)\} \, de(t) \quad (4.52)$$

where $P_1\{e(t)\}$ is the first probability density function of the input element

To find the value of K_{eq} which minimises the mean square error M is differentiated with respect to K_{eq}.

$$\frac{dM}{dK_{eq}} = -2 \int_{-\infty}^{\infty} f\{e(t)\}e(t)P_1\{e(t)\} \, de(t)$$

$$+ 2K_{eq} \int_{-\infty}^{\infty} e^2(t) P_1\{e(t)\} \, de(t) \quad (4.53)$$

Slope overload noise in linear d.m. systems

When dM/dK_{eq} is equated to zero the value of K_{eq} is found which gives the minimum value of M.

$$K_{eq} = \frac{\int_{-\infty}^{\infty} f\{e(t)\}e(t)P_1\{e(t)\}\,de(t)}{\int_{-\infty}^{\infty} e^2(t)P_1\{e(t)\}\,de(t)}$$

$$= \frac{1}{\sigma_e^2}\int_{-\infty}^{\infty} f\{e(t)\}e(t)P_1\{e(t)\}\,de(t) \qquad (4.54)$$

where σ_e^2 is the mean square value (variance) of the input signal $e(t)$, and

$$\sigma_e^2 = \int_{-\infty}^{\infty} e^2(t)P_1\{e(t)\}\,de(t) \qquad (4.55)$$

The probability density function of the input signal $P_1\{e(t)\}$ is

$$P_1\{e(t)\} = \frac{1}{\sigma_e\sqrt{2\pi}}\exp\left\{-\frac{e^2(t)}{2\sigma_e^2}\right\} \qquad (4.56)$$

Substituting Equation 4.56 in Equation 4.54

$$K_{eq} = \frac{1}{\sigma^3\sqrt{2\pi}}\cdot\int_{-\infty}^{\infty} f\{e(t)\}e(t)\exp\left\{-\frac{e^2(t)}{2\sigma_e^2}\right\}de(t) \qquad (4.57)$$

and this is the coefficient b_1 when $n = 1$ in Equation 3.10.
For the non-linear element shown in Figure 4.7

$$f\{e(t)\} = \zeta, \qquad e(t) \geq 0$$
$$= -\zeta, \qquad e(t) < 0$$

Hence

$$K_{eq} = \frac{\zeta}{\sigma_e^3\sqrt{2\pi}}\left[\int_0^{\infty} e(t)\exp\left\{-\frac{e^2(t)}{2\sigma_e^2}\right\}de(t)\right.$$
$$\left. - \int_{-\infty}^{0} e(t)\exp\left\{-\frac{e^2(t)}{2\sigma_e^2}\right\}de(t)\right]$$

$$= \frac{2\zeta}{\sigma_e^3\sqrt{2\pi}}\int_0^{\infty} e(t)\exp\left\{-\frac{e^2(t)}{2\sigma_e^2}\right\}de(t)$$

Putting

$$u = \frac{e^2(t)}{2\sigma_e^2}$$

$$K_{eq} = \frac{2\zeta}{\sigma_e\sqrt{2\pi}}\int_0^{\infty}\exp(-u)\,du$$

Since

$$\int_0^\infty \exp(-u)\,du = 1$$

$$K_{eq} = \frac{2\zeta}{\sigma_e \sqrt{2\pi}} \tag{4.58}$$

Equation 4.58 gives the equivalent gain K_{eq} of an isolated, memoryless, amplitude sensitive non-linear network when a Gaussian input is applied to its input terminals. A Gaussian signal is defined here as a signal whose amplitude has a probability density function which is Gaussian.

K_{eq} depends on the characteristic of the non-linearity and the r.m.s. value of the input signal.

Having discussed the basis for replacing an isolated non-linear element by a linear one having a gain K_{eq}, it will now be shown that the non-linear element in the encoder in Figure 4.7 can be replaced by a linear amplifier.

Suppose that the closed loop system is a linear one. For a Gaussian input signal $x(t)$, every variable throughout the system is also a Gaussian signal. However, in Figure 4.7 the error signal is not strictly Gaussian due to the presence of the non-linearity. The closed loop system endeavors to make the output of the local decoder $y(t) = x(t)$. When it achieves this condition $y(t)$ approximates to a Gaussian signal, and the difference between the two Gaussian signals is another Gaussian signal $e(t)$. In these circumstances the quasi-linearisation method is applicable. When the error signal becomes large the $y(t)$ and $e(t)$ signals have statistical properties which differ from those of a Gaussian signal and there is an error in using the quasi-linearisation technique. Over a long period of time the experimental evidence shows that $e(t)$ is a reasonable approximation to a Gaussian signal and therefore the quantizer shown in Figure 4.7 is replaced by the linear amplifier as displayed in Figure 4.8. The distortion terms $e_H(t)$, defined by Equation 4.50, are severely attenuated by the local decoder in the feedback network. This attenuation aids the assumption that $e_H(t)$ can be neglected.

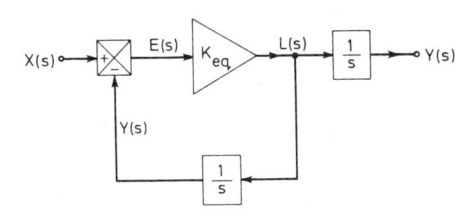

Fig. 4.8. The quantizer in Fig. 4.7 replaced by a linear amplifier

A second equation must now be sought which relates K_{eq} to σ_e. This is because the response of the non-linear element effects its input due to the presence of the feedback path.

Writing down the equations for the system in Figure 4.8 where s is the Laplacian variable

$$E(s) = X(s) - Y(s)$$
$$L(s) = E(s)K_{eq} = s \cdot Y(s)$$

Hence

$$\frac{E(s)}{X(s)} = \frac{1}{1 + \frac{K_{eq}}{s}} \qquad (4.59)$$

Replacing s by $j2\pi f$ in order to find the transfer function

$$\frac{E(j2\pi f)}{X(j2\pi f)} = \frac{1}{1 + \frac{K_{eq}}{j2\pi f}}$$

The power spectrum of the error signal is

$$S_{ee}(f) = \left| \frac{1}{1 + \frac{K_{eq}}{j2\pi f}} \right|^2 S_{xx}(f) \qquad (4.60)$$

The mean square value of the error signal $e(t)$ is

$$\langle \{e(t)\}^2 \rangle = \sigma_e^2 = \int_{-\infty}^{\infty} S_{ee}(f) \, df$$

and from Equation 4.60

$$\sigma_e^2 = 2 \int_0^{\infty} \left| \frac{1}{1 + \frac{K_{eq}}{j2\pi f}} \right|^2 S_{xx}(f) \, df \qquad (4.61)$$

σ_e^2 cannot be determined until the shape of the input energy density spectrum $S_{xx}(f)$ is specified.

Band-limited white noise input signal

The spectral density of the error signal from Equation 4.60 is

$$S_{ee}(f) = \frac{S_{xx}(f)}{1 + \left(\frac{K_{eq}}{2\pi f}\right)^2} \qquad (4.62)$$

Slope overload noise in linear d.m. systems 137

Let
$$\mu = K_{eq}/2\pi \quad (4.63)$$
then
$$S_{ee}(f) = \left\{\frac{f^2}{f^2 + \mu^2}\right\} S_{xx}(f) \quad (4.64)$$

The term $[f^2/(f^2 + \mu^2)]$ increases monotonically with frequency from zero to infinity. The integral of Equation 4.61 can be analytically evaluated[7] for a wide variety of input density functions, $S_{xx}(f)$. However, a frequently used test signal is band-limited white noise, and in this case the integral of Equation 4.61 enables σ_e^2 to be simply calculated.

Let the two-sided spectral density $S_{xx}(f)$ be

$$S_{xx}(f) = N_0/2 \text{ W/Hz}, \quad -f_{c2} \le f \le f_{c2}$$
$$= 0, \text{ elsewhere} \quad (4.65)$$

Then $S_{ee}(f)$ from Equations 4.64 and 4.65 becomes

$$S_{ee}(f) = \frac{f^2}{f^2 + \mu^2} \cdot \frac{N_0}{2}, \quad -f_{c2} \le f \le f_{c2} \quad (4.66)$$

Hence the mean square error power from Equations 4.61 and 4.66 is

$$\sigma_e^2 = N_0 \int_0^{f_{c2}} \frac{f^2}{f^2 + \mu^2} \cdot df$$

$$\sigma_e^2 = N_0 \left\{f_{c2} - \mu \tan^{-1}\left(\frac{f_{c2}}{\mu}\right)\right\} \quad (4.67)$$

Then the mean square value of the input signal is

$$\langle\{x(t)\}^2\rangle = \sigma_x^2 = 2 \int_0^{f_{c2}} S_{xx}(f) \, df = N_0 f_{c2} \quad (4.68)$$

From Equations 4.63, 4.67 and 4.68

$$\frac{\sigma_e^2}{\sigma_x^2} = 1 - \frac{K_{eq}}{2\pi f_{c2}} \tan^{-1}\left(\frac{2\pi f_{c2}}{K_{eq}}\right) \quad (4.69)$$

Equation 4.69 is the second equation which relates σ_e^2 with K_{eq}, and when considered with Equation 4.58 enables σ_e to be eliminated and hence the value of K_{eq} to be determined.

Jump phenomena

Before solving Equations 4.69 and 4.58 to find the equivalent gain of the quantizer an examination will be made to determine if jump phenomena can exist in the system. Jump phenomena can arise if the mean square value of the error signal σ_e^2 is multivalued for a given set of conditions. For example, if for a particular value of σ_x^2, σ_e^2 has more than one value, then when it jumps between these values K_{eq} will jump, and the transfer function of the equivalent linear filter will be unstable.

Substituting for K_{eq} from Equation 4.58 in Equation 4.69

$$\frac{\sigma_e^2}{\sigma_x^2} = 1 - \frac{\zeta}{\sigma_e \pi \sqrt{2\pi} f_{c2}} \tan^{-1}\left(\frac{\sigma_e \pi \sqrt{2\pi} f_{c2}}{\zeta}\right) \qquad (4.70)$$

Let

$$A = \frac{\zeta}{\pi f_{c2} \sqrt{2\pi}} \qquad (4.71)$$

then

$$\frac{\sigma_e^2}{\sigma_x^2} = 1 - \frac{A}{\sigma_e} \tan^{-1}\left(\frac{\sigma_e}{A}\right) \qquad (4.72)$$

If both sides of Equation 4.72 are plotted as a function of σ_e^2, then provided the graphs do not intersect more than once, σ_e^2 and hence K_{eq}, will not jump. These graphs must be plotted for the range of values of the constant A.

For a certain range of A it is possible to specify that jump phenomena will not occur without actually plotting the graphs.

Representing $\tan^{-1}(\sigma_e/A)$ in Equation 4.72 by its series expansion

$$\frac{\sigma_e^2}{\sigma_x^2} = \left(\frac{\sigma_e}{A}\right)^2 \left\{\frac{1}{3} - \frac{1}{5}\left(\frac{\sigma_e}{A}\right)^2 + \cdots\right\} \qquad (4.73)$$

If $A \gg \sigma_e$

$$A^2 \doteq \sigma_x^2 \left\{\frac{1}{3} - \frac{1}{5}\left(\frac{\sigma_e}{A}\right)^2\right\}$$

hence

$$\sigma_e^2 \doteq 5A^2 \left(\frac{1}{3} - \frac{A^2}{\sigma_x^2}\right) \qquad (4.74)$$

As σ_e^2 is always positive $A^2 < \sigma_x^2/3$. Thus provided the inequality

$$\sigma_e^2 \ll A^2 < \frac{\sigma_x^2}{3} \qquad (4.75)$$

holds, Equation 4.74 indicates that jump phenomena cannot occur and σ_e^2 is specified by this equation.

Slope overload noise in linear d.m. systems

From Equation 4.58 the equivalent gain of the non-linear element is

$$K_{eq} \doteq \frac{2\zeta}{\sqrt{2\pi}} \cdot \frac{1}{A\left[5\left(\frac{1}{3} - \frac{A^2}{\sigma_x^2}\right)\right]^{1/2}} \qquad (4.76)$$

Substituting for A from Equation 4.71 in Equation 4.76 gives

$$K_{eq} \doteq \frac{2\pi f_{c2}}{\left[\frac{5}{2}\left\{2/3 - \frac{1}{\pi^3}\left(\frac{\zeta}{f_{c2}\sigma_x}\right)^2\right\}\right]^{1/2}} \qquad (4.77)$$

Transfer function of the linear filter

The equivalent gain K_{eq} of the quantizer shown in Figure 4.7 has been established by Equation 4.77 and shown not to exhibit jump phenomena. It is now a simple process to determine the transfer function of the equivalent linear filter.

Writing down the equations for the closed-loop system in Figure 4.8

$$L(s) = [X(s) - Y(s)]K_{eq} = sY(s)$$

$$G(s) = \frac{Y(s)}{X(s)} = \frac{1}{1 + \frac{s}{K_{eq}}} \qquad (4.78)$$

The generalised transfer function $G(s)$ can be changed to the complex transfer function by putting $s = j2\pi f$. Thus

$$G(j2\pi f) = \frac{1}{1 + j\frac{f}{\mu}} \qquad (4.79)$$

where

$$\mu = \frac{K_{eq}}{2\pi} \qquad (4.80)$$

Equations 4.79 and 4.80 show that at d.c. $G(j2\pi f)$ is unity, break frequency is μ, and $|G(j2\pi f)|$ falls off at 6 dB per octave. The break angular frequency is K_{eq}, and is a function of the input power σ_x^2. The overload characteristic of the delta modulator also falls off at 6 dB per octave where the break frequency is $1/(2\pi RC)$.

4.3.3. Mean square value of quantization noise

The signal $n(t)$ is that part of $e(t)$ which occurs when the encoder is correctly tracking the input signal. If the encoder is never overloaded then $n(t)$ can be calculated as described in Section 3.10. However, because the situation considered here relates to the encoder being frequently overloaded the value of $n(t)$ is hard to determine analytically, even though it relates to the quantization noise when the encoder is not overloaded. If slope overload occurs to such an extent that the amount of in-band noise due to quantization effects is trivial compared to that resulting from slope overloading then $n(t)$ can be neglected. However, in cases where $n(t)$ constitutes a significant proportion of the noise in the message band its value must be found by computer simulation or measurements on the actual codec.

The measurement technique together with the relevant theory is described in Section 12.2.

4.3.4. Spectral density function of error signal

The error signal $e(t)$ is the difference between the input signal $x(t)$ and the signal $y(t)$ at the output of the local decoder. Using the representation of $y(t)$ given by Equation 4.31 the error signal becomes

$$e(t) = x(t) - \{n(t) + g(t) * x(t)\}$$

Taking the Fourier transform

$$E(j2\pi f) = X(j2\pi f) - N(j2\pi f) - G(j2\pi f) \cdot X(j2\pi f)$$
$$= X(j2\pi f)[1 - G(j2\pi f)] - N(j2\pi f) \qquad (4.81)$$

Note that convolution in the time domain corresponds to multiplication in the frequency domain.

Now the spectral density of the error signal $e(t)$ is

$$S_{ee}(f) = E(j2\pi f)E^c(j2\pi f) = |E(j2\pi f)|^2$$

Taking the square of the modulus of Equation 4.81

$$S_{ee}(f) = [X(j2\pi f)\{1 - G(j2\pi f)\} - N(j2\pi f)]$$
$$\times [X(j2\pi f)\{1 - G(j2\pi f)\} - N(j2\pi f)]^c$$

where as before c indicates the complex conjugate of a symbol. Writing

$$X(j2\pi f)\{1 - G(j2\pi f)\} = a + jb \qquad (4.82)$$

and

$$N(j2\pi f) = c + jd \qquad (4.83)$$

Slope overload noise in linear d.m. systems

then
$$S_{ee}(f) = |a+jb|^2 + |c+jd|^2 - (c+jd)(a-jb) - (c-jd)(a+jb) \quad (4.84)$$

Substituting in Equation 4.84
$$\begin{aligned}S_{ee}(f) &= |X(j2\pi f)\{1 - G(j2\pi f)\}|^2 + |N(j2\pi f)|^2 \\ &\quad - N(j2\pi f)\{X(j2\pi f)\{1 - G(j2\pi f)\}\}^c \\ &\quad - N^c(j2\pi f)\{X(j2\pi f)\{1 - G(j2\pi f)\}\} \end{aligned} \quad (4.85)$$

Since the noise signal $n(t)$ and the input signal $x(t)$ are uncorrelated the last two terms in Equation 4.85 are zero. Hence
$$\begin{aligned}S_{ee}(f) &= |X(j2\pi f)\{1 - G(j2\pi f)\}|^2 + |N(j2\pi f)|^2 \\ &= S_{xx}(f)|1 - G(j2\pi f)|^2 + S_{nn}(f)\end{aligned} \quad (4.86)$$

where
$$S_{xx}(f) = |X(j2\pi f)|^2 \quad (4.87)$$
$$S_{nn}(f) = |N(j2\pi f)|^2 \quad (4.88)$$

4.3.5. Noise at the output of the decoder

The decoder is composed of the local decoder, an integrator and a filter which may be a band-pass type having critical frequencies f_{c1} and f_{c2}. If frequencies are identical to the frequency band occupied by the input signal $x(t)$ and assuming that no errors are introduced in the transmission link and that the encoder is perfectly symmetrical, i.e. there is no idle channel noise, then the decoded power is

$$N_e^2 = 2\int_{f_{c1}}^{f_{c2}} S_{ee}(f)\, df$$

Substituting the expression for $S_{ee}(f)$ from Equation 4.86 in the above equation

$$N_e^2 = 2\int_{f_{c1}}^{f_{c2}} S_{xx}(f)|1 - G(j2\pi f)|^2\, df + 2\int_{f_{c1}}^{f_{c2}} S_{nn}(f)\, df \quad (4.89)$$

For an encoder with
$$\begin{aligned}S_{xx}(f) &= N_0/2 \ \text{V}^2\text{s}, & 0 \le |f| \le f_{c2} \\ &= 0, & |f| > f_{c2}\end{aligned}$$

and $G(j2\pi f) = 1/[1 + (jf/\mu)]$, from Equation 4.79 the noise power at the output of the decoder is

$$N_e^2 = N_0 \int_0^{f_{c2}} \frac{f^2}{\mu^2 + f^2} df + 2 \int_0^{f_{c2}} S_{nn}(f) df \quad (4.90)$$

The first term on the right side of Equation 4.90 is σ_e^2 (Equation 4.67). The second term is the mean square value of the component in the error waveform due to quantization effects, i.e. due to $n(t)$. Its value is found experimentally as described in Section 12.2.

The mean square value of the error waveform $e(t)$ is

$$N_e^2 = \sigma_e^2 + \langle \{n(t)\}^2 \rangle \quad (4.91)$$

4.3.6. Signal-to-noise ratio

When the slope overload noise is dominant $N_e^2 \doteqdot \sigma_e^2$ and the signal-to-noise ratio as defined by Equation 3.90, where N_q^2 is replaced by σ_e^2, is given by

$$\text{s.n.r.} = \frac{\sigma_x^2}{\sigma_e^2} = \frac{1}{1 - \frac{K_{eq}}{2\pi f_{c2}} \tan^{-1}\left(\frac{2\pi f_{c2}}{K_{eq}}\right)} \quad (4.92)$$

using Equations 4.67 and 4.63.

4.3.7. Conclusions

Representing a delta modulator by an equivalent linear filter with additive noise at its output when the input signal is band-limited Gaussian noise gives good agreement with computer simulated and measured results, particularly if Equation 4.4 is used.

For example, by employing the linear filter model to determine the variation of the ratio of the mean square value of the input signal to the mean square value of the error signal as a function of the step height γ in the $y(t)$ waveform, a curve is found with a single maximum of 12 dB at $\gamma = 1$. This applies for a clock rate of 8 times the highest frequency in the band-limited signal. When this curve is obtained both experimentally and by computer simulation and compared with the theoretical curve, then over a range of the abscissa γ from 0·25 to 2·0 the discrepancy never exceeds 3 dB between the ordinates of these three curves.

Slope overload noise in linear d.m. systems 143

4.4. SLOPE LIMITER MODEL

This model[4] aims to produce a signal at its output which has a spectral density function over the message band which is a close replica of the spectral density function of the signal emanating from the filter in the linear d.m. decoder. The model is only applicable when the input signal overloads the codec to such an extent that the noise in the message band is mainly due to slope overload rather than quantization noise.

The slope-limiter model is basically simpler than the linear filter model and gives an accurate representation of the filtered $y(t)$ signal when $x(t)$ is narrow-band white noise.

When $x(t)$ has a slope which exceeds the maximum slope ζ of the encoder the latter produces an output signal $L(t)$ which is a sequence of all ones if

$$dx(t)/dt = x'(t) > \zeta$$

This sequence will cause the feedback signal at the output of the local decoder, i.e. integrator, to produce a ramp if $L(t)$ is composed of pulses of duration T, or a staircase function if $L(t)$ is a sequence of impulses spaced T seconds apart. The ramp will have a negative slope if $x'(t)$ is more negative than $-\zeta$. Because the effects of quantization are ignored, when

$$|x'(t)| < \zeta$$

$$y(t) = x(t)$$

The slope-limiter model which is shown in Figure 4.9 is proposed for the case when the encoder is overloaded, and when it is not overloaded and

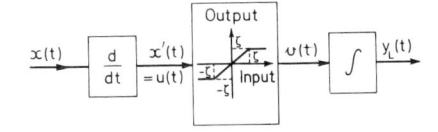

Fig. 4.9. Slope-limiter model
(Courtesy of I.E.R.E.)

quantization effects are ignored. The model is composed of a differentiator, followed by a non-linear element whose output is integrated to give $y_L(t)$. Let the differentiated value of $x(t)$ be $u(t)$. Suppose that $u(t)$, the differentiated value of $x(t)$, had always been insufficient to overload the encoder, then because the non-linear element is linear for values of $u(t) \leq \zeta$, $v(t) = u(t)$ and $y_L(t) = x(t)$ ($v(t)$ is the signal at the output of the non-linear element and the suffix L in $y_L(t)$ denotes that this is the output signal from the model). For the perfect model $y_L(t)$ would always equal $y(t)$. When $|u(t)| > \zeta$ for large periods, $v(t)$ remains at ζ and $y_L(t)$ is ramp shaped, a condition which would occur in a delta modulator.

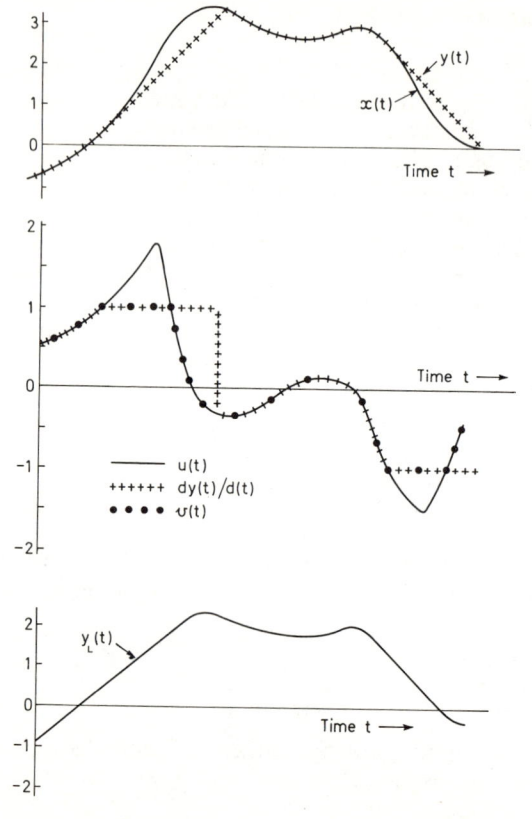

Fig. 4.10. Comparison between the behaviour of d.m. system and slope-limiter model in the time domain for the same arbitrary input signal x(t) (Courtesy of I.E.R.E.)

However, the model does not accurately represent the d.m. system in slope overload for when $|u(t)| > \zeta$ and subsequently $|u(t)| \leq \zeta$ the output signal $y_L(t)$ is different from $x(t)$; in other words $y_L(t)$ does not track $x(t)$ because there is no closed loop in the model. A further discrepancy between the model and the d.m. system is that the latter can be in slope overload when $u(t)$ (which is $x'(t)$) is zero. This can be seen from Figures 4.1 and 4.4. From the model it can be seen that when $u(t) = 0$ the nonlinearity is operating at the origin and not in its saturation state. Figure 4.10 shows an arbitrary input signal and the corresponding waveforms $u(t)$, $v(t)$ and $y_L(t)$ associated with the model. For comparison the $y(t)$ signal for the d.m. system is presented and although it is similar in shape to $y_L(t)$ it has different amplitude values.

Slope overload noise in linear d.m. systems 145

4.4.1. Frequency domain

Having commented upon the behaviour of the model in the time domain, the spectral density function of $y_L(t)$ will now be calculated.

Figure 4.11(a) shows the normalised spectral density function $\phi_x(f)$ of the input signal $x(t)$, the same as that shown in Figure 3.4. The actual spectral density function of $x(t)$ is

$$S_x(f) = \sigma^2 \phi_x(f) \qquad (4.93)$$

where

$$\sigma^2 = \langle \{x(t)\}^2 \rangle$$

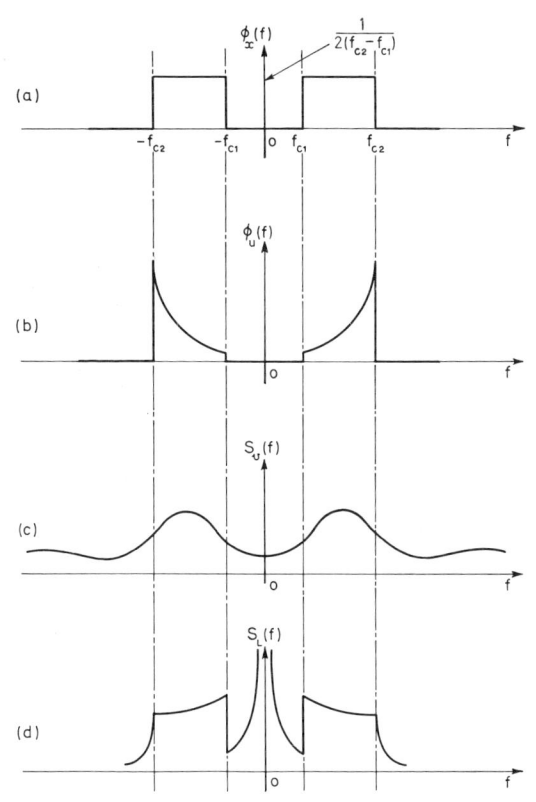

Fig. 4.11. *Power spectral density functions of the signals in the slope-limiter model (Courtesy of I.E.R.E.)*

Because the power transfer function of a differentiator is $(2\pi f)^2$, the spectral density function $S_u(f)$ of the signal $u(t)$ at the output of the differentiator is

$$S_u(f) = 4\pi^2 f^2 \phi_x(f) \cdot \sigma^2 \qquad (4.94)$$

The normalised spectral density function of $u(t)$ is

$$\phi_u(f) = 4\pi^2 f^2 \phi_x(f) \qquad (4.95)$$

and is sketched in Figure 4.11(b).

Proceeding through the model, the signal $u(t)$ is distorted by the non-linear device and the signal $v(t)$ results. From Section 3.4, and Equation 3.12,

$$S_v(f) = \sum_{n=0}^{\infty} b_{n,L}^2 N_n \phi_u^n(f) \qquad (4.96)$$

The determination of $b_{n,L}$ involves the use of Equation 3.10, where $z(x)$ becomes $v(x)$ in this case. Now $b_{1,L}$ is

$$b_{1,L} = \frac{1}{N_1} \int_{-\infty}^{\infty} H_1(u) \cdot v(u) f(u) \, du(t)$$

From Equations 3.3 and 3.6, $H_1 = v(t)$ and $N_1 = \sigma^2$ respectively. The non-linearity is defined by

$$v(t) = u(t), \qquad -\zeta \le u(t) \le \zeta$$
$$= \zeta, \qquad u(t) > \zeta$$
$$= -\zeta, \qquad u(t) < \zeta$$

and the expression for $f(u(t))$ is given by Equation 3.5. Combining this information

$$b_{1,L} = \frac{1}{\sigma^3 \sqrt{2\pi}} \left[\int_{-\infty}^{\zeta} u(t)(-\zeta) \exp\left\{-\frac{u^2(t)}{2\sigma^2}\right\} du(t) \right.$$
$$+ \int_{-\zeta}^{\zeta} u(t) u(t) \exp\left\{-\frac{u^2(t)}{2\sigma^2}\right\} du(t)$$
$$\left. + \int_{\zeta}^{\infty} u(t)(\zeta) \exp\left\{-\frac{u^2(t)}{2\sigma^2}\right\} du(t) \right] \qquad (4.97)$$

Integrating by parts the second integral in Equation 4.97

$$-\frac{2\zeta}{\sigma\sqrt{2\pi}} \exp\left\{-\frac{\zeta^2}{2\sigma^2}\right\} + \mathrm{erf}\left(\frac{\zeta}{\sigma\sqrt{2}}\right)$$

Slope overload noise in linear d.m. systems   147

The remaining two integrals in Equation 4.97 become

$$\frac{2\zeta}{\sigma\sqrt{2\pi}}\exp\left\{-\frac{\zeta^2}{2\sigma^2}\right\}$$

and consequently

$$b_{1,L} = \operatorname{erf}\left(\frac{\zeta}{\sigma\sqrt{2}}\right) \qquad (4.98)$$

By proceeding in a similar manner for higher values of n the high order coefficients become

$$b_{n,L} = \frac{1}{N_n}\sigma\left(\frac{2}{\pi}\right)^{1/2}\left\{-H_{n-2}(u)\exp\left\{-\frac{\sigma^2}{2}\right\}\right\} \qquad (4.99)$$

for $n \geq 3$, where n is integer and odd.

Having determined $b_{n,L}$, the spectral density function of the signal $v(t)$ at the output of the non-linear element is found by using Equation 4.96. This involves the nth order autoconvolution of $\phi_u(f)$ which produce normal spectra which overlap each other. The result is that $S_v(f)$ is a relatively smooth curve as the stylised one in Figure 4.11(c) indicates.

The signal $v(t)$ is integrated to produce $y_L(t)$. The power function of an integrator is $1/(4\pi^2 f^2)$ and consequently the spectral density function of $y_L(t)$, obtained with the aid of Equations 4.96 and 4.95, is

$$S_L(f) = \frac{S_v(f)}{4\pi^2 f^2} = \sum_{n=0}^{\infty} b_{n,L}^2 \frac{N_n}{4\pi^2 f^2}\{4\pi^2 f^2 \phi_x(f)\}^n \qquad (4.100)$$

Figure 4.11(d) indicates the effect that integrating $v(t)$ has on its spectral density function. $S_L(f)$ depends critically on the shape of the spectral density function applied to the input of the non-linear element. The theory is applicable for any shape provided that the signal remains a Gaussian one. The coefficients $b_{n,L}$ do not depend on the shape of $\phi_u(f)$ but on its mean square value.

4.4.2. Signal-to-noise ratio

The required output spectral density function is the first term in Equation 4.100.

$$S_{1,L}(f) = b_{1,L}^2 \frac{N_1}{(2\pi f)^2}(2\pi f)^2 \phi_x(f) = b_{1,L}^2 \sigma^2 \phi_x(f) = b_{1,L}^2 S_x(f) \qquad (4.101)$$

$N_1 = \sigma^2$ from Equation 3.6.

The mean square power of the signal is

$$P_{so} = b_{1,L}^2 \cdot 2 \int_{f_{c1}}^{f_{c2}} S_x(f)\,df = b_{1,L}^2 \cdot P_{si} \qquad (4.102)$$

where P_{si} is the mean square value of $x(t)$, i.e. $P_{si} = \sigma_x^2$.

The $b_{n,L}$ coefficients are zero for even values of n due to the non-linear element having an odd characteristic. Consequently the distortion spectral density function at the output of the model is

$$S_L(f) - S_{1,L}(f) = \sum_{n=3}^{\infty} b_{n,L}^2 \frac{N_n}{(2\pi f)^2} \{(2\pi f)^2 \phi_x(f)\}^n \qquad (4.103)$$

and the distortion power in the message band is

$$P_D = 2 \int_{f_{c1}}^{f_{c2}} S_L(f)\,df - b_{1,L}^2 P_{si} \qquad (4.104)$$

The signal-to-noise ratio is

$$\text{s.n.r.} = \frac{P_{so}}{P_D} = \frac{b_{1,L}^2 P_{si}}{P_d} \qquad (4.105)$$

4.4.3. Performance

The spectral density functions $S_L(f)$ and $S_A(f)$ of the slope-limiter model and the d.m. system, respectively are displayed in Figure 4.12 for the same band-limited white noise input signal $x(t)$. The frequency scale has been normalised relative to f_{c2}, the highest frequency in $x(t)$. In Figure 4.12(a) to (e), f_p/f_{c2} has a constant value of 32, and f_{c2} is fixed. The only difference between the sub-figures is the bandwidth of the input signal $x(t)$ which is different in each case due to f_{c1}, the lowest frequency in $x(t)$,

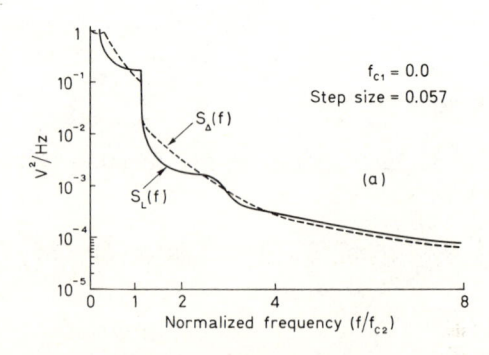

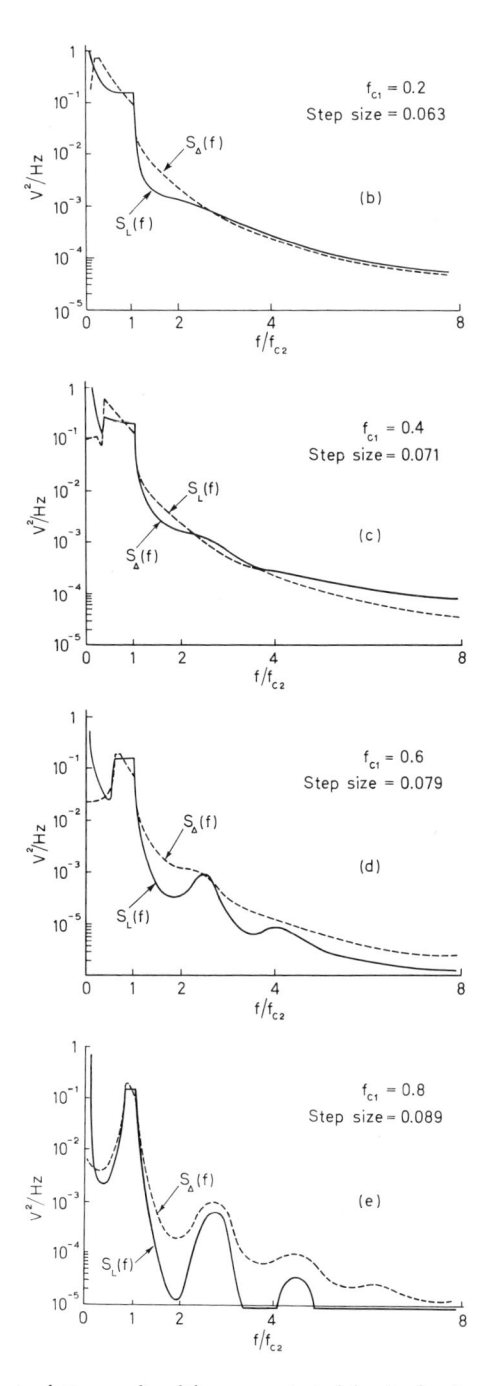

Fig. 4.12. Computer simulation results of the power spectral density function of the waveforms at the output of the d.m. system and slope limiter model (Courtesy of I.E.R.E.)

being changed. In order to avoid differing amounts of slope overloading for each value of f_{c1} the ratio of $\zeta/u(t)$ is made a constant.

The $S_L(f)$ and $S_\Delta(f)$ functions show relatively good agreement for frequencies outside the message band f_{c1} to f_{c2} when the input signal has a wide bandwidth. This is illustrated in Figure 4.12(a) and (b). As the bandwidth of $x(t)$ is reduced, i.e. f_{c1} is increased, the discrepancy between $S_L(f)$ and $S_\Delta(f)$ reduces in the message band and increases elsewhere. Because the final filter in the decoder only accepts those frequencies which lie between f_{c1} and f_{c2} it follows that the slope limiter model is more representative of the d.m. system in slope overload when the bandwidth is reduced. For example in Figure 4.12(e) the values of $S_L(f)$ and $S_\Delta(f)$ only differ by approximately 1 dB.

The spectral density functions contain peaks. In the case of the d.m. system they are the odd high order spectra caused by slope overload. It has already been argued that the slope overload noise correlates with the input signal and this is evident from Figure 4.12(e). The same situation occurs in the model due to saturation effects in $v(t)$ when $|v(t)| > \zeta$. These peaks are less apparent when f_{c1} is small as the distortion spectra overlap, in Figures 4.12(a) and (b) they are just perceptible.

The model has more low frequency distortion than the d.m. system below f_{c1}. The reason for this is that every time $u(t)$ overloads the non-linear element, i.e. forces it to operate in its non-linear region, there is a constant difference between $x(t)$ and $y_L(t)$. This difference changes at each new overload condition and the frequency of these changes is lower than f_{c1}.

4.5. HERMITE POLYNOMIAL MODEL

The Normal spectrum technique discussed in Section 3.4 relates to memory-less non-linear devices. Although the d.m. system is not memory-less its output signal $y(t)$ in slope overload can be approximately represented by a signal $y_h(t)$. From Equation 3.2

$$y_h(t) = \sum_{n=0}^{\infty} b_{n,h} H_n(x) \qquad (4.106)$$

where $H_n(x)$ is the Hermite polynomial of order n.

The Hermite polynomial model of the d.m. system is consequently a 'black box' having a Gaussian signal $x(t)$ at its input and $y_h(t)$ at its output. From Equation 3.12 the spectral density function of $y_h(t)$ is

$$S_h(f) = \sum_{n=0}^{\infty} b_{n,h}^2 N_n \phi_x^n(f) \qquad (4.107)$$

If the $b_{n,h}$ coefficients for the d.m. system in slope overload can be found it can be represented by an infinite power series. However, $b_{n,h}$ cannot be found analytically and so $S_h(f)$ is first determined by computer simulation when a Gaussian input is applied to the d.m. system. The inverse discrete Fourier transform enables the autocorrelation function $R_h(\tau)$ of $y_h(t)$ to be found, and from Equation 3.11, $R_h(\tau)$ can be equated to a power series

$$R_h(\tau) = \sum_{n=0}^{\infty} b_{n,h}^2 N_n \rho_x^n(\tau) \qquad (4.108)$$

In Section 3.10.2 we saw that for odd non-linearities the $b_{n,h}$ coefficients are zero when n is even. The first terms in the $R_h(\tau)$ expansion are the most significant, and the first three relevant coefficients $b_{1,h}$, $b_{3,h}$ and $b_{5,h}$ can be found from three values of $R_h(\tau)$, i.e. three equations with three unknown values.

The required output spectral density function is obtained when $n = 1$ in Equation 4.107.

$$S_{1,h}(f) = b_{1,h}^2 \cdot \sigma^2 \phi_x(f) = b_{1,h}^2 S_x(f) \qquad (4.109)$$

For the Hermite polynomial model to represent the d.m. system in slope overload, Equations 4.101 and 4.109 must be nearly equivalent, i.e. $b_{1,L} \doteq b_{1,h}$. When these coefficients are plotted over a wide range of input power to step-size they agree very closely, i.e. $b_{1,h}$ has the shape of an error function. This close agreement indicates that a d.m. system in slope overload can be represented by a memory-less non-linearity.

REFERENCES

1. Protonotarios, E. N., 'Slope overload noise in differential pulse code modulation systems,' *Bell Systems Tech. J.*, **46**, No. 9, 2119–2162, November (1967)
2. O'Neal, J. B. Jr., 'Delta modulation quantizing noise analytical and computer simulation results for Gaussian and television signals', *Bell Systems Tech. J.*, **45**, No. 1, 117–141, January (1966)
3. Aaron, M. R., Fleischam, J. S., McDonald, R. A. and Protonotarios, E. N., 'Response of delta modulation to Gaussian signals, *Bell Systems Tech. J.*, **48**, 1167–1195, May–June (1969)
4. Steele, R. and Passot, M., 'Slope-limiter model of a delta modulator in slope overload with Gaussian input signals,' *Joint Conference Digital Processing of Signals in Communications*, Loughborough University of Technology, England, 285–303, April (1972)
5. Baikovskii, V. M., 'Application of the describing function method to the study of automatic unitary code systems'. *Automatic and Remote Control*, **26**, No. 9, 1494–1504, September (1965)
6. Booton, R. C., 'The analysis of non-linear control systems with random inputs', *Symposium Non-Linear Circuit Analysis*, Polytechnic Institute of Brooklyn, April 23–24 (1953)
7. Booton, R. C., Jr., Mathews, Max V. and Seifert, W. W., 'Non-linear servo mechanisms with random inputs,' *Report No. 70*, Dynamic Analysis and Control Lab., M.I.T., Cambridge, Mass., 38–42, August 20 (1953)
8. Slepian, D., 'On delta modulation', Bell System Tech. J. **51**, No. 10, 2101–2137, December (1972)
9. Greenstein, L. J., 'Slope overload noise in linear delta modulators with Gaussian inputs'. *Bell Systems Tech. J.*, **52**, No. 3, 387–421, March (1973)

Chapter 5

IDLE CHANNEL NOISE AND TRANSMISSION ERRORS IN LINEAR d.m. SYSTEMS

5.1. IDLE CHANNEL NOISE IN D.M. CODECS

When the input signal to the linear delta modulator is zero it is said to be in an idle state and the binary pattern of the signal $L(t)$ at the output of the encoder consists of a sequence of alternate logical ones and zeros, i.e. ... 1 0 1 0 1 0 1 0 .... This idle channel pattern is transmitted to the receiver and after regeneration is conveyed to the d.m. decoder where it is integrated and then rejected by the final filter. This rejection occurs because the fundamental frequency of the received binary signal $L(t)$ is $1/2f_p$; this is the lowest frequency in the signal and none of its frequency components will be able to pass through a filter whose upper critical frequency f_{c2} is much less than $1/2f_p$. Practical filters do not have infinite attenuation at $1/2f_p$ and consequently the output signal from the decoder in the idle channel condition will be negligible.

However, if the encoding and decoding processes are asymmetrical the idle channel pattern will occasionally have two adjacent logical 'ones' or 'zeros' inserted into the transmitted ... 1 0 1 0 1 ... pattern and under certain circumstances an unwanted signal $i(t)$ will emanate from the output of the filter in the decoder. This $i(t)$ signal is called the idle channel noise.[1]

Figure 5.1 shows the linear d.m. system and the channel terminal equipment. It will be assumed that the integrators are ideal and that there are no channel errors. The asymmetry of the codec is considered to result

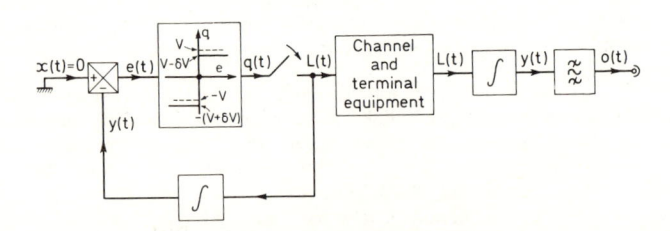

Fig. 5.1. Asymmetrical delta modulator with channel and terminal equipment and symmetrical decoder

Idle channel noise and transmission errors 153

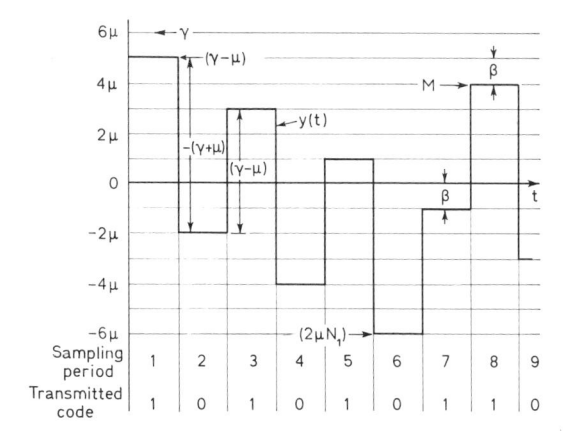

Fig. 5.2. *Waveform at the output of the integrators due to an asymmetrical quantizer,* $\gamma = 6\mu$

from the quantization characteristic, although the same waveforms at the output of the integrators can result if the quantization characteristic is symmetrical and the integrators are sufficiently and identically asymmetrical. The signal $q(t)$ at the output of the quantizer is sampled for τ seconds every sampling period, where $\tau \ll T$. For simplicity, the $L(t)$ waveform will be considered to be composed of impulses of strength $(V - \delta V)\tau$ or $-(V + \delta V)\tau$ respectively depending on whether $e(t) \geq 0$ or < 0 at a sampling instant. V is assumed always to be considerably greater than δV. The positive and negative impulses in the $L(t)$ sequence produce steps of $(\gamma - \mu)$ and $-(\gamma + \mu)$ respectively.

The discussion which follows is based on the asymmetry shown in Figure 5.1. However, if the asymmetry is such that the positive impulses are larger than the negative ones, the remarks still apply providing the polarity of μ is changed.

Assume that at $t = 0$

$$y(t) = Y_0 = \gamma - \mu$$

and this value of $y(t)$ is maintained for one clock period T. As there is no input signal the error signal $e(t) = -Y_0$ throughout the period and at the next sampling instant $q(t)$ is negative and consequently $L(t)$ is a negative impulse which produces a negative step in $y(t)$ of

$$Y_1 = Y_0 - (\gamma + \mu) = -2\mu$$

The error signal is now $+2\mu$ and at the next sampling instant a positive impulse is produced at the output of the encoder and Y_2 becomes $\gamma - 3\mu$. The situation is illustrated in Figure 5.2 for the first nine sampling periods when $\gamma = 6\mu$. After the third negative impulse has been produced $y(t) = -6\mu$ and two positive impulses must be generated in order for the error to change sign, i.e. the ideal ... 1 0 1 0 1 0 1 0 ... idling output pattern has been disturbed by the asymmetry in the encoder. The same waveform $y(t)$ is of course produced at the encoder and the decoder as shown in Figure 5.1.

The following observations are made with reference to the waveform $y(t)$ shown in Figure 5.2 which applies for an encoder having a particular asymmetry. While the ... 1 0 1 0 0 1 0 ... pattern persists $y(t)$ decreases by 2μ after every negative impulse, and after N negative impulses $y(t) = -2\mu N$. If after N_1 negative impulses $y(t) = -2\mu N_1$ and the subsequent positive impulse causes $y(t)$ to increase to $-2\mu N_1 + (\gamma - \mu) = -\beta$, a second positive impulse is produced. The value of β is dependent on the magnitude of the asymmetry. If β is a rational number the $y(t)$ signal is periodic having a fundamental frequency determined by $(\gamma - \mu)/2\mu$ and the clock frequency f_p.

Figure 5.3 shows some simple $y(t)$ waveforms where the relevant values of the parameters are integers. Figure 5.3(a) illustrates the case where $y(t)$ can be considered to be the sum of a sawtooth waveform and the ideal idling waveform. The decoded idling signal $i(t)$ is the sawtooth waveform filtered by the final band-pass filter. In Figure 5.3(b) the asymmetry is increased, but the fundamental frequency f_y is nearly the same as in 5.3(a). There is a second peak which is less than the maximum one. The asymmetry is increased in 5.3(c) but the fundamental frequency is considerably increased when compared to the values in 5.3(a) and (b) because the waveform has no subsidiary peaks. Of the 4 waveforms 5.3(d) has the greatest asymmetry. There are three subsidiary peaks which produce two envelopes. The fundamental frequency f_y of the idling patterns in Figure 5.3 is

$$f_y = \left(\frac{1}{\lambda}\right)\left(\frac{\gamma-\mu}{\mu}+1\right)^{-1} f_p$$

where λ is the number of positive peaks in the periodic pattern of $y(t)$.

Irrespective of the complexity of $y(t)$ it is possible to calculate the next peak in $y(t)$ provided the last peak is known. In Figure 5.3(d) the peak in the tenth sampling period can be expressed as

$$M_1 = 2(\gamma-\mu) - 2\mu N_1 = (\gamma-\mu-\beta_1) \tag{5.1}$$

The encoder now generates a zero and $y(t)$ reduces to

$$y(t) = M_1 - (\gamma+\mu) = -(2\mu+\beta_1)$$

After N_2 negative impulses

$$y(t) = -(2\mu N_2 + \beta_1)$$

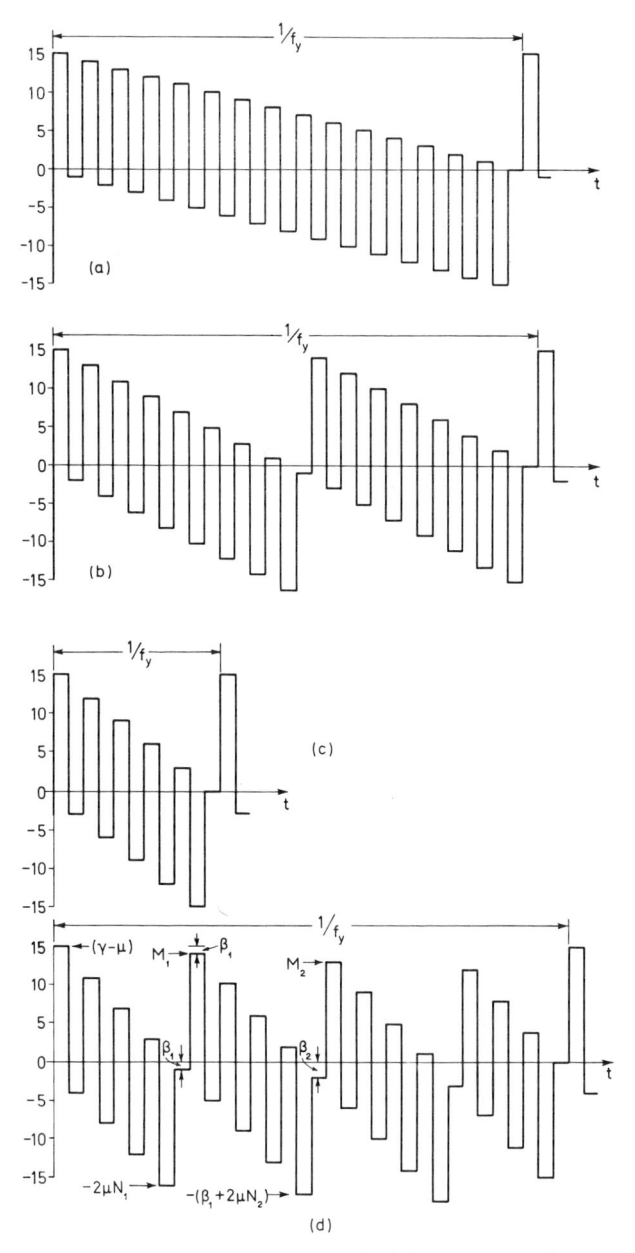

Fig. 5.3. *Waveform at the output of the integrator for different degrees of asymmetry:* (a) $\gamma = 15.5$, $\mu = 0.5$, (b) $\gamma = 16$, $\mu = 1$, (c) $\gamma = 16.5$, $\mu = 1.5$, (d) $\gamma = 17$, $\mu = 2$

and this is followed by two positive steps to give the next maximum M_2

$$M_2 = 2(\gamma - \mu) - (2\mu N_2 + \beta_1)$$
$$= (\gamma - \mu) - 2\mu N_2 + M_1 \qquad (5.2)$$

N_2 can be found as it is the value of the first integer greater than $M_1/2\mu$: a double one will occur if $M_1 - 2\mu N_2$ is just negative.

The value of β, i.e. the value of $y(t)$ when a second consecutive one is about to occur, must lie in the range

$$2\mu < \beta_k \leq 0, k = 1, 2, 3, \ldots$$

and as

$$2\mu N_{k+1} + \beta_k \leq \gamma - \mu$$

$2\mu N_{k+1}$ is bounded

$$\gamma - \mu < 2\mu N_{k+1} < \gamma + \mu$$

The differences between successive maxima $M_{i+1} - M_i$ must lie in the range $0 < |M_{i+1} - M_i| < 2\mu$.

Iwersen[2] analyses the effect of asymmetry in linear d.m. He represents the behaviour of a linear d.m. system having a positive asymmetry by a symmetrical linear d.m. encoding an appropriate ramp input signal. The theoretical results obtained by this approach which have been verified experimentally[3] are given here without proof.

The discrete lines f_l in the waveform $y(t)$ are

$$f_l = f_p \left| \frac{l}{2}\left(1 - \frac{\mu}{\gamma}\right) - N(\alpha) \right| \qquad (5.3)$$

where $N(\alpha)$ is the integer nearest to $(l/2)\{1 - (\mu/\gamma)\}$. The mean square value associated with a frequency line f_l is

$$p_l = (2\gamma^2)/(\pi^2 l^2) \qquad (5.4)$$

The equations hold for non-integer ratios of γ/μ. The noise power in the message band depends on whether the asymmetry is such as to cause a low order frequency line, i.e. one having low values of l, to appear in this band.

Although the line f_1 has more power than lines corresponding to higher values of l, the frequency f_1 may be a high or low frequency. For example, if the values of γ and μ are 12 mV and 2 mV respectively, and the clock rate is 50 kbits/s, then with the aid of Equations 5.3 and 5.4 the mean square value of the discrete frequency $f_1 = 20.8$ kHz is 29.2 μW, whereas when $l = 7$ the line frequency is only 4.17 kHz although its mean square value is 0.6 μW.

Idle channel noise and transmission errors 157

5.1.1. Effect of asymmetry on tracking

If the encoder and decoder have the same asymmetry due to different charging currents in their integrators when positive and negative impulses are applied, then the $y(t)$ waveform at the encoder has the same shape as that at the decoder, assuming no transmission errors. Observe that in the first part of Section 5.1 the asymmetry in the encoder is considered to reside in the quantizer. This approach has been made as a matter of convenience. The important point is that asymmetry manifests itself in the $y(t)$ waveform, i.e. at the output of the integrators.

The overload characteristic is asymmetrical because the $y(t)$ waveform can increase at $(\gamma + \mu)f_p$ V/s in one direction or a reduced rate of $(\gamma - \mu)f_p$ V/s in the other.

The discrete line spectrum produced by the asymmetrical encoder in the idle channel state become phase modulated[2] by the input signal $x(t)$, an effect which can be explained with reference to the discrete pulse phase modulation in Section 1.6.1. When the input signal is Gaussian the discrete lines change to a Gaussian shape having an effective width proportional to the r.m.s. slope of $x(t)$. Increasing the input power results in the Gaussian spectra increasing until a large Gaussian spectrum associated with a powerful discrete frequency in the idling condition makes a significant contribution to the noise power in the message band.

For a sinusoidal input signal $E_s \sin 2\pi f_s t$ each discrete idle channel line f_l is phase modulated such that

$$\Delta f \doteq 2\pi f_s E_s / \text{Hz}$$

If the asymmetry occurs at the encoder, and the decoder is symmetrical, then the decoded signal differs from the required $x(t)$ to a greater extent than if the asymmetry was identical in both encoder and decoder. Figure 5.4 illustrates the argument for an arbitrary input signal. The decoder experiences an effect similar to slope overload, i.e. a breakdown

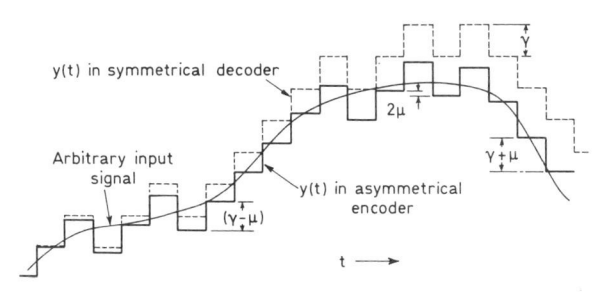

Fig. 5.4. The $y(t)$ waveforms at the encoder and decoder in response to an arbitrary input signal $x(t)$. The asymmetry occurs in the integrator in the encoder

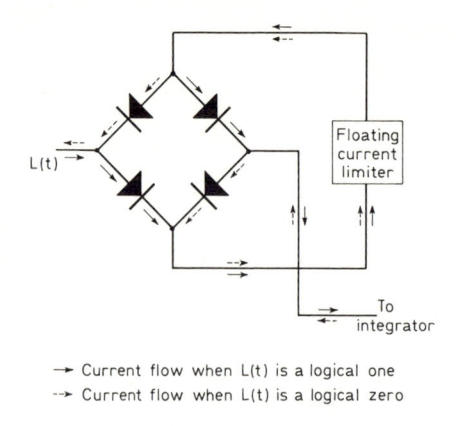

→ Current flow when L(t) is a logical one
--→ Current flow when L(t) is a logical zero

Fig. 5.5. Circuit for eliminating idle channel noise. (After Schindler[4]*)*

in tracking, but this occurs where the encoder is tracking the input signal over a wide range.

The presence of idle channel noise degrades the performance of the codec. Fortunately this noise can be reduced to zero or negligible proportions by careful design. For example, Figure 5.5 shows a circuit which when placed between the binary waveform $L(t)$ at the output of the delta modulator and the input to the integrator in the feedback loop eliminates idle channel noise. The floating current limiter ensures that the magnitudes of the positive and negative current applied to the integrator are identical.

5.2. TRANSMISSION ERRORS IN D.M. SYSTEMS

The received binary signal $L_R(t)$ will suffer from distortion due to imperfections in the terminal equipments and communication channel. $L_R(t)$ is regenerated into a binary waveform $L_D(t)$ by deciding at each sampling instant whether $L_R(t)$ is a logic one or zero. This decision process enables $L_D(t)$ to be identical with the binary waveform $L(t)$ at the output of the encoder in spite of $L_R(t)$ being considerably distorted. However, if the distortion in $L_R(t)$ is sufficiently large the decision system will make errors[5,6,7] and some transmitted logical ones will be identified as logical zeros and vice versa. The waveform $L_D(t)$ now contains digital errors and the effect of these errors on the decoded signal-to-noise ratio will now be investigated for an exponential d.m. system.[5] This type of system is considered in order to see what effects a leaky integrator has on the performance of the system in the presence of noise.

Idle channel noise and transmission errors 159

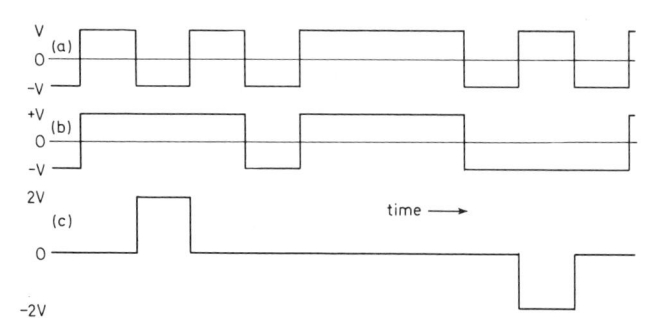

Fig. 5.6. $L(t)$ at the encoder, (b) regenerated waveform $L_D(t)$ at receiver, (c) the error component $L_e(t)$ of $L_D(t)$

Figure 5.6(a) shows an arbitrary waveform $L(t)$ which originates at the output of an exponential delta modulator. Due to the effects of distortion the regenerated signal $L_d(t)$ is seen to contain two errors. $L_D(t)$ can be separated into the sum of two waveforms, the errorless data waveform $L(t)$ and an error waveform $L_e(t)$. Observe that the errors in $L_R(t)$ at sampling times manifest themselves as error pulses having a duration of one bit period and an amplitude which is double those of $L(t)$.

Assume that the probability of any regenerated digit being in error is P_e, and that these errors are randomly distributed in time. The decoder is composed of two linear components, an RC circuit and a sharp cut-off filter F_0. The data signal $L(t)$ and the error signal $L_e(t)$ are statistically independent. Consequently the decoded output signal is the sum of the signal $O(t)$ which is due to $L(t)$ and contains errors due to the encoding process, and a transmission error signal $O_e(t)$ due to $L_e(t)$, i.e. due to transmission errors. The noise power of $O_e(t)$ is found by considering the decoding of $L_e(t)$ only.

If the error waveform $L_e(t)$ is considered over a number of clock periods, and provided that the error pulses are relatively infrequent, for example $< 10\%$, then the autocorrelation function $R_e(\tau)$ will be a maximum when τ is zero, and will be effectively zero for $\tau > T$ and $\tau < -T$. This is because for $|\tau| > T$ the product of $L_e(t) \cdot L_e(t+\tau)$ is very close to zero. Thus $R_e(\tau)$ shown in Figure 5.7(a) has an autocorrelation function which is similar to that of an isolated error pulse, but a magnitude dependent on the probability P_e.

The Fourier transform of $R_e(\tau)$ is the spectral density function $S_e(f)$ shown in Figure 5.7(b). As the message band, shown shaded in Figure 5.7(b), occurs at frequencies which are close to d.c., i.e. on the nearly flat part of the $S_e(f)$ function it is necessary to calculate $S_e(0)$ the d.c. value of $S_e(f)$. Having calculated $S_e(0)$ the effect of the passive decoder on the

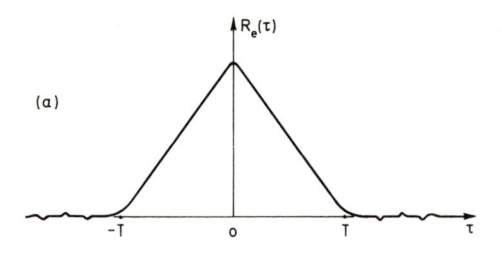

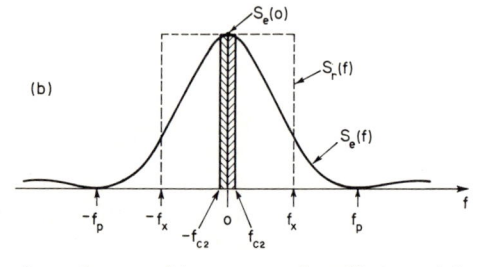

160   Idle channel noise and transmission errors

Fig. 5.7. (a) autocorrelation function of the error waveform, (b) spectral density function $S_e(f)$ of the error waveform and the equivalent rectangular spectral density function $S_r(f)$

shaded part of the $S_e(f)$ function can be found, and hence the mean square value of the decoded error signal.

Proceeding to calculate $S_e(0)$, the mean square value associated with the error waveform $L_e(t)$ is $(2V)^2 P_e$, i.e.

$$4V^2 P_e = 2 \int_0^\infty S_e(f) \, df \qquad (5.5)$$

Consider a rectangular spectral density function $S_r(f)$ shown in Figure 5.7(b), and defined by

$$S_r(f) = S_e(0), \quad |f| < f_x \qquad (5.6)$$
$$= 0, \text{ elsewhere}$$

Suppose f_x has a value such that

$$2 \int_0^\infty S_e(f) \, df = 2 \int_0^{f_x} S_r(f) \, df \qquad (5.7)$$

Because of the shape of $R_e(\tau)$, $S_e(f)$ is a sin squared function of the form

$$S_e(f) = S_e(0) \left| \frac{\sin(\pi f T)}{(\pi f T)} \right|^2 \qquad (5.8)$$

Idle channel noise and transmission errors 161

$$2\int_0^\infty S_e(f)\,df = \frac{2S_e(0)}{\pi T}\int_0^\infty \left|\frac{\sin(\pi f T)}{(\pi f T)}\right|^2 d(\pi f T)$$

and

$$\int_0^\infty \left|\frac{\sin(\pi f T)}{(\pi f T)}\right|^2 d(\pi f T) = \frac{\pi}{2} \tag{5.9}$$

Hence

$$2\int_0^\infty S_e(f)\,df = A^2 T \tag{5.10}$$

From Figure 5.7(b)

$$2\int_0^{f_x} S_r(f)\,df = 2S_e(0)f_x \tag{5.11}$$

Equating 5.7, 5.10 and 5.11 gives

$$f_x = f_p/2 \tag{5.12}$$

From Equations 5.5 and 5.10

$$S_e(0) = 4V^2 P_e/f_p \tag{5.13}$$

Thus the power spectral density function $S_e(f)$ of the error signal $L_e(t)$ can be replaced over the message band (shown shaded in Figure 5.7(b)) by the rectangular function $S_r(f)$.

The spectral density at the output of the RC integrator is $S_e(f).|H(j2\pi f)|^2$, where $H(j2\pi f)$ is the transfer function of the RC integrator expressed by Equation 2.32.

If the final filter has a transfer function $H_d(j2\pi f)$ then the spectral density at the output of the decoder due to channel errors is

$$O_e(f) = S_e(f)|H(j2\pi f)|^2 |H_d(j2\pi f)|^2 \tag{5.14}$$

The output noise due to transmission errors is

$$N_e^2 = 2\int_{f_{c1}}^{f_{c2}} O_e(f)\,df = 2\int_{f_{c1}}^{f_{c2}} \left(\frac{4V^2 P_e}{f_p}\right)\left(\frac{f_1^2}{f_1^2+f^2}\right)|H_d(j2\pi f)|^2\,df \tag{5.15}$$

If $H_d(j2\pi f)$ is an ideal filter having a magnitude of unity over the frequency bands $\pm(f_{c2}-f_{c1})$ and zero elsewhere, and a linear phase shift characteristic, then

$$N_e^2 = \frac{8V^2 P_e f_1}{f_p}\left\{\tan^{-1}\left(\frac{f_{c2}}{f_1}\right) - \tan^{-1}\left(\frac{f_{c1}}{f_1}\right)\right\} \tag{5.16}$$

162 Idle channel noise and transmission errors

A typical arrangement is $f_{c1} = f_1$ and $f_{c2} \gg f_1$, when

$$N_e^2 \doteqdot \frac{8V^2 P_e f_1}{f_p} \cdot \frac{\pi}{4} \tag{5.17}$$

We now consider whether the noise due to transmission errors be curtailed by a judicious choice of f_1. It would appear from Equation 5.17 that N_e^2 can be reduced by simply lowering f_1, and of course this is so, but it has serious consequences for the signal, as it affects the overload characteristic of the delta modulator shown in Figure 2.28 for a fixed message band f_{c1} to f_{c2}. If the overload characteristic is to be the same in the message band then lowering f_1 must be compensated by an increase in V. If f_1 is increased by G, say, where $G > 1$ and V decreased by G, the quantization noise and s.q.n.r. given by Equations 2.44 and 2.45, respectively are unaltered but N_e^2 is reduced by G. Thus in practical systems which often make $f_1 = f_{c1}$, the noise due to transmission errors can be diminished without effecting the capability of the system to encode, transmit and decode an analogue signal by reducing the RC time constant. However, the amplitude range, represented by Equation 2.43, becomes unacceptably small if f_1 is increased too much.

The fact that a lower RC time constant gives a smaller N_e^2 is expected because it allows the error pulse to decay relatively rapidly. Errors have a more accumulative, i.e. long term effect, for longer time constants. A perfect integrator would retain for all time the step at its output due to the advent of an error pulse.

When $f_1 \ll f_{c1}$ there is no advantage to be gained in trading f_1 against V in order to reduce N_e^2.

If the encoder is tracking correctly thereby avoiding slope overload the total decoded noise N^2 is the sum of N_e^2 and N_q^2. From Equations 5.17 and 2.44

$$N^2 \doteqdot N_e^2 \left\{ 1 + 2\pi K_q \cdot \frac{f_c f_1}{P_e f_p^2} \right\} \tag{5.18}$$

where $f_c = f_{c2} - f_{c1}$.

For the quantization noise to be equal to the noise due to transmission errors

$$P_e \doteqdot 2\pi K_q f_c f_1 / f_p^2$$

For $f_1 = f_{c1} = 300$ Hz, $f_{c2} = 3{,}400$ Hz and $K_q = \frac{1}{3}$ the probability of error P_e (in per cent when f_p is in kbits/s) is approximately $200/f_p^2$. When $f_p = 56$ kbits/s, $P_e = 0\cdot064\%$ and when $f_p = 20$ kbits/s, $P_e = 0\cdot5\%$.

These figures illustrate that when $f_p = 20$ kbits/s and the error rate is as high as 1 in 200 the noise produced at the output of the decoder is of the same magnitude as the quantization noise, which is of course small compared to the signal power. The linear d.m. system can therefore operate in the presence of substantial random errors.

5.2.1. Transmission errors in d.s.m. systems

The d.s.m. decoder following the decision system is simply a band pass filter. $H_d(j2\pi f)$. The RC integrator is not used. Equation 5.15 becomes

$$N_e^2 = \frac{8V^2 P_e}{f_p} \int_{f_{c1}}^{f_{c2}} |H_d(j2\pi f)|^2 \, df \qquad (5.19)$$

and assuming that $H_d(j2\pi f)$ is an ideal band-pass filter

$$N_e^2 = \frac{8V^2 P_2}{f_p}(f_{c2} - f_{c1}) \qquad (5.20)$$

For a d.s.m. system the RC integrator used in the encoding process does not affect the decoded noise due to transmission errors. When f_{c1} is zero, i.e. $H_d(j2\pi f)$ is a low-pass filter, then from Equations 5.16 and 5.20

$$P_e = \frac{\pi^2}{6} \cdot K_{eq} \left(\frac{f_{c2}}{f_p}\right)^2$$

for $N_e^2 = N_q^2$.

If $K_q = \frac{1}{3}$ and $f_{c2} = 3{,}400$ Hz, then from the above equation $P_o \doteq 630/f_p \%$, where f_p is in kbits/s. Consequently the d.s.m. system performs better in the presence of channel errors than the d.m. system.

REFERENCES

1. Wang, P. P., 'Idle channel noise of delta modulation', *I.E.E.E. Trans. on Com. Tech.*, **COM-16**, No. 5, 737, October (1968)
2. Iwersen, J. E., 'Calculated quantization noise of single integration delta modulation coders', *Bell Systems Tech. J.*, 2359–2389, September (1969)
3. Laane, R. R., 'Measured quantization noise spectrum for single integration delta modulation coders', *Bell Systems Tech. J.*, 191–195, February (1970)
4. Schindler, H. R., 'Delta modulation', *I.E.E.E. Spectrum*, 69–78, October (1970)
5. Johnson, F. B., 'Calculating delta modulation performance', *I.E.E.E. Trans.*, **AU-16**, 121–129 (1968)

6. Butter, L. S. and Kiddle, L., 'The rating of delta-sigma modulation systems, with constant errors, burst errors and tandem links in free conversation test using the reference speech link', Report No. 69014, *Signals Research & Development Establishment*, Christchurch, Hants., February (1969)
7. Wolf, J. K., 'Effects of channel errors on delta modulation', *I.E.E.E. Trans. Com. Tech.*, **COM-14**, 2–7, February (1966)

Chapter 6

ASYNCHRONOUS DELTA MODULATION SYSTEMS

6.1. ASYNCHRONOUS DELTA MODULATION

Asynchronous delta modulation systems have a digital output which is quantized in amplitude but not in time. Consequently no sampling processes are involved.

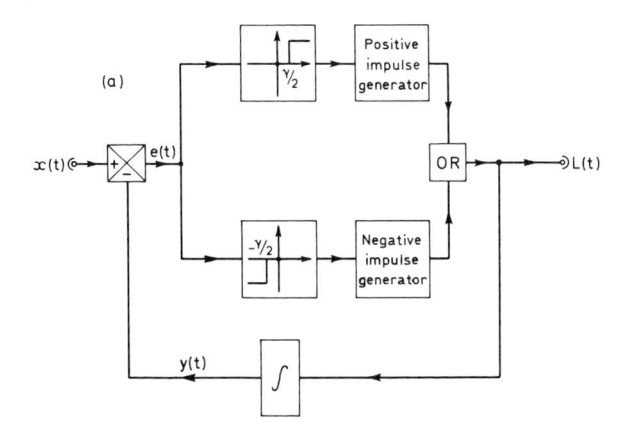

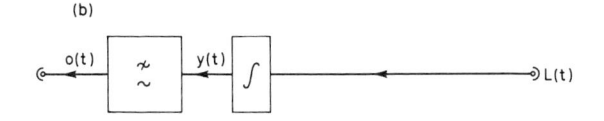

Fig. 6.1. Asynchronous d.m. codec: (a) encoder, (b) decoder

The system arrangement is shown in Figure 6.1. The encoder[1,2] generates an output waveform $L(t)$ consisting of impulses of magnitude $\pm V$. Each impulse is generated directly when the output from a quantizer changes from zero to its high magnitude state. The polarity is determined

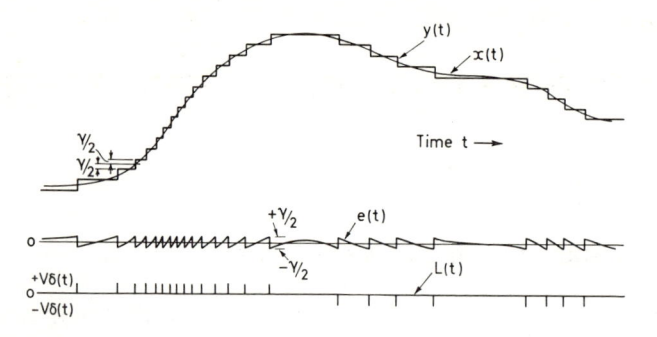

Fig. 6.2. Asynchronous d.m. waveforms when encoding an arbitrary input signal x(t)

from the magnitude of the error signal $e(t)$ by applying the following rules:

$$L(t) = +V\delta(t), \quad \text{if} \quad e(t) \geq \gamma/2$$
$$= -V\delta(t), \quad \text{if} \quad e(t) \leq -(\gamma/2) \quad (6.1)$$
$$= 0, \quad \text{if} \quad -\gamma/2 < e(t) < \gamma/2$$

These output impulses are integrated by a perfect integrator having an impulse function

$$h_i(t) = \gamma/V, \quad \text{for} \quad t \geq 0$$
$$= 0, \quad t < 0$$

The convolution of the $L(t)$ sequence $\{L_r\}$ with $h_i(t)$ produces the step-like waveform $y(t)$ shown in Figure 6.2. The amplitude of these steps is $\pm\gamma$ corresponding to L_r being an impulse of $\pm V\delta(t)$, respectively. The error signal $e(t)$ is the difference between $x(t)$ and $y(t)$ and as can be seen from Equation 6.1 cannot exceed $|\gamma/2|$. Figure 6.2 shows the behaviour of the encoder for an arbitrary input signal $x(t)$. The $y(t)$ waveform is identical to that obtained by uniformally quantizing $x(t)$. Although the excursions of $e(t)$ are limited to $\pm\gamma/2$, the duration $D(t)$ between these maxima in $e(t)$ is a variable and depends on the slope of $x(t)$.

The decoder shown in Figure 6.1(b) consists of an integrator to recover $y(t)$ from the reformed $L(t)$ impulses followed by a filter to remove the steps in $y(t)$. The filter has the same pass-band as the filter which band-limits the input signal $x(t)$.

6.1.1. Zero crossings

Consider a sinusoid $x(t)$ equal to $E_s \sin 2\pi f_s t$. The number of steps in the $y(t)$ waveform required to reach a given value of $x(t)$ is a constant, independent of the frequency. However, the rate of these steps is a function of the slope of $x(t)$.

During the first quarter cycle of $x(t)$ the number of steps in the $y(t)$ waveform is E_s/γ. Over a complete cycle of $x(t)$ the number of steps in $y(t)$ is $4E_s/\gamma$. The number of steps per second is

$$n = \frac{4E_s f_s}{\gamma} \qquad (6.2)$$

and the number of zero crossings per second in the error waveform $\mu = 2n$.

The average slope of $x(t)$ is

$$\langle |x'(t)| \rangle = \frac{1}{1/(2f_s)} \int_{-1/(4f_s)}^{1/(4f_s)} E_s 2\pi f_s \cos 2\pi f_s t \, dt = 4f_s E_s \qquad (6.3)$$

where x' represents differentiation with respect to time and the $\langle \ \rangle$ brackets denotes time averaging. From Equations 6.2 and 6.3 the number of zero crossing in the $e(t)$ waveform is

$$\mu = \frac{2}{\gamma} \langle |x'(t)| \rangle \qquad (6.4)$$

This equation provides a useful result, and it will now be used to estimate the long term average number of zero crossings in the error waveform when the input signal $x(t)$ has zero mean, mean square value σ_x^2, band-limited to f_{c2} Hz, and a Gaussian amplitude probability density function $f(x)$ defined by Equation (3.5).

Rice[3] shows that the slope of $x(t)$ also has a Gaussian probability density function $f(x'(t))$.

$$f(x'(t)) = \frac{1}{2\pi f_{c2} \sigma} \sqrt{\frac{3}{2\pi}} \cdot \exp\left[-\frac{3}{2}\left\{\frac{x'(t)}{2\pi f_{c2}\sigma}\right\}^2\right] \qquad (6.5)$$

To calculate the number of zero crossings μ in the error waveform, $\langle |x'(t)| \rangle$ must be determined. Now

$$\langle |x'(t)| \rangle = \int_{-\infty}^{\infty} |x'(t)| f\{|x'(t)|\} \, d|x'(t)| \qquad (6.6)$$

Substituting for $f\{x'(t)\}$ from Equation 6.5 in Equation 6.6 and rearranging

$$\langle |x'(t)| \rangle = 4\sqrt{\frac{2\pi}{3}} \cdot f_{c2}\sigma \int_0^\infty u \exp(-u^2) \, du \qquad (6.7)$$

where

$$u = \sqrt{\frac{3}{2}} \cdot \frac{|x'(t)|}{2\pi f_{c2}\sigma}$$

Now

$$\frac{d}{du}(e^{-u^2}) = -2u\, e^{-u^2}$$

hence

$$\langle |x'(t)| \rangle = 4\sqrt{\frac{2\pi}{3}} \cdot f_{c2}\sigma \left[-\frac{\exp(-u^2)}{2} \right]_0^\infty$$

$$\langle |x'(t)| \rangle = \sqrt{\frac{8\pi}{3}} \cdot f_{c2}\sigma \tag{6.8}$$

The number of zero crossings μ is found by substituting the value of $\langle x'(t) \rangle$ in Equation 6.4

$$\mu = 4\sqrt{\frac{2\pi}{3}} \left(\frac{f_{c2}\sigma}{\gamma} \right) \tag{6.9}$$

The average duration $D(t)$ between the $L(t)$ impulses is related to the slope of the input signal $x(t)$, illustrated in Figure 6.2.

$$D(t) \doteqdot \frac{\gamma}{\langle |x'(t)| \rangle} \tag{6.10}$$

This result gives an alternative expression $\mu = 2/D(t)$.

In a practical system the $L(t)$ impulses are transmitted in pulse form where every pulse has a width equal to the minimum value of $D(t)$.

6.1.2. Minimum channel bandwidth

The minimum channel bandwidth W for the digital transmission of $x(t)$ is equal to the reciprocal of the minimum value of $D(t)$, i.e.

$$W = \frac{|x'(t)_{max}|}{\gamma} \tag{6.11}$$

6.1.3. Overload characteristic

This is defined in terms of a sinusoidal input signal $E_{sm} \sin \pi f_{sm} t$ which satisfies Equation 6.11.

$$W = 2\pi f_{sm} E_{sm}/\gamma$$

or

$$E_{sm} = W\gamma/(2\pi f_{sm}) \tag{6.12}$$

The graph of E_{sm} against f_{sm} is the overload characteristic. E_{sm} falls with increasing f_{sm} at 6 dB/octave. The overload characteristic is the same shape as that for the synchronous linear d.m. although it includes the clock rate f_p in the numerator instead of the channel bandwidth W. If the amplitude of the input sinusoid $E_s > E_{sm}$ the encoder will generate adjacent pulses whose spacing is $< D(t)_{min}$ over part of the cycle. Thus although the encoder is functioning satisfactorily with $y(t)$ tracking $x(t)$ the $L(t)$ signal requires a bandwidth in excess of the channel bandwidth, with the result that the decoded signal will be distorted.

6.1.4. Overload condition

Equation 6.11 indicates the limitations on a sinusoidal input signal to avoid overload. When $x(t)$ is band-limited white noise, as defined when establishing Equation 6.9, $x'(t)$ has a mean square value

$$\langle \{x'(t)\}^2 \rangle = (4/3)\pi^2 f_{c2}^2 \sigma^2$$

This equation was derived in Section 3.3.

The encoder is overloaded with a probability $< 4 \times 10^{-5}$ if the r.m.s. value of the slope of $x(t)$ is less than a quarter of the maximum slope that the encoder can accommodate. Hence for negligible overloading

$$(2/\sqrt{3})\pi f_{c2}\sigma < \tfrac{1}{4}W\gamma \qquad (6.13)$$

6.1.5. Amplitude range

Defining the amplitude range AR of the encoder as E_{sm}/E_{st}, where E_{sm} is given by Equation 6.12 and E_{st} is the smallest amplitude of the sinusoidal input which causes an impulse to appear at the output of the encoder

$$\text{AR} = W/(\pi f_{sm}) \qquad (6.14)$$

6.1.6. Quantization noise

The $y(t)$ waveform shown in Figure 6.2 is identical to the waveform obtained by passing $x(t)$ through a linear quantizer having an infinite staircase characteristic. Consequently the analysis of the quantization noise can be achieved by using the simple open-loop model of the infinite staircase quantizer, rather than considering the actual closed-loop coder.

Provided the input signal $x(t)$ is a band-limited white noise signal specified in Section 3.4.1, then the Normal spectrum technique can be applied to calculate the spectral density function $S_{em}(f)$ of the error signal $e(t)$.

$$S_{em}(f) = \frac{1}{2\pi^3}\sqrt{\frac{3}{2\pi}} \cdot \frac{\gamma^3}{\sigma f_{c2}} \sum_{n=1}^{\infty} \frac{1}{n^3} \exp\left\{-\frac{3}{8\pi^2}\left(\frac{\gamma f}{n\sigma f_{c2}}\right)^2\right\} \qquad (6.15)$$

where $S_{em}(f)$ is a single-sided spectral density function and f_{c2} is the highest frequency in $x(t)$.

Bennett[4] in his classical paper on the spectra of quantized signals arrives at an identical result by using Equations 3.7 and 3.8 but without recourse to Mehler's formula of Equation 3.9. Bennett expresses the summation term in the equation for $S_e(f)$ as

$$B(z) = \sum_{n=1}^{\infty} \frac{1}{n^3} \exp\left(-\frac{z^2}{n^2}\right) \qquad (6.16)$$

where

$$z = \frac{3}{8\pi^2}\left(\frac{\gamma f}{\sigma f_{c2}}\right)^2 \qquad (6.17)$$

Figure 6.3 shows the graph of $B(z)$ against z. The curve is a good approximation to that obtained for the function

$$B(z) = \frac{B(0)z_1}{(z_i^2 + z^2)^{1/2}} \qquad (6.18)$$

where z_1 is the value of z when $B(z)$ is 3 dB down from its value $B(0)$; i.e. the value of $B(z)$ when z is zero

$$B(0) = \sum_{n=1}^{\infty} \frac{1}{n^3} = 1 \cdot 202 \qquad (6.19)$$

The value of z for the highest frequency in $x(t)$ is from Equation (6.17).

$$z_{c2} = \frac{3}{8\pi^2}\left(\frac{\gamma}{\sigma}\right)^2$$

As the r.m.s. value σ of the input signal $x(t)$ is much greater than the step-size γ in the $y(t)$ waveform

$$\gamma/\sigma \ll 1$$

and from Figure 6.3 $z_{c2} \ll z_1$ with the result that when $f = f_{c2}$, $B(z) \simeq B(0)$. Consequently $S_{em}(f)$ from Equations 6.15 and 6.16 is given by

$$S_{em}(f) = \frac{1}{2\pi^3}\left(\frac{3}{2\pi}\right)^{1/2} \cdot \frac{\gamma^3}{\sigma f_{c2}} \cdot B(0) \qquad (6.20)$$

and because this expression is independent of frequency the quantization

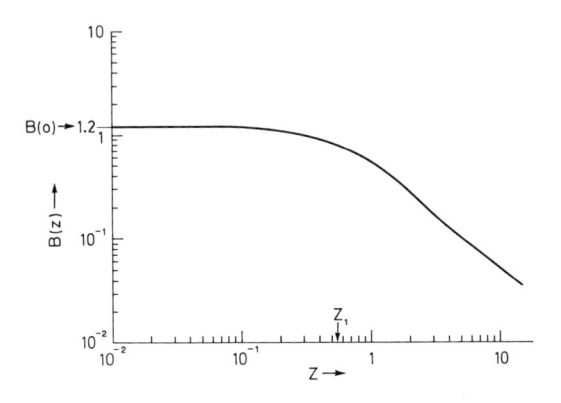

Fig. 6.3. B(z) as a function of z

noise is

$$N_q^2 = S_{em}(f) \cdot f_{c2} = \gamma^3/(75\sigma) \tag{6.21}$$

The error signal at the output of the integrator in the decoder is the same as the difference signal between the input and output of an infinite staircase function. This error signal is merely filtered in the asynchronous d.m. decoder to give the decoded distortion signal having a spectral density function $S_{em}(f)$. This function is smaller than the corresponding p.c.m. spectral density function because in p.c.m. the error signal is sampled at the Nyquist rate which produces severe aliasing. This effect is discussed in Section 9.2.

The performance of asynchronous d.m. is poor when it has to operate through channels which cause intersymbol interference and where there is sufficient additive noise to cause the receiver to make errors in specifying the time location of the pulses.

6.1.7. Quantization noise as a function of zero crossings

The error signal $e(t)$ has a simple and useful relationship between its spectral density function over the message band and the number of its zero crossings. Eliminating the ratio $\gamma/f_{c2}\sigma$ in Equation 6.20 using Equation 6.9

$$S_{em}(f) = 0.077\gamma^2/\mu, \quad 0 \leq f \leq f_{c2} \tag{6.22}$$

This is a simple formula which expresses the spectral density function of the error signal in terms of its zero crossings and the step-size in the feedback waveform.

The quantization noise is

$$N_q^2 = 0.077\gamma^2 f_{c2}/\mu \tag{6.23}$$

The break frequency f_n, i.e. the value of f when $z = z_1$, and the number of zero crossings μ given by Equations 6.17 and 6.9, respectively have the same dimensions. The relationship is

$$\mu = 1.54 f_n \tag{6.24}$$

6.1.8. Signal-to-quantization noise ratio

The ratio of the input signal power σ^2 and the quantization noise power given by Equation 6.21 is

$$\begin{aligned}\text{s.q.n.r.} &= 75(\sigma/\gamma)^3 \\ &= 18.8 + 30 \log_{10}(\sigma/\gamma) \text{ dB}\end{aligned} \tag{6.25}$$

If $\sigma = 5\gamma$ the s.q.n.r. = 40 dB. An alternative expression for s.q.n.r. can be obtained using Equation 6.23, i.e.

$$\text{s.q.n.r.} = 13 \frac{\mu}{f_{c2}} \left(\frac{\sigma}{\gamma}\right)^2 \tag{6.26}$$

6.2. ASYNCHRONOUS PULSE DELTA MODULATION

This modulation has been called asynchronous binary slope quantized p.c.m., and pulse-interval modulation,[1,5,6] but it will be known here as asynchronous pulse delta modulation as it is the asynchronous version of the pulse delta modulator described in Chapter 2. The schematic diagram of the encoder and decoder is shown in Figure 6.4.

Whenever $e(t) \geq +\gamma/2$, $q(t)$ becomes a logic one and causes the monostable to be triggered producing a pulse of amplitude E and period τ. This pulse is applied to the RC integrator in the feedback loop effecting an increase in the feedback signal $y(t)$. The value of $y(t)$ is given by Equation 2.53

$$y_i(t) = E\{1 - \exp(-\tau/RC)\} + Y_i \exp(-\tau/RC) \tag{6.27}$$

where Y_i is the value of $y(t)$ prior to the advent of the $L(t)$ pulse.

When the $L(t)$ pulse is completed $y(t)$ has the value given by the above equation. The $y(t)$ signal now decays, and from Section 2.5.1

$$y(t) = y_i(t) \exp(-t/RC) \tag{6.28}$$

Asynchronous delta modulation systems

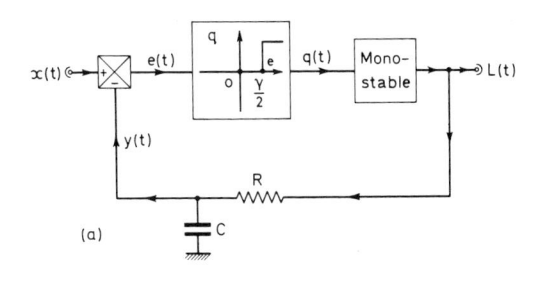

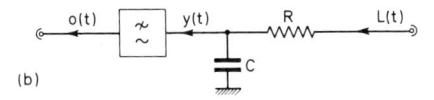

Fig. 6.4. Asynchronous pulse d.m. system: (a) encoder, (b) decoder

By making $\tau \ll RC$ Equations 6.30 and 6.31 can be approximated as

$$y_i(t) \doteqdot \gamma + Y_i(1-(\tau/RC)) \quad (6.29)$$

$$y(t) \doteqdot y_i(t)(1-(\tau/RC)), \; t = \tau \quad (6.30)$$

where

$$\gamma = E\tau/RC. \quad (6.31)$$

To a first approximation, the arrival of a pulse in the RC integrator causes its output to increase by γ. Subsequently the voltage at the output of the integrator, i.e. the voltage on the capacitor, decays with a time constant RC and an aiming potential of zero volts.

It is essential to use a leaky integrator in the feedback path. For example, if the integrator is ideal its output voltage $y(t)$ could only assume positive values because $L(t)$ has only positive pulses. The result would be that the

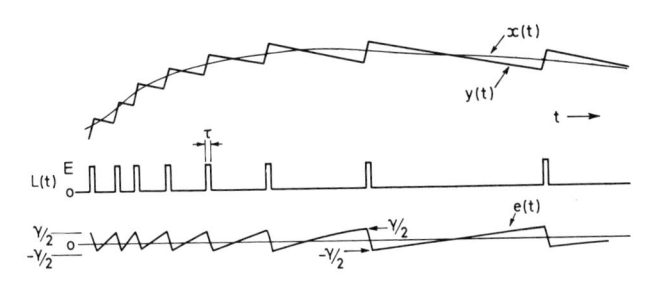

Fig. 6.5. Waveforms of an asynchronous pulse d.m. codec in response to an arbitrary input signal x(t)

encoder could not function, being incapable of tracking input signals having negative slope.

The $y(t)$ signal is subtracted from the input signal $x(t)$ to form the error signal $e(t)$. When an output pulse arrives $y(t)$ rapidly changes level, by approximately $+\gamma$, and $e(t)$ is negative. $y(t)$ decays until $e(t) \geq \gamma/2$ when another output pulse is generated. When $x(t)$ has a negative slope the error may be nearly constant for a long period because $y(t)$ also has a decaying negative slope. However, when $x(t)$ has a positive slope the error signal reaches or exceeds $\gamma/2$ in a shorter time as the slope is increased. The behaviour of the encoder for an arbitrary input is shown in Figure 6.5 where the proximity of adjacent output pulses is seen to be highly dependent on the slope of $x(t)$.

The decoder is simply an integrator, for recovering $y(t)$, followed by a low-pass filter for removing the transients in the waveform.

6.2.1. Zero crossings

The $y(t)$ waveform for $x(t)$ with positive and negative slope is shown in Figure 6.6. Between points C and E in Figure 6.6(a), $y(t)$ increases by

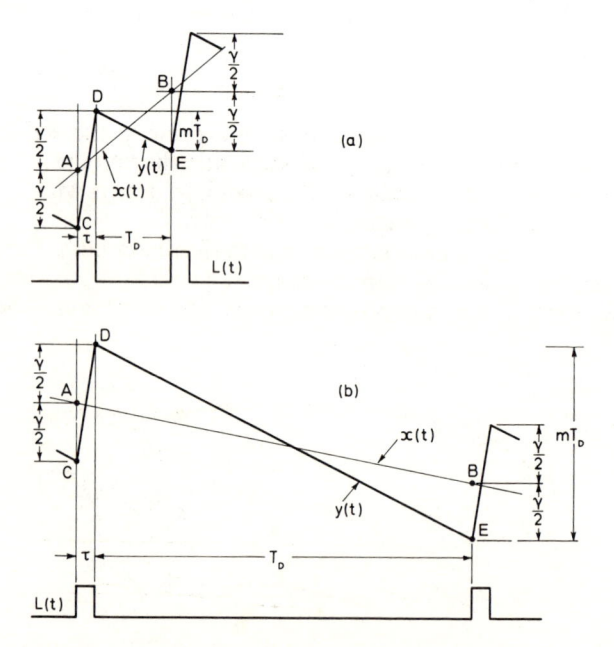

Fig. 6.6. $y(t)$ and $L(t)$ waveforms of an asynchronous pulse d.m. codec when the input signal $x(t)$ has (a) a positive slope and (b) a negative slope

$\gamma - mT_D$, where m is the decaying slope of $y(t)$; T_D is the time interval between pulses and is a function of $x'(t)$. From Equation 6.28

$$m = \frac{dy(t)}{dt} = -\frac{y(t)_i}{RC}\exp\left(-\frac{t}{RC}\right) \qquad (6.32)$$

m is clearly a function of $y(t)_i$, which is constrained to be very small compared with E. The diagram implies that $x(t)$ must pass through the points A and B and that $e(t) < \gamma/2$ between these points. The average slope of $y(t)$ is approximately equal to $x'(t)$, assuming $x(t)$ is a straight line connecting points A and B, i.e.

$$x'(t) = \frac{\gamma - mT_D}{T_D + \tau}$$

$$T_D = \frac{\gamma - x'(t)\cdot\tau}{m + x'(t)} \qquad (6.33)$$

For Figure 6.6(b) the above equation for T_D applies but with the sign of $x'(t)$ reversed. The minimum value of T_D is

$$T_{D\,min} = \frac{\gamma - x'(t)_{max}\cdot\tau}{m + x'(t)_{max}} \qquad (6.34)$$

and

$$\gamma \gg |x'(t)_{max}\cdot\tau|$$

The encoder is overloaded when the negative slope of $x(t)$ is greater than m. When $x(t)$ has a large positive slope which causes overloading $y(t)$ increases at its fastest rate of $\gamma/(\tau + t_r)$ approximately, where t_r is the minimum time between the end of the monostable pulse and the initiation of another pulse. Generally, the main limitation on overloading is when encoding signals with negative slopes.

When encoding a sinusoid $E_s \sin 2\pi f_s t$ over the range $-E_s$ to $+E_s$ the $y(t)$ waveform increases where the slope of the sinusoid is positive by the following increments for successive $L(t)$ pulses, $(\gamma - mT_{D1})$, $(\gamma - mT_{D2})$, $(\gamma - mT_{D3})$, etc. By the time the sinusoid arrives at $+E_s$

$$2E_s = n_p\gamma - m\sum_{i=1}^{n_p} T_{Di} \qquad (6.35)$$

where n_p is the number of $L(t)$ pulses.

Due to the symmetry of the sinusoid over its region of positive slope,

$$m\sum_{i=1}^{n_p} T_{Di} \doteq n_p \cdot mT_{D\,av\,+} \qquad (6.36)$$

where T_{Dav+} is the average value of T_D. From Equations 6.33, 6.4 and 6.2

$$T_{Dav+} = \frac{\gamma - 4f_s E_s \tau}{m + 4f_s E_s} \tag{6.37}$$

When the region where the sinusoid has a negative slope is considered, i.e. from E_s to $-E_s$, then by applying similar arguments

$$-2E_s = n_n \gamma - n_n m T_{Dav-}$$

where

$$T_{Dav-} = \frac{\gamma + 4f_s E_s \tau}{m - 4f_s E_s} \tag{6.38}$$

Then over a complete cycle, the number of $L(t)$ pulses

$$n_s = n_n + n_p$$

and the number of $L(t)$ pulses per second is $(n_s + n_p)f_s$. During each $L(t)$ pulse the $y(t)$ waveform exceeds that of $x(t)$ as shown in Figure 6.5. Consequently each $L(t)$ pulse causes two zero crossings in the error waveform. The number of zero crossings per second μ is

$$\mu = 2(n_n + n_p)f_s$$

Combining the above equations yields

$$\mu = 2m/(\gamma + m\tau) \tag{6.39}$$

Unlike the asynchronous delta modulator μ is independent of the input signal and dependent only on the parameters of the encoder.

For a band-limited Gaussian signal having zero mean the number of zero crossings is approximately the same as for the sinusoidal input over a long time, since over a long period of time T_{Dav} is independent of $\langle |x'(t)| \rangle$.

6.2.2. Signal-to-quantization noise ratio

The asynchronous pulse delta modulator has an error waveform $e(t)$ which tends to be triangular (see Figure 6.5). The number of zero crossings μ in $e(t)$ is much larger when the slope of the input signal $x(t)$ is positive than when $x'(t)$ is negative. When compared to the delta modulator version the pulse type has a smaller μ when $x'(t) < 0$ and a larger μ when $x'(t) > 0$. This can be readily appreciated from Figure 6.2, where $y(t)$ remains constant following an output impulse, whereas the asynchronous pulse delta modulator has a waveform $y(t)$ which declines exponentially. A consequence of this difference in the distribution of the zero crossings is that the pulse encoder has an error signal whose duration between adjacent zero-crossings cover a wider range than the corresponding ones in the asynchronous delta modulator.

An increase in the r.m.s. value of the input signal, having zero mean, to the asynchronous pulse delta modulator does not effect the number of zero crossings μ over a long time period, although there are instants when bunching and dispersion of the zero crossings occur. A similar increase in the input signal in the case of the asynchronous delta modulator results in an increase in μ as revealed by Equations 6.9 and 6.39.

It has been suggested[1,8] that Equation 6.22 can be applied to the asynchronous pulse delta modulator. An estimate for the s.q.n.r. when $x(t)$ is a Gaussian input signal is then given by using Equations 6.26 and 6.39 and assuming $\gamma \gg m\tau$

$$\text{s.q.n.r.} = 26 \frac{m}{\gamma f_{c2}} \left(\frac{\sigma}{\gamma}\right)^2 \tag{6.40}$$

The graph of s.q.n.r. against σ^2, the mean square value of $x(t)$, is a straight line. Experimental results[6,11] confirm this statement, and the shape of the graph over a wide range of input levels covering tens of decibels is similar to the one obtained for linear d.m., see Figure 3.18. The overload condition in asynchronous pulse d.m. occurs when $|x'(t)|$ is sufficiently large that the correct time locations of the $L(t)$ pulses are indeterminate at the receiver.

6.2.3. Minimum channel bandwidth

The minimum channel bandwidth W is the reciprocal of T_{Dmin}. From Equation 6.34

$$W = \frac{m + x'(t)_{max}}{\gamma - x'(t)_{max}\tau} \tag{6.41}$$

For a sinusoid, $x'(t)_{max} = 2\pi f_s E_s$ giving

$$E_s = \frac{W\gamma - m}{2\pi f_s(1 + W\tau)} \tag{6.42}$$

As the minimum input amplitude which will produce an output pulse is $\gamma/2$ the amplitude range AR defined in Section 6.2.5 is

$$\text{AR} = \frac{2(W\gamma - m)}{2\pi f_s \gamma(1 + W\tau)} \tag{6.43}$$

6.2.4. Idle channel state

The asynchronous delta modulator does not transmit when in its idle channel state, unlike the pulse encoder which produces an output pulse whenever the decaying waveform $y(t) = -\gamma/2$. The duration between successive pulses is found from Equation 6.37 when $\langle |x'(t)| \rangle = 0$, and the idle channel pulse rate $f_i = m/\gamma$. The s.q.n.r. is improved if f_i is increased. This result is expected from Equation 6.40, although in practical systems the rate of improvement decreases as f_i is increased.[5]

6.3. RECTANGULAR WAVE MODULATION

When the memory-less quantizer of the delta modulator shown in Figure 1.1 is replaced by another non-linear element whose characteristic is that of an hysteresis loop and the sampler is permanently closed the resulting encoder is called a rectangular wave modulator, abbreviated r.w.m. (Sharma and Das[8,9]). The modulator is shown in Figure 6.7 together with its decoder.

The binary waveform $L(t)$ at the output of the encoder has levels $\pm V$. Suppose $L(t) = +V$, then the waveform $y(t)$ has a slope of gradient $+V$. When the error signal $e(t)$ is reduced to $-\gamma/2$ the $L(t)$ voltage changes to $-V$. The $y(t)$ signal continues to decrease until $e(t) \geq \gamma/2$ when $L(t)$ reverts to $+V$. Thus the $L(t)$ waveform consists of alternate levels, where the time duration between these levels depends on the slope of the input signal $x(t)$. Figure 6.8 shows the $y(t)$ and $L(t)$ waveforms for an arbitrary input signal. The rate of zero crossings in $e(t)$ is greatest when the encoder

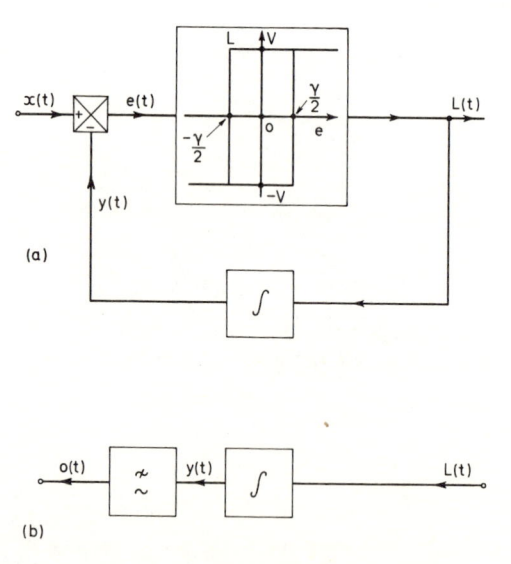

Fig. 6.7. Rectangular wave modulation system: (a) encoder, (b) decoder

Asynchronous delta modulation systems

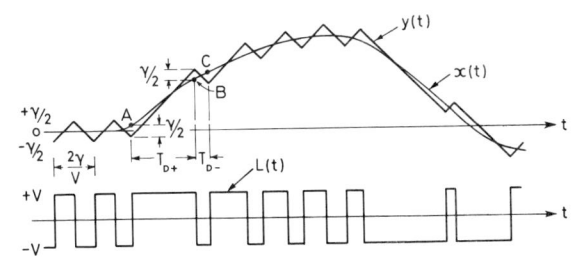

Fig. 6.8. R.W.M. waveforms when encoding an arbitrary input signal x(t)

is in the idle channel condition, or when $x'(t)$ is zero. In this condition $L(t)$ is a square wave and $y(t)$ is a triangular waveform whose peak values are $\pm \gamma/2$. As the slope m of $y(t)$ is V then the period of the waveform is $2\gamma/V$ and the frequency of the idle channel waveform is $V/2\gamma$. From Figure 6.9,

$$T_{D+} = \frac{\gamma}{m - x'(t)_{AB}} \qquad (6.44)$$

where $x'(t)_{AB}$ is the slope of $x(t)$ between A and B. Similarly for BC

$$T_{D-} = \frac{\gamma}{m + x'(t)_{BC}} \qquad (6.45)$$

The duration between successive positive (or negative) binary levels in the $L(t)$ waveform is

$$T_D = \frac{\gamma}{m - x'(t)} + \frac{\gamma}{m + x'(t)} = \frac{2\gamma m}{m^2 - \{x'(t)\}^2} \qquad (6.46)$$

where $\langle x'(t) \rangle$ is the long term average slope of $x(t)$. The period of the square waveform $L(t)$ in the idle channel condition is $2\gamma/m$, where $m = V$, and is found by putting $\langle x'(t) \rangle$ to zero in Equation 6.46.

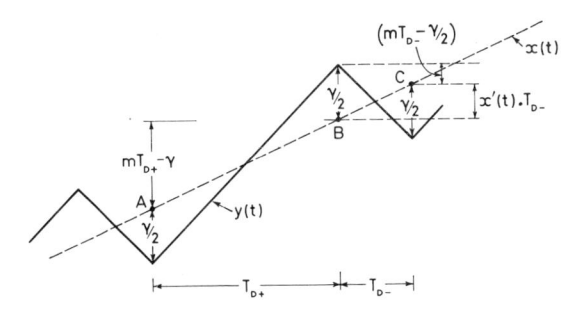

Fig. 6.9. Duration $T_{D+} + T_{D-}$ between successive positive binary levels in $L(t)$

The signal-to-noise ratio can be improved[9] by applying $r(t)$ to the r.w.m., not $x(t)$. The $r(t)$ signal is obtained by passing $L(t)$ through a double integrator adding the resulting signal to $x(t)$ and then applying suitable amplification. This in effect is identical to representing the integrator in the encoder in Figure 6.7 by a network having a transfer function

$$\frac{K}{\{(1+s\tau_1)(1+s\tau_2)\}}$$

where K is a gain constant and τ_1 and τ_2 are time constants. The modulator is of the bang-bang type and analytical techniques[10] are available to calculate its performance; no difficulty is experienced in producing a stable idling pattern, and the modulator behaves as described when tracking input signals.

The signal $y(t)$ can be estimated by replacing the hysteresis loop by its describing function and analysing the resulting linear system.

The similarity between r.w.m. and frequency modulation is apparent. When these systems are compared, for the same bandwidths,[11] their output versus input signal-to-noise ratio characteristics have similar shapes. For high input s.n.r. the output s.n.r. is up to 3 dB higher for f.m. than r.w.m. but the threshold of the characteristic for r.w.m. is extended by 3 dB compared to the f.m. system. Consequently r.w.m. can be used as an alternative to f.m. Observe that r.w.m. has a self-generated carrier.

A delta-sigma version of the rectangular wave modulator[9,12] is produced by removing the integrator from the feedback loop and placing it after the error point in the feed-forward path. The decoder is simply a low-pass filter.

6.3.1. Decoder

In the r.w.m. of Figure 6.7 slope overload occurs when the slope of the input signal becomes too large for $y(t)$ to track. This condition can be overcome by increasing the quantization level V. Unfortunately, increasing V raises the idling frequency which is related to the channel bandwidth W by a factor of 2.5, approximately. Consequently for fixed W the level V is fixed and thereby the probability of slope overload for a given input signal.

The received signal is filtered by F_W, to remove any noise frequencies in excess of W, and then passed through a zero crossing detector to produce a binary waveform $L_D(t)$ having levels of $\pm V$. Ideally this waveform should be identical to $L(t)$, but generally it will contain errors which may manifest themselves as differences between the positions of the zero crossings when compared to $L(t)$. Even in the absence of channel noise $L_D(t)$

will differ from $L(t)$ due to the characteristics of F_W and the zero crossing detector. For example, narrow $L(t)$ pulses may not reach their full magnitude $\pm V$ due to the impulse response of F_W. Provided there is no channel distortion the discrepancy between $L(t)$ and $L_D(t)$ can be reduced by either restricting the bandwidth of the message signal $x(t)$ or by altering the feedback loop of the encoder. If the latter is allowed to contain a network which is representative of the channel, F_W, the zero crossing detector and finally the integrator, then the feedback signal $y(t)$ will be a close approximation to the signal at the output of the integrator in the decoder.

6.4. DISCUSSION

The asynchronous delta modulator does not produce a digital output in the absence of an input signal, whereas the asynchronous pulse delta modulator and r.w.m. have idling frequencies of m/γ and $m/(2\gamma)$, respectively. The asynchronous pulse delta modulation and r.w.m. systems have the same channel bandwidths W and amplitude range AR for a given input signal $x(t)$. The asynchronous delta modulator requires a lower W, and generally has a greater AR for the same $x(t)$, than the asynchronous pulse delta modulation and r.w.m. codecs.

The signal-to-quantization noise ratio for asynchronous d.m. and asynchronous pulse d.m. are equal for Gaussian input signals having the same σ and γ values, if

$$\frac{f_i}{f_{c2}} = 2 \cdot 9 \left(\frac{\sigma}{\gamma}\right)$$

where f_i is the idle channel frequency of the asynchronous pulse d.m. codec. For $f_{c2} = 3 \cdot 4$ kHz, $\gamma = 10$ mV and an input signal having a power level of -20 dBm (i.e. the power of the input signal in dB relative to a power of 1 mV across a 600 Ω resistor) $f_i = 76$ kHz and $m = 760$ V/s.

The channel bandwidth for asynchronous d.m. is less than that required by linear d.m. This is particularly so in the case of asynchronous d.m. which does not need to generate pulses as long as the slope of the input signal is less than $|\gamma/2|$, a situation which occurs when encoding d.c. and speech with conversational pauses.

The maximum signal-to-noise ratio is greater by about 6 dB for asynchronous pulse d.m. than linear d.m.[13] for the same channel bandwidth. The asynchronous modulation systems having the best signal-to-noise ratio for a given channel bandwidth[8] are in order of merit, r.w.m., asynchronous d.m. and finally asynchronous pulse d.m. However, the differences are no more than a few dB. The shape of the characteristics of signal-to-noise ratio as a function of input signal power is similar for both synchronous and asynchronous systems, and are typified by the curves shown in Figure 3.18.

When analogue signals have to be conveyed over relatively short distances, as in a factory, it is frequently necessary to buffer these signals in order to reduce the impedance levels prior to transmission. This buffering reduces the noise picked by the signals when they arrive at their destination. Rather than use an analogue amplifier buffer, a delta modulator can be used to encode and transmit the signal in a binary waveform thereby improving the received signal-to-noise ratio. If asynchronous d.m. or r.w.m. is used a clock is not required.

The r.w.m. is cheaply built and can operate at a high idling rate without bandwidth difficulties in this type of environment.

The above example illustrates a particular use of an asynchronous d.m. system. However, in communications it is much more likely that a synchronous d.m. system would be used and that it would be companded. The next chapter deals with syllabically companded d.m. for speech transmission.

REFERENCES

1. Sharma, P. D., 'Characteristics of asynchronous delta-modulation and binary-slope-quantized p.c.m. systems', *Electronic Engineering*, 32–37, January (1968)
2. Inose, H., Aoki, T. and Watanabe, K., 'Asynchronous delta modulation system', *Electronics Letters*, **2**, No. 3, 95–96, March (1966)
3. Rice, S. O., 'Mathematical analysis of random noise', *Bell Systems Tech. J.*, **23**, No. 282 (1944) and **24**, No. 706 (1945)
4. Bennett, W. R., 'Spectra of quantized signals', *Bell Systems Tech. J.*, **27**, 446–472, (1948)
5. Sharma, P. D., 'Pulse interval modulation—a new method of signal approximation', *Electronic Engineering*, 582–587, October (1968)
6. Das, J. and Sharma, P. D., 'Pulse interval modulation', *Electronic Letters*, **3**, No. 6, 288–289, June (1967)
7. Steele, R., 'Pulse delta modulators-inferior performance but simpler circuitry', *Electronic Engineering*, **42**, 75–79, (1970)
8. Das, J. and Sharma, P. D., 'Some asynchronous pulse modulation systems', *Electronic Letters*, **3**, No. 6, 284–286, June (1967)
9. Sharma, P. D., 'Signal characteristics of rectangular-wave modulation', *Electronic Engineering*, **40**, 103–107, February (1968)
10. Graham, D. and McRuer, D., *Analysis of Nonlinear Control Systems*, Wiley (1961)
11. Das, J. and Sharma, P. D., 'Rectangular wave modulation—a hybrid p.l.m.-f.m. system', *Electronic Letters*, **2**, No. 1, 7–9, January (1966)
12. Sharma, P. D., 'Companding and slicing in r.w.m. systems', *Electronic Engineering*, 372–378, July (1968)
13. Das, J. and Sharma, P. D., 'Optimised hybrid unidigit p.c.m. system', *Electronic Letters*, **2**, No. 1, 9–10, January (1966)

Chapter 7

SYLLABICALLY COMPANDED DELTA MODULATION

7.1. INTRODUCTION

Linear d.m., linear d.s.m., exponential d.m., double integration d.m. and the asynchronous d.m. systems described in the last chapter all have the undesirable feature that there is one input level which maximises the signal-to-noise ratio, as typified by the curves in Figure 3.18 which apply to the linear d.s.m. codec.

A method of overcoming this deficiency is to compress the large amplitude levels in the signals relative to the smaller ones prior to encoding using a compressor circuit.[1,2] In this way the input level to the encoder can be maintained close to the value which gives the maximum signal-to-noise ratio. The receiver decodes the d.m. binary stream and passes the analogue signal through an expander which compensates for the distortion in the input signal caused by the compressor. The graph of decoded s.q.n.r. as a function of input level is no longer sharply peaked but is relatively flat over a large range of input levels. The effect of using a compressor and its complimentary expander, known as companding on a linear d.s.m. system can be seen in Figure 7.5 for the same transmitted bit rate.

Most of the d.m. systems described in this chapter do not achieve companding by the method just described. The companding is accomplished[6,7,30] solely by the delta modulator which unlike the encoders described in the previous six chapters has a *non*-linear active network in its feedback loop.

In the companded encoder the height of the pulses fedback to the error point are adapted by the local decoder in the feedback loop to enable a wider range of input amplitudes to be tracked for the same bit rate compared to the linear d.m. encoder. By making the d.m. encoder adaptive a simpler circuit arrangement is achieved and the need to produce complementary compressor and expander analogue circuits is avoided.

The companding characteristic is a graph whose abscissa is the amplitude of the sinusoid which is applied to the input of the encoder, and whose ordinates are the corresponding amplitudes of the sinusoid at the output of a linear decoder, i.e. an integrator followed by a low-pass filter in d.m. and a low-pass filter in d.s.m. This should not be confused with the p.c.m. characteristics which are well defined and independent of the input signal, see Section 9.4.1. In some types of d.m. and d.s.m. systems the companding characteristic depends on the frequency of the applied sinusoid, which necessitates the specification of this frequency. The value of this characteristic is that it gives an indication of the amount of compression being achieved by the encoder. The slope of the characteristic is much less than unity over the region where companding occurs and this means that the amplitude of the signal obtained by decoding the binary waveform without the presence of an 'expander' is nearly constant with changes in the amplitude of the applied sinusoid.

The type of expander which must be employed to recover the original signal at the decoder is clearly to be found in the feedback loop of the encoder. This is because the local decoder contained in the feedback loop operating on the output binary waveform produces an analogue signal which results in a small error signal at the input to the quantizer. Thus the companded encoder is adaptive in that it adapts the binary signal at the output of the encoder to produce a small error signal, and in so doing 'expands' the amplitudes of the baseband signals which reside in a latent form in the output binary waveform. The decoder therefore consists of an adaptive local decoder followed by a low-pass filter for removing out-of-band quantization noise.

For example, if speech is being encoded then filtering the binary output signal by a filter which only passes speechband signals will produce the original speech modified by the companding law. The expander having the inverse companding law will operate on the output binary signal to expand the baseband signal in this waveform to produce a signal whose baseband components differ only from the input speech from the microphone by the effects of quantization. This is due to the forward path having a quantizer and sample and hold circuit.

The adaptive processes of syllabically companded d.m. will be discussed in subsequent sections of this chapter. It should be mentioned, however, that the adaptive algorithms used by d.m. encoders adjust the magnitude of the pulses feedback to the error point at a much slower rate than the instantaneous variations in speech signal. This rate is, generally speaking, the pitch rate of the speech signal, i.e. of the order of 10 ms, and not the syllabic rate which is generally in excess of 100 ms. However, these companded d.m. systems have, in the opinion of the Author, been mis-named 'syllabically' companded rather than 'pitch' companded d.m., and because they are well known by this erroneous title it will be continued to be used here. The meaning of pitch rate is made apparent in Section 7.6.

7.2. SYLLABICALLY COMPANDED DELTA-SIGMA MODULATION SYSTEMS

In some types of noisy environments, for example, certain military ones, it is desirable to use microphones which pre-emphasise the high frequencies of the speech signal such that the spectrum of the signal at the output of the microphone is substantially flat.[4] This requires an encoder with a flat overload characteristic; this section concerns itself with an investigation into such an encoder which together with its decoder is known as a syllabically companded delta-sigma modulation system, abbreviated s.c.d.s.m.[3] The d.m. as distinct from the d.s.m. version of this system has been investigated,[5] but will not be considered here because of the similarities of the companding arrangements. The essential differences between the systems are the inherent differences between d.m. and d.s.m. In particular, the s.q̂.n.r. is dependent on the frequency of the input signal, unlike the s.c.d.s.m.

7.2.1. Preamble

The block diagram of an analogue s.c.d.s.m. system[3] is shown in Figure 7.1. The difference between this block diagram and that for the simple delta-sigma modulation system described in Section 2.4 is the presence of a local decoder. The function of this local decoder is to feedback to the error point, not a binary signal $L(t)$ having values $\pm V$, as in the case of a delta-sigma modulator, but a multi-level signal $y(t)$ having magnitudes $\pm H$ which are controlled by the syllabic amplitude of the signal extracted from the binary output $L(t)$. This signal can consequently vary

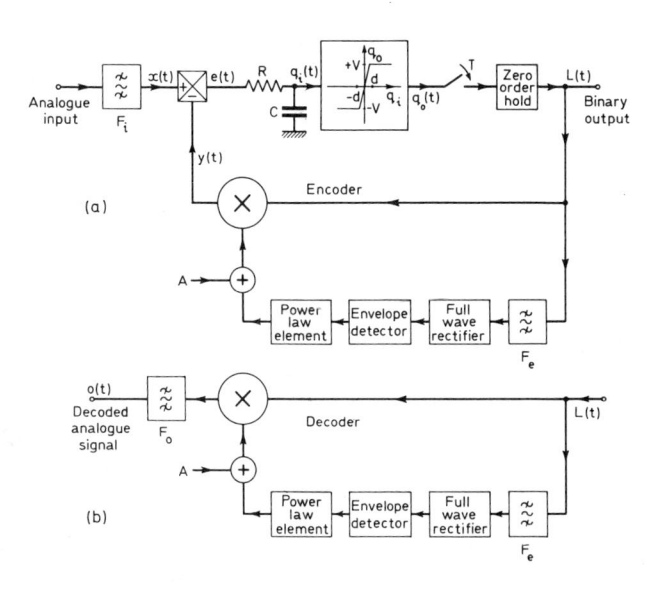

Fig. 7.1. Analogue s.c.d.s.m. system: (a) encoder, (b) decoder

continuously over a wide range of values always adopting a polarity identical to that of $L(t)$. As the syllabic rate of change is much slower than the clock rate f_p it follows that for tens of clock periods s.c.d.s.m. encoder has an approximately constant magnitude H and hence can be considered to be a simple delta-sigma modulator having a voltage $y(t) = \pm H$, rather than $\pm V$, fedback to the error point.

If a sinusoidal input is applied to the encoder it produces a constant value H because a sinusoid has a constant envelope. A speech signal generates a constant value H for tens or hundreds of clock periods, the number depends on the actual speech signal, the syllabic time constant and the clock rate used in the encoder. The reasonable constancy of H over many clock periods for two very different types of signals enables the performance of the system to be calculated by using a simple sinusoidal input signal.

These calculations show very good agreement with experimental results using narrow-band white noise input signals, and this is significant because objective evaluation of speech frequently requires this type of input signal. Additional justification of why narrow-band white noise inputs give good agreement with sinusoidal inputs can be appreciated from the following argument. Band-limited white noise can be represented by a sinusoid having a frequency equal to its mid-band value, a phase angle which can have any value between 0 and 2π with equal probability, and an amplitude which has a Rayleigh probability density function. Further the envelope variations of this signal are slow compared to its frequency. Thus although the test signal is narrow-band white noise it can be represented by a sine wave whose amplitude is varying very slowly compared to the clock rate. Hence an analysis of the encoder can be satisfactorily accomplished by applying a simple sine wave, whose amplitude is approximately equal to $\sqrt{2} \cdot \sigma$, where σ is the r.m.s. value of the Gaussian test signal. The following theory although useful for estimating the performance of the codec when narrow band white noise input signals are used, is invalid for wide band input signals.

7.2.2. Operating principle of the codec

The speech signal from a pre-emphasised microphone is band-limited by a filter F_i to produce the signal $x(t)$ which is subsequently encoded in a binary signal $L(t) = \pm V$. This $L(t)$ signal, suitably filtered, amplified and perhaps used to modulate an analogue carrier, is transmitted to its destined receiver/s. However, $L(t)$ is also applied at two points to the system in the feedback loop, referred to as the local decoder. One connection is to a low pass filter and the other to a multiplier. When $L(t)$ is passed through a filter F_e (Figure 7.1) having identical characteristics to filter F_i, it produces an analogue signal which is then full-wave rectified before being passed to an envelope detector which essentially behaves as a low pass filter having a time constant approximating to the pitch rate of the speech signal, i.e. of the order of 10 ms. The output of the envelope detector is next raised to the power n, added to a voltage A, and the sum is multiplied with the signal $L(t)$ to produce the signal $y(t)$. As $L(t)$ is being multiplied by a signal which is always positive, it follows that the $y(t)$ waveform consists of pulses having the same polarity as those of the $L(t)$ waveform. This multiplier behaves as a pulse height modulator.

The $y(t)$ waveform is subtracted from the band-limited speech waveform $x(t)$ to produce an error signal which is integrated before being amplitude quantized to realise the signal $q_0(t)$. Finally the polarity of $L(t)$ is the polarity of $q_0(t)$ at the sampling instants. The system is clearly behaving as a delta-sigma modulation system with $y(t)$ pulses rather than $L(t)$ pulses feedback. The behaviour of the forward-loop is fully described in Sections 1.4 and 2.4.

As the idling pattern of $L(t)$ is a square waveform, i.e. . . . 1 0 1 0 1 0. . . then as $x(t)$ is zero, by the definition of idling, there is no output from the envelope extractor. Consequently $y(t)$ has an identical pattern to that of $L(t)$ with amplitudes $\pm I$, where $I = AV$. The sole function of A is to establish the correct idling conditions.

The s.c.d.s.m. decoder consists of the system used in the feedback loop of the encoder, i.e. the local decoder plus a filter F_0 which is assumed identical to F_i and F_e. Ignoring errors in the channel, and assuming identical local decoders at the encoder and decoder, it follows that the $y(t)$ waveform is the input to F_0. Filtering $y(t)$ by F_0 to remove the high-frequency effects produces $0(t)$ which is a close replica of the original speech waveform $x(t)$.

7.2.3. Overload characteristic

It has been shown in Section 2.4 that if the maximum sinewave amplitude E_{sm} is equal to the magnitude of the binary level of the signal at the output of the encoder, i.e. V volts, then overload of the d.s.m. encoder will not occur. Further this value of $E_{sm} = V$ is independent of frequency.

In the s.c.d.s.m. codec the overload situation is more complex as the feedback voltage is not a constant voltage V, but a voltage H which has a minimum value I and a maximum value H_M. It is shown in Equation 7.3 that the maximum decoded signal-to-quantization noise ratio, abbreviated s.q.n.r., occurs when E_s reaches a value E_{sm} equal to H, say H_m. A value of $E_s > E_{sm}$ will reduce the signal-to-noise ratio indicating an overload condition. The s.c.d.s.m. system therefore has an overload characteristic defined by the following equation,

$$E_{sm} = H_m \qquad (7.1)$$

Observe that H_m has replaced V in Equation 2.49.

7.2.4. Signal to quantization noise ratio

The quantization noise N_q^2 in a linear d.s.m. system is given by Equation 2.48. This equation assumes that the frequency components of the input signal are large compared to the break frequency

$$f_1 = 1/(2\pi RC)$$

of the RC integrator circuit and that the decoding filter F_0 is of the low-pass type having a pass band which extends to f_{c2}. When F_0 has critical frequencies f_{c1} and f_{c2}, where the pass-band is

$$f_c = f_{c2} - f_{c1}$$

and f_1 is of the order f_{c1}, Equation 2.48 becomes

$$N_q^2 = \frac{K_q 4\pi^2 V^2}{3f_p^3}(3f_1^2 f_c + f_{c2}^3 - f_{c1}^3)$$

In the s.c.d.s.m. codec V is replaced by H enabling N_q^2 to be expressed as

$$N_q^2 = \frac{H^2}{2R} \qquad (7.2)$$

where

$$R = \frac{3f_p^3}{8\pi^2 K_q(3f_1^2 f_c + f_{c2}^3 - f_{c1}^3)}$$

R is a factor which depends on the cube of the clock rate, the RC time constant and the critical frequencies of the filter F_0.

For s.c.d.s.m. having a sinusoidal input

$$x(t) = E_s \sin 2\pi f_s t$$

which does not overload the encoder, the s.q.n.r. is

$$\text{s.q.n.r.} = R\left(\frac{E_s}{H}\right)^2, \qquad E_s \leq H \qquad (7.3)$$

7.2.5. Basic equations

Suppose a sinewave is applied to the encoder shown in Figure 7.1.

$$x(t) = E_s \sin 2\pi f_s t \tag{7.4}$$

and this sinusoid is encoded into the binary waveform $L(t)$. If $L(t)$ is fed through a low-pass filter, which is the procedure for decoding in an uncompanded d.s.m. system, a sinewave is produced having the same frequency but an amplitude $E_{sr} \neq E_s$. This situation is expected because the s.c.d.s.m. encoder has compressed the input sinusoid. Thus the output signal from filter F_e, in Figure 7.1, is $E_{sr} \sin 2\pi f_s t$ which after full-wave rectification and ideal envelope detection, is converted into a constant signal of magnitude E_{sr}.

This voltage is amplified by a power-law element to give GE_{sr}^n, where G is a constant and n is the order of the power-law element. A d.c. voltage A is necessary to establish the idling pattern and is added to the amplified voltage. The sum is then multiplied by the waveform $L(t)$ of magnitude $\pm V$ to produce $y(t)$ (which has the same polarity as $L(t)$ since $A + GE_{sr}^n$ is always positive) with magnitude H volts, i.e.

$$H = AV + GVE_{sr}^n \tag{7.5}$$

We now establish a relationship between E_s and H.

The encoder operates by adjusting the average value of $y(t)$ to equal the average value of the amplitude of the input signal $x(t)$ over a short period of time. As the value of H is a constant for a given E_s, i.e. for a given envelope E_{sr}, the $y(t)$ and $L(t)$ waveforms have identical binary patterns but differing magnitudes $\pm H$ and $\pm V$, respectively. When the input sinewave is in the vicinity of its zero crossings the number of logical ones and zeros corresponding to $\pm H$, $\pm V$ in $y(t)$ and $L(t)$, respectively are nearly equal and these waveforms approximate to square waves. When the input sinewave is near its positive maximum the number of ones in $y(t)$ and $L(t)$ greatly exceed the number of zeros and vice versa when the input sinewave is near its negative maximum. If the $y(t)$ or $L(t)$ waveforms are viewed on an oscilloscope over a number of cycles of the input sinusoid, it will be seen that the ones and zeros are sinusoidally distributed at the sinusoid frequency.

When $L(t)$ is applied to filter F_e the sinusoid $E_{sr} \sin 2\pi f_s t$ is extracted and when $y(t)$ is applied to filter F_0 in the decoder, then to a good approximation the original signal $E_s \sin 2\pi f_s t$ is recovered.

As both filters F_e and F_0 are identical and as the sinusoidal outputs of these filters are proportional to the amplitude of their input binary waveforms

$$\frac{H}{V} = \frac{E_s}{E_{sr}} \tag{7.6}$$

Eliminating E_{sr} from Equations 7.5 and 7.6

$$H^{n+1} - AVH^n - GV^{n+1}E_s^n = 0$$

or

$$H^{n+1} - IH^n - BE_s^n = 0 \qquad (7.7)$$

where $B = GV^{n+1}$. I is the value of H when $x(t)$ is zero and the encoder is idling.

$$I = AV \qquad (7.8)$$

For a simple d.s.m. encoder the companding is zero, i.e. $n = 0$. From Equation 7.7

$$H_0 = I + B \qquad (7.9)$$

Equation 7.9 shows that in a d.s.m. encoder the feedback voltage H_0 is a constant, independent of E_s.

For $n = 1$, i.e. no power law element in the system, from Equation 7.7

$$H_1 = \frac{I}{2} + \left[\left(\frac{I}{2}\right)^2 + BE_s\right]^{1/2} \qquad (7.10)$$

For higher values of n the solution of Equation 7.7 to produce a practical value of H becomes more difficult, but before tackling this problem the values of I and B will be considered.

7.2.6. Codec parameters

Selection of I

The height of the feedback pulse is a variable quantity. It will be demonstrated in Section 7.2.8 that the wide range of H values are responsible for the vastly improved dynamic range of this modulator compared to the linear one. To maximize the range of H it is necessary to arrange for the encoder to idle with H equal to the smallest value of I. In order to find this value the minimum value of $q_i(t)$ is now sought which can operate the quantizer shown in Figure 7.1.

When $x(t)$ is zero, the encoder idles with $L(t)$ and $y(t)$ having a ...1 0 1 0 1 0... pattern. The error waveform $e(t)$ is simply the inverted $y(t)$ and is integrated by the RC circuit to produce the triangular waveform $q_i(t)$ shown in Figure 7.2. The RC circuit because its time constant T_1 is very much greater than T transforms the square wave form $e(t)$ at its input to a triangular waveform. From Figure 7.2,

$$\frac{I}{T_1} = \frac{2D}{T}$$

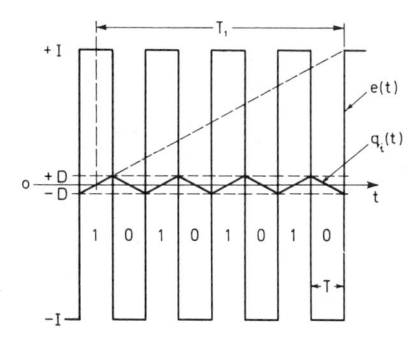

Fig. 7.2. *Idling error waveform and corresponding $q_c(t)$.*

or

$$D = \frac{\pi I f_1}{f_p} \qquad (7.11)$$

where

$$f_1 = \frac{1}{2\pi T_1} \qquad (7.12)$$

If the minimum value of D is $\pm d$ to operate the quantizer and produce $\pm V$ then the minimum value of I is

$$I_{min} = \frac{df_p}{\pi f_1} \qquad (7.13)$$

Selection of B

The constant B is a function of the gain of the local decoder, and its value decides the performance of the encoder in the vicinity of overload. As an example consider the first order system where $n = 1$. Figure 7.3 shows the variation of the magnitude H_1 of the feedback signal $y(t)$ as a function of the amplitude E_s of the input sinusoid for three different values of B. The curves have been plotted using Equation 7.10. The maximum signal-to-quantization noise ratio, s.q.n.r. = R, occurs when $E_s = H_1$, as indicated by Equation 7.3. When $E_s > H_1$ tracking deteriorates and the signal-to-noise ratio decreases. When $B = 2$, $E_s = H_1$, when H_1 is below the maximum value of H, namely, $H_M = 10$ V say. This causes a peak in the s.n.r. well below $E_s = 10$ V. Increasing E_s further causes the signal-to-noise ratio to decrease, but this decrease is not as great as in d.s.m. encoders as H_1 is still capable of becoming larger in response to the increase in E_s.

Syllabically companded delta modulation

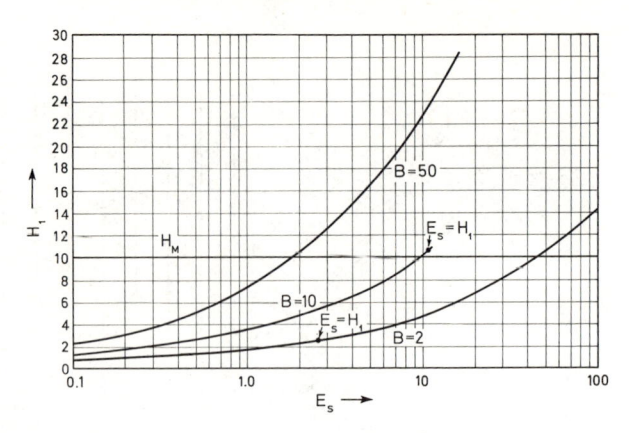

Fig. 7.3. Variation of H_1 as a function of E_s for different values of B, $n = 1$

For the dynamic range of the multiplier to be fully used H_1 should be capable of reaching H_M. However, for this low value of $B = 2$ it can be seen from Figure 7.3 that for E_s to equal H_M, E_s must be increased to a prohibitively large value which the input circuits will be unable to accommodate. Figure 7.4 shows the sketch of signal-to-noise ratio, normalised by the factor R, for $B = 2$.

If B is large, say 50, then by the time E_s reaches 1·8 V, $H_1 = H_M$. When this occurs the s.q.n.r. $\ll R$. For $E_s > 1\cdot 8$ V the encoder behaves as a linear d.s.m. having a fixed step height of H_M and the s.n.r. reduces relatively quickly.

From Figure 7.3, when $B = H_M = 10$, $E_s = H_1$ when H_1 is just greater than H_M. If the minimum magnitude I (from Equation 7.10) of the signal at the output of the multiplier is made to approach zero, $E_s = H_1$ when $H_1 = H_M$. Provided I is very small compared to H_M the optimum choice of B is $B = H_M$. This is because over the range of signal amplitudes E_s, from zero to H_M, the value of E_s is always less than the magnitude of the H_1 pulses in the feedback signal $y(t)$. As the encoder is overloaded if $E_s > H_1$, the encoder is able to track the input signal over the full amplitude range of the multiplier. Consequently the dynamic range of the input signal amplitudes are maximised relative to some arbitrary signal-to-noise ratio. The dynamic range is referred to here as the range of amplitudes of E_s in which the encoder is not overloaded. Note that if some overload distortion is acceptable then values of $B < H_M$ can give a larger dynamic range than the optimum value of H_M.

Figure 7.4 shows the s.q.n.r. normalised by R for the three values of B discussed, together with the curve relating to a linear d.s.m. codec having binary levels of $\pm H_M$. Observe that the dynamic range for $B = 10$

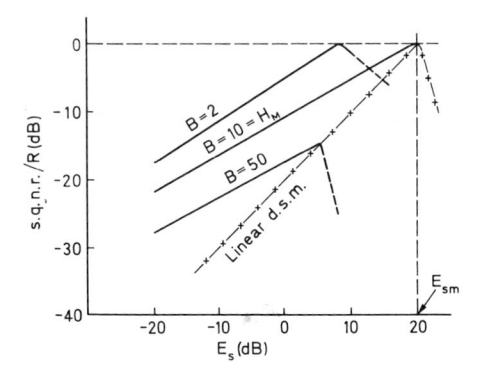

Fig. 7.4. *Normalised s.q.n.r. as a function of E_s for different values of B where $n = 1$, and for linear d.s.m.*

measured from the curve to the value $E_s = E_{sm}$ for any s.q.n.r. is approximately double that of the linear d.s.m. The extension of the dynamic range would be exactly double if $I = 0$ rather than $I = 0.5$ V. It is a practical proposition to make I smaller, i.e. below 0·1 volt.

The best choice of values for B and I, based on the above comments, is for I to be as small as possible and $B = H_M$. When H approaches H_M, Equation 7.7 can be expressed as

$$H = B^{1/(n+1)} \cdot E_s^{n/(n+1)} \qquad (7.14)$$

When $E_s = H$, at $H = H_M$

$$B = H_M \qquad (7.15)$$

irrespective of the value of n.

Equation 7.15 gives the optimum value of the parameter B, i.e. equal to the maximum output of the multiplier. For $B = H_M$, the overload characteristic defined by Equation 7.1 replaces H_m by H_M. For $B \neq H_M$ the encoder overload occurs at H_m which does not coincide with the overload of the multiplier, i.e. a wrong choice of B means that the system may overload with $H_m < H_M$; if $H_m > H_M$ the maximum s.q.n.r. cannot occur.

7.2.7. Dynamic range

The dynamic range of the encoder is defined here as

$$DR = 20 \log_{10} \left(\frac{E_{sm}}{E_{st}} \right) \mathrm{dB} \qquad (7.16)$$

where E_{sm} is the value of E_s which results in the maximum signal-to-quantization noise ratio, and E_{st} is the value of E_s which gives a subjectively acceptable decoded signal-to-quantization noise ratio χ. The suffix t will be used to denote subjective threshold of encoding. From Equation 7.3

$$\chi = R\left(\frac{E_{st}}{H_t}\right)^2$$

giving

$$H_t = \left(\frac{R}{\chi}\right)^{1/2} \cdot E_{st} \qquad (7.17)$$

Substituting for H_t in Equation 7.7

$$\left(\frac{R}{\chi}\right)^{(n+1)/2} \cdot E_{st}^{n+1} - I\left(\frac{R}{\chi}\right)^{n/2} E_{st}^n - BE_{st}^n = 0$$

Hence

$$E_{st} = \frac{B + I(R/\chi)^{n/2}}{(R/\chi)^{(n+1)/2}} \qquad (7.18)$$

From Equation 7.1

$$E_{sm} = H_m \qquad (7.19)$$

where Equation 7.19 gives the value of E_s for maximum signal-to-noise ratio.

Substituting the value of E_{st} and E_{sm} from Equations 7.18 and 7.19 in Equation 7.16 the DR for the encoder is

$$DR = 20 \log_{10} \frac{H_m(R/\chi)^{(n+1)/2}}{B + I(R/\chi)^{n/2}} \qquad (7.20)$$

For the optimum s.c.d.s.m. encoder, $H_m = H_M = B$ and $I = I_{min}$

$$DR = 20 \log_{10} \frac{H_M(R/\chi)^{(n+1)/2}}{H_M + I(R/\chi)^{n/2}} \qquad (7.21)$$

In the linear d.s.m. encoder the feedback binary signal has a magnitude of V volts at all times. Equations 7.1 and 7.17 can be used for this encoder if H and H_t are replaced by V. Hence $E_{sm0} = V$

$$E_{st0} = V\left(\frac{\chi}{R}\right)^{1/2}$$

and the dynamic range for the uncompanded encoder is from Equation 7.16

$$DR_0 = 20 \log_{10} \left(\frac{R}{\chi}\right)^{1/2} \text{dB} \qquad (7.22)$$

If

$$\left(\frac{H_M}{I}\right) \ll \left(\frac{R}{\chi}\right)^{(n/2)} \qquad (7.23)$$

then Equation 7.21 reduces to

$$DR = 20 \log_{10} \left\{\frac{H_M}{I}\left(\frac{R}{\chi}\right)^{1/2}\right\} \text{dB} \qquad (7.24)$$

As the dynamic range of the multiplier is

$$DR_\otimes = 20 \log_{10} (H_M/I) \text{ dB} \qquad (7.25)$$

the dynamic range in dB of the s.c.d.s.m. can be expressed as

$$DR = DR_\otimes + DR_0 \qquad (7.26)$$

In order for the Inequality 7.23 to apply

$$n \gg DR_\otimes/DR_0$$

Suppose the multiplier has a minimum pulse height $I = 0.1$ V and its maximum output $H = 10$ V, i.e., $DR_\otimes = 40$ dB. If the peak s.q.n.r. $R = 500$ and the dynamic range is measured for a s.q.n.r. $\chi = 250$ then from Equation 7.22 $DR_0 = 3$ dB; from the simplified Equation 7.26, $DR = 43$ dB. If $n = 20$, from Equation 7.21, $DR = 42$ dB indicating that for this value of n Equation 7.26 is valid. However, Equation 7.26 is increasingly in error for low values of n, for example, if $n = 10$, $DR = 31$ dB. Equation 7.26 represents the maximum attainable dynamic range.

7.2.8. High order companding

As the input amplitude E_s is increased from zero the value of H remains approximately constant at its idling value I, and the system behaves like an uncompanded d.s.m. system.

Consider the situation when E_s increases to a value greater than I volts. Equation 7.7 can be rearranged to give

$$H = \left(\frac{H}{H-I}\right)^{1/(n+1)} \cdot B^{1/(n+1)} \cdot E_s^{n/(n+1)} \qquad (7.27)$$

The series expansion of the first term in Equation 7.27 gives

$$\left(\frac{H}{H-I}\right)^{1/(n+1)} = 1 - \frac{1}{(n+1)}\left(\frac{I}{H}\right) + \cdots \quad (7.28)$$

If $n > 10$ say, then when H is only 10% greater than the idling feedback voltage I, the right-hand side of Equation 7.28 equals 1 ± 0.1. Thus Equation 7.27 becomes

$$H \doteqdot H_M^{1/(n+1)} \cdot E_s^{n/(n+1)} \quad (7.29)$$

where the optimum value of $B = H_M$ is used. Further as H_M is limited by circuit restraints then for $H_M \not> 10$ V, say, $H_M^{1/(n+1)}$ is close to unity and

$$E_s^{n/(n+1)} \doteqdot E_s$$

As E_s increases from zero, H is approximately equal to I. For values of E_s up to I the amount of compression is small. When E_s is approximately 10% greater than I Equation 7.29 applies, i.e. the magnitude of H becomes close to but less than E_s, which means that the s.q.n.r. is almost at its maximum value of R as can be seen by referring to Equation 7.3. As E_s is further increased H remains greater than E_s but by a diminishing amount and consequently the s.q.n.r. although close to R becomes even closer. A high s.q.n.r. is maintained over a wide dynamic range of E_s by the encoder keeping H just greater than E_s over the amplitude range of the multiplier. Eventually E_s becomes as large as H, when $H = H_M$, and the s.q.n.r. is exactly equal to R. The encoder is now on the point of overload; increasing E_s above this value results in the encoder behaving as a linear delta-sigma modulator in overload where V has been replaced by H_M.

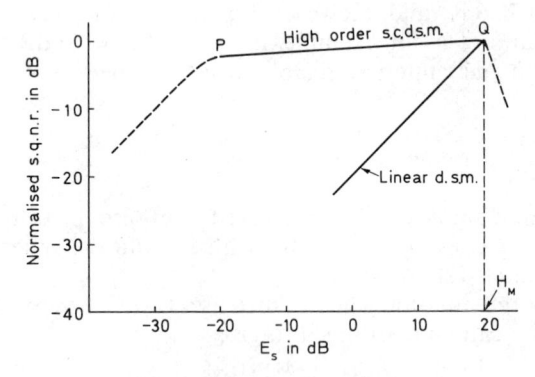

Fig. 7.5. *Normalised s.q.n.r. as a function of E_s for high order s.c.d.s.m. and linear delta-sigma modulation systems*

Syllabically companded delta modulation 197

These remarks are translated into graphical form in Figure 7.5 which shows the characteristic of the normalised s.q.n.r. versus input signal amplitudes for a high order s.c.d.s.m. system. The curve from P to Q was calculated using Equation 7.29 for $H_M = 10$ V and $n = 20$. The characteristic for linear d.s.m. is also shown where its binary level has for convenience been equated to H_M. The effect of companding the linear d.s.m. codec is to break the single peak characteristic of the codec (Figure 7.5) at the overload point and to insert the nearly horizontal portion PQ. The range of E_s corresponding to PQ is the dynamic range of the multiplier. The effects of high order companding are discussed in Section 9.5.2.

7.2.9. Effect of transmission errors

The noise due to transmission errors N_e^2 in a s.c.d.s.m. system is obtained from Equation 5.20 by replacing V by H

$$N_e^2 = \frac{8H^2 P_e f_c}{f_p} \tag{7.30}$$

where P_e is the probability of wrongly deciding at the decoder input that a logic zero is a logic one or vice versa, and f_c is the difference between f_{c2} and f_{c1}.

Assuming that the noise due to quantization and that due to transmission errors are statistically independent, then these noise powers can be added to give the combined noise N_T^2 at the decoder output.

$$N_T^2 = N_q^2 + N_e^2$$

Substituting from Equations 7.2 and 7.30

$$N_T^2 = \frac{H^2}{2R} + \frac{8 P_e f_c H^2}{f_p} \tag{7.31}$$

and for a sinusoid of peak value E_s the signal power

$$S^2 = E_s^2/2$$

hence

$$\frac{S^2}{N_T^2} = \frac{R}{J}\left(\frac{E_s}{H}\right)^2 \tag{7.32}$$

where

$$J = 1 + \frac{16 P_e f_c R}{f_p} \tag{7.33}$$

Comparing Equations 7.3 and 7.32 it can be seen that J is the transmission error factor.

From Equation (7.32) when $E_s = H$

$$\text{s.q.n.r.} = R/J$$

This value is independent of the order of companding. The effect of transmission errors is to reduce the maximum s.q.n.r. by J dB.

Proceeding as in Section 7.2.8 the dynamic range in the presence of transmission errors is

$$DR_e = 20 \log_{10} \left\{ \frac{H_M(R/J\chi)^{(n+1)/2}}{H_M + I(R/J\chi)^{n/2}} \right\} \quad (7.34)$$

The DR for the uncompanded system becomes

$$DR_{0e} = 20 \log_{10} \left\{ \frac{R}{J\chi} \right\}^{1/2}$$
$$= DR_0 - 10 \log_{10} J \quad (7.35)$$

where DR_0 is the DR in the absence of channel errors, specified by Equation 7.22. For the high order companded system Equation 7.34 reduces to

$$DR_e = DR_\otimes + DR_0 - 10 \log_{10} J \quad (7.36)$$

Thus for a given error-rate, i.e. given J, the DR of the companded and uncompanded systems are reduced by the same factor $10 \log J$. Consequently the effect of channel errors on the DR of the high order companded system is less serious than on the uncompanded one. However, both codecs have their maximum signal-to-noise ratio degraded by the same amount in the presence of channel errors.

7.2.10. s.c.d.s.m. system having n syllabic expanders

A s.c.d.s.m. encoder of order n can be realised by using n syllabic expanders in the feedback loop of a delta-sigma modulator. Each syllabic expander consists of an envelope extractor and a multiplier, i.e. the local decoder shown in Figure 7.1, where the output of the envelope detector is connected directly to the input of the multiplier.

Figure 7.6 shows the arrangement of the n syllabic expanders. It will now be shown that the encoders in Figures 7.1 and 7.6 are the same.

If a filter F_0 is connected to the output of the final multiplier, Figure 7.6, $x(t)$ will be produced together with some quantization noise which can be ignored for the same reasons as stated in the establishment of Equation 7.6. If $x(t)$ is a sinusoid $E_s \sin 2\pi f_s t$ the feedback signal at the error point in the encoder is a binary waveform $y(t)$ having levels $\pm H_n$. As described

Syllabically companded delta modulation

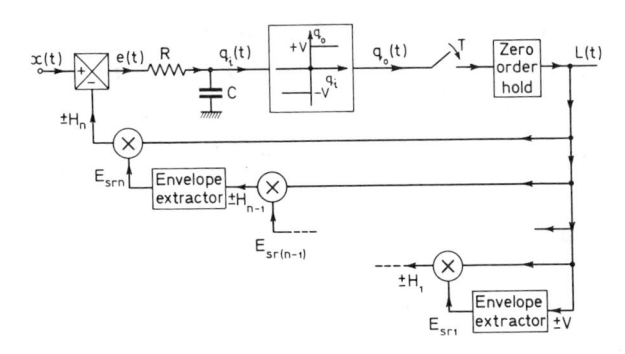

Fig. 7.6. High order s.c.d.s.m. encoder having n envelope extractions and multipliers

in Section 7.2.2 the $x(t)$ signal is recovered from $y(t)$ by passing it through the filter F_0. Similarly filtering binary waveforms having values

$$\pm H_{n-1}, \pm H_{n-2}, \pm H_{n-3}, \ldots, \pm V$$

will produce sinusoids having amplitudes

$$E_{srn}, E_{sr(n-1)}, E_{sr(n-2)}, \ldots E_{sr1}$$

Hence

$$\frac{H_n}{H_{n-1}} = \frac{E_s}{E_{srn}}; \frac{H_{n-1}}{H_{n-2}} = \frac{E_{srn}}{E_{sr(n-1)}}; \ldots \frac{H_2}{H_1} = \frac{E_{sr3}}{E_{sr2}}; \frac{H_1}{V} = \frac{E_{sr2}}{E_{sr1}}$$

Thus

$$\frac{V}{E_{sr1}} = \frac{H_1}{E_{sr2}} = \frac{H_2}{E_{sr3}} = \cdots = \frac{H_{n-1}}{E_{srn}} = \frac{H_n}{E_s} \quad (7.37)$$

From Equation 7.5 the value of H for a first order syllabic expander ($n = 1$) is

$$H = AV + GVE_{sr} = I + GVE_{sr}$$

and over most of the dynamic range the idling value of H can be ignored, i.e.

$$H \simeq GVE_{sr} \quad (7.38)$$

Thus the magnitude of the voltages at the output of the multipliers in Figure 7.6 is given by

$$H_k = GVE_{srk} \quad (7.39)$$

where k has integer value between 1 and n corresponding to the n multipliers.

Combining Equation 7.39 with Equation 7.37

$$\frac{V}{H_1} = \frac{H_1}{H_2} = \frac{H_2}{H_3} = \cdots = \frac{H_{n-1}}{H_n} = \frac{H_n}{GVE_s} \qquad (7.40)$$

From Equation 7.40 $H_1 = V^{1/2}H_2^{1/2}$ and $H_2 = H_1^{1/2}H_3^{1/2}$ hence

$$H_2 = V^{1/3}H_3^{2/3}$$

Proceeding in a similar manner the following generalised equation is arrived at

$$H_{n-1} = V^{1/n}H_n^{(n-1)/n} \qquad (7.41)$$

Dividing Equation 7.41 by H_n and using the result of Equation 7.40

$$\frac{V^{1/n} \cdot H_n^{(n-1)/n}}{H_n} = \frac{H_n}{GVE_s}$$

i.e.

$$H_n = VG^{n/n+1} \cdot E_s^{n/n+1}$$

From Equation 7.7, $B = GV^{n+1}$ and hence

$$H_n = G^{(n-1)/(n+1)} \cdot B^{1/(n+1)} \cdot E_s^{n/(n+1)} \qquad (7.42)$$

This equation applies when $H_n \gg I$ and it has the same form as H for the encoder shown in Figure 7.1 When $n = 1$

$$H_1 = B^{1/2} \cdot E_s^{1/2} \qquad (7.43)$$

and when n is large

$$H_n = \text{constant} \times B^{1/(n+1)} \cdot E_s^{n/n+1} \qquad (7.44)$$

These last two equations agree with Equations 7.10 and 7.29 respectively when B is replaced by its optimum value H_M.

This analysis indicates that by using an nth-law power element, $(n-1)$ fewer syllabic expanders of the type shown in Figure 7.6 are required.

7.3. CONTINUOUS D.M.

The continuous d.m. codec designed by Greefkes and de Jager[8] employs an encoder whose block diagram is shown in Figure 7.7. This delta modulation instead of extracting the envelope of the analogue signal which resides in $L(t)$, as is the case for the encoders described in Sections 7.2 and 7.3 passes $L(t)$ through an RC circuit having a cut-off frequency of about 100 Hz to form the slowly varying signal $f(t)$. This signal represents the mean number of logic 'one' pulses in $L(t)$. $f(t)$ is multiplied by $L(t)$ to produce a pulse modulated waveform.

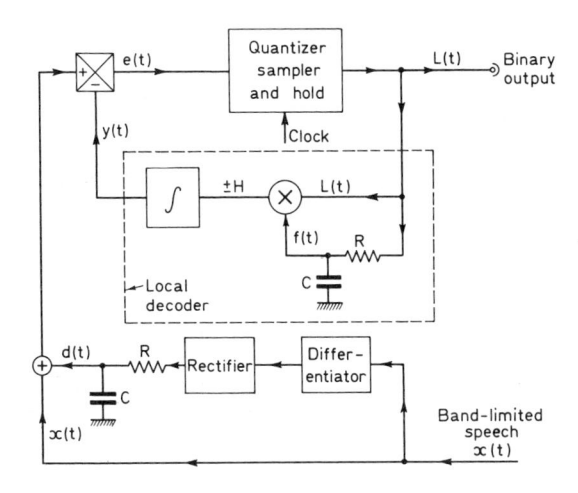

Fig. 7.7. *Continuous d.m. encoder.* (*After Greefkes and de Jager*[8])

The name of this system derives from the way the magnitude of the pulses at the multiplier output vary in an almost continuous fashion, as they do in the s.c.d.s.m. codec. The encoder is arranged such that for very small input signals the mean number of logic 'one' pulses at the output of the encoder is $\frac{1}{3}$, and for large input signals $\frac{1}{2}$, i.e. $f(t)$ is 1·5 times larger for large inputs than for small ones.

The multiplier characteristic can be chosen such that this variation of $f(t)$ can cause the feedback pulses $\pm H$ to change by a large factor M, say 20 to 100. The corresponding change in the d.c. voltage of $y(t)$ is 1·5 M, i.e. the d.c. component of $y(t)$ is proportional to the product of the pulse amplitude and the pulse density of logic ones.

The speech signal band limited between 300 Hz to 3400 Hz is unsuitable for encoding by the continuous d.m. encoder because it compands by responding to frequency components below 300 Hz. In order to perform the encoding of speech signals these signals must be modified to produce the necessary low frequency components. This modification is achieved by differentiating and then rectifying the speech input signal $x(t)$, before smoothing the resulting waveform by an *RC* network having a cut-off frequency of about 100 Hz, to produce a slowly varying signal $d(t)$ with a d.c. component. It is necessary to use a differentiator because excessive slope causes the encoder to be overloaded. The frequency band occupied by $d(t)$ is below the lowest frequency in $x(t)$ and consequently the signal applied to the encoder, $x(t) + d(t)$, has frequency components similar to $y(t)$. This is because $y(t)$ is composed of pulses which change at the clock rate in an attempt to follow the instantaneous variations in the speech input and a slowly changing component due to the mean level changes in the slope of the input signal which occur at the pitch-rate, i.e. of the order of 10 ms.

The decoder consists of the local decoder used in the encoder followed by a band-pass filter whose highest cut-off frequency is approximately equal to the highest frequency in $x(t)$, and whose lowest frequency is greater than the highest frequency in $d(t)$. By using a band-pass filter the signal $d(t)$ is rejected and only $x(t)$ is recovered.

The response of a continuous d.m. system to a sinusoidal input

$$x(t) = E_s \sin 2\pi f_s t$$

will now be investigated. The frequency f_s will be taken to by 800 Hz as a sinusoidal input of this frequency is similar to the speech condition discussed in Section 2.2.6.

Now the d.c. components in the $y(t)$ signal are proportional to the product of H and the density of 'ones' in $L(t)$, and approximates to the low frequency components in the input signal, i.e. to the signal $d(t)$. The high frequency components in $y(t)$ are used to track the input signal $x(t)$ but the companding is achieved by the low frequency components in $y(t)$ adjusting to $d(t)$. When a sinusoidal input of constant frequency f_s is applied the signal $d(t)$ is reasonably constant having a value kE_s, where k is a constant associated with the differentiator, rectifier, pitch filter and the value of f_s.

The signal-to-quantization noise ratio will now be calculated as a function of E_s. Let n be the relative number of positive binary pulses in the output signal $L(t)$, where $0 \leq n \leq 1$. When E_{sm}, the largest amplitude of the input sinusoid which does not cause the encoder to be overloaded is applied,

$$d(t) = kE_{sm}$$

$n = n_2$, and the magnitude of the pulses at the output of the multiplier $H = H_M$. When the encoder is idling, $x(t) = 0$, $n = n_1$, say and $H = I$, its minimum value.

From the previous discussion, the low frequency components $d(t)$ at the input to the encoder are proportional to H and the pulse density of the binary output signal. Consequently

$$d(t) = kE_s = a_1 nH - a_2$$

where a_1 is a constant of proportionality and a_2 is a constant which satisfies the idling condition, since $H = I$ when $E_s = 0$.

Introducing the boundary conditions stated above

$$kE_{sm} = a_1 n_2 H_M - a_2$$
$$0 = a_1 n_1 I - a_2$$

the constants a_1 and a_2 are determined. Substituting in the above equation giving E_s in terms of n and H

$$E_s = \frac{E_{sm}}{H_M}\left(\frac{nH - n_1 I}{n_2 - n_1(I/H_M)}\right) \quad (7.45)$$

If a linear multiplier is used H and n are linearly related, i.e.

$$H = b_1 n + b_2$$

where b_1 and b_2 are constants. Applying the boundary conditions once again

$$H_M = b_1 n_2 + b_2$$
$$I = b_1 n_1 + b_2$$

The constants b_1 and b_2 are now found and thus the equation relating H to n, i.e.

$$H = \frac{H_M}{n_2 - n_1}\left\{\left(1 - \frac{I}{H_M}\right)n - \left(n_1 - \frac{I}{H_M}n_2\right)\right\} \quad (7.46)$$

If the dynamic range of the multiplier, as defined in Section 7.2.7, is very large Equations 7.45 and 7.46 may be approximated by

$$\frac{E_s}{E_{sm}} = \frac{nH}{n_2 H_M} - \left(\frac{I}{H_M}\right)\frac{n_1}{n_2} \quad (7.47)$$

and

$$\frac{H}{H_M} = \frac{n - n_1}{n_2 - n_1} \quad (7.48)$$

respectively. Eliminating n from Equations 7.47 and 7.48 yields

$$\left(\frac{H}{H_M}\right)^2 + \frac{n_1}{n_2 - n_1}\left(\frac{H}{H_M}\right) - \frac{n_2}{n_2 - n_1}\left(\frac{E_s}{E_{sm}} + \frac{I}{H_M}\cdot\frac{n_1}{n_2}\right) = 0 \quad (7.49)$$

For $n_1 = \frac{1}{3}$, $n_2 = \frac{1}{2}$ and $(E_s/E_{sm}) \gg \{(I/H_M)(\frac{2}{3})\}$

$$\frac{H}{H_M} = -1 + \left(1 + 3\cdot\frac{E_s}{E_{sm}}\right)^{1/2} \quad (7.50)$$

Figure 7.8 displays the plot of Equation 7.50 and shows that E_s/E_{sm} is just less than H/H_M over its entire range, and $H = H_M$ when $E_s = E_{sm}$. The variation of H_1 with E_s for $B = H_M$ for an s.c.d.s.m. having $n = 1$ is also shown. A high order s.c.d.s.m. system obeying Equation 7.29 is similar to a continuous d.m. system, i.e., close to the ideal.

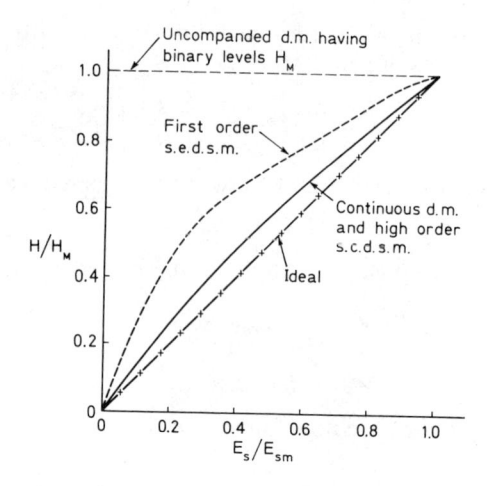

Fig. 7.8. *Variation of normalised H with normalised input amplitude for various d.m. systems*

If $E_{sm} = H_M$ then Equation 7.3 applies. For the codec in Figure 7.7, $E_{sm} \neq H_M$, and Equation 7.3 is rearranged to give

$$\text{s.q.n.r.} = R\left(\frac{E_s}{E_{sm}} \cdot \frac{H_M}{H}\right)^2 \tag{7.51}$$

H can be found from Equation 7.55 and for any value of E_s. When $E_s = E_{sm}$, $H = H_M$ and the peak s.q.n.r. $= R$.

When a single integrator is used in the feedback network the value of R is given by Equation 2.46. If the single integrator is replaced by a double integrator with prediction then R is represented by Equation 2.30 and

$$\text{s.q.n.r.} = \frac{f_p^{4\cdot 3}}{630}\left(\frac{E_s H_M}{E_{sm} H}\right)^2 \tag{7.52}$$

The procedure for calculating the dynamic range is similar to that used in Section 7.2.7 and yields the same result, i.e. it is largely dependent on the dynamic range of the multiplier. Finally, the equations are approximately valid for narrow band Gaussian input signals if E_s is replaced by $\sqrt{2}\,\sigma$.

7.4. EXTERNALLY COMPANDED D.M.

The encoder for the d.m. system shown in Figure 7.9(a) has an integrator in both its forward and feedback paths. If the integrator in the forward

Syllabically companded delta modulation

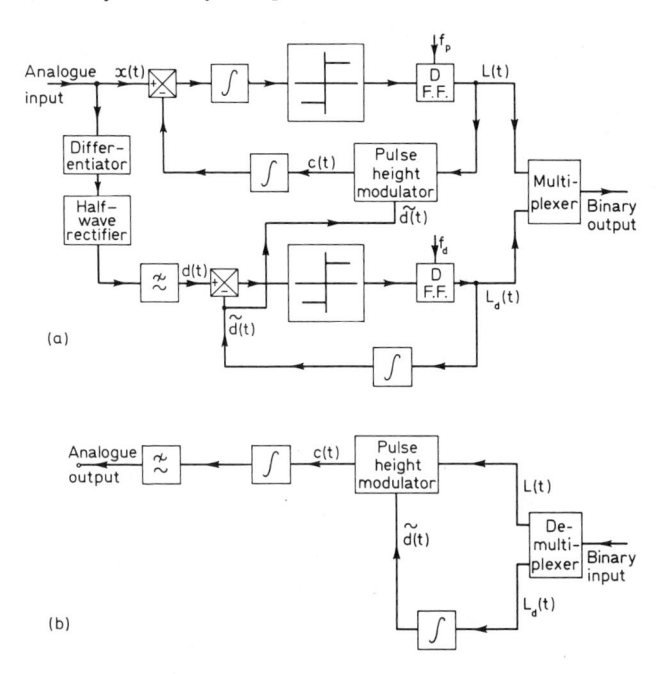

Fig. 7.9. Externally companded d.m. (a) encoder, (b) decoder. (After Brolin and Brown[9])

path is removed and placed on either side of the error point the encoder is a double integration type whose input signal has been de-emphasised. This arrangement results in an overload characteristic which is the same as for a single integration d.m.

The d.m. encoded is made adaptive by controlling the step height of the signal $c(t)$ by variations in $\tilde{d}(t)$. The signal $\tilde{d}(t)$ is an analogue one derived from an integrator in a low bit rate d.m. encoder which is tracking an input signal $d(t)$. $\tilde{d}(t)$ is a close representation of $d(t)$ and is derived from the speech signal in a similar manner to that used in continuous d.m., and for other reasons. However, because $\tilde{d}(t)$ does not derive from $L(t)$ the decoder cannot decode the speech signal unless information concerning $\tilde{d}(t)$ is transmitted. Clearly as the output $L(t)$ of the encoder is in a binary form it is an advantage for $d(t)$ to be encoded in the same way so that time division multiplexing can be used. The encoding of $d(t)$ is achieved in a linear d.m. encoder operating at a bit rate $f_d \ll f_p$ as, compared to $x(t)$, $d(t)$ contains only low frequency components.

The output binary waveform $L(t)$ is then multiplexed with $L_d(t)$ prior to transmission. The decoder de-multiplexes the incoming signal and operates on $L_d(t)$, to produce $\tilde{d}(t)$, which when multiplied by $L(t)$ gives

$C(t)$. This latter signal is integrated and filtered to recover the speech signal.

An adaptive d.m. system using this strategy is due to Brolin and Brown.[9] The external signal $\tilde{d}(t)$ is used to adjust the dynamic range of the encoder in preference to using a linear delta modulator to encode a compressed audio signal. It is for this reason that it has been called here 'externally companded d.m.'

The consequence of controlling $C(t)$ by a signal produced from the input signal $x(t)$ is the requirement of an additional delta modulator, and the incorporation in the system of a multiplexer and de-multiplexer.

7.5. SCALE

SCALE is an acronym of Syllabically Companded All Logic Encoder.[10,11] The SCALE system is designed to transmit speech signals in a binary form at digit rates of from 10 to 40 kbits/s. It is an attractive system in that it behaves similarly to the s.c.d.s.m. codec, discussed in the previous sections, but it is implemented in a digital form thereby improving reliability and reducing cost.

The encoder is shown in Figure 7.10. The logic stores the previous two $L(t)$ values then together with the current $L(t)$ value performs the simple logic operation

$$G_L = L_r L_{r-1} L_{r-2} + \bar{L}_r \bar{L}_{r-1} \bar{L}_{r-2} \qquad (7.53)$$

where L_r, L_{r-1}, L_{r-2} refer to the values of $L(t)$ at the rth, $(r-1)$ and $(r-2)$ sampling instants, respectively. $\bar{L}_r$ means the logical inversion of L_r,

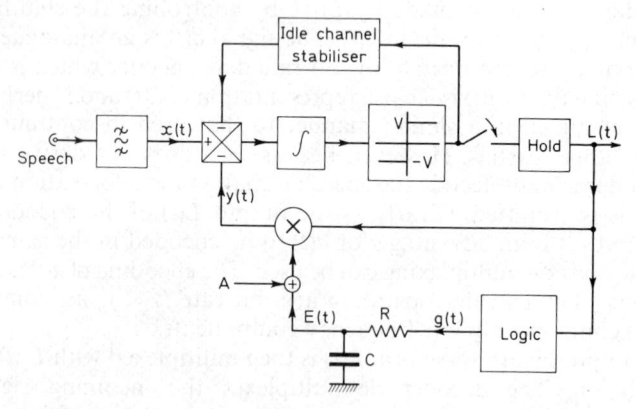

Fig. 7.10. SCALE encoder

Syllabically companded delta modulation

namely if $L_r = 1$, $\bar{L}_r = 0$ and vice versa. Hence the signal $g(t)$ at the output of the logic is only a logic 'one' if there are sequences of 3 or more digits of the same polarity at the output of the encoder.

If $x(t)$ is insufficient to cause $g(t)$ to go to its 'one' state, the encoder is uncompanded, and behaves as a linear delta-sigma modulator where $y(t)$ consists of pulses having amplitudes

$$H = \pm I = \pm AV$$

When the encoder is companding, $g(t)$ occasionally has positive pulses of amplitude V_L which charge up the capacitor C via the resistor R. This RC circuit is known as the syllabic filter; its time constant is made 10 ms because this value results in a good subjective performance when encoding speech signals. If the bit rate is 20 kbits/s, then in 10 ms there are 200 bits, hence the signal $E(t)$ at the output of the filter changes slowly when compared to the bit rate. Further, $E(t)$ is always positive as $g(t)$ has levels of either $+V_L$ or zero. Consequently the waveform $E(t)$ has some similarities with the waveform at the output of the power-law element in Figure 7.1.

In order to ensure the symmetry of the encoder, thereby guaranteeing the correct ... 1 0 1 0 1 0 ... idling pattern, an idle channel stabiliser is introduced (Figure 7.10). It consists of an integrator having a time constant many times greater than that used in the forward path of the encoder, and usually a voltage level shifter which depends on the particular circuit configuration employed. The subtractor, idle channel stabiliser and integrator are usually implemented using only resistors and capacitors.

The s.c.d.s.m. system of Figure 7.1 and the SCALE encoder shown in Figure 7.10 can be seen to differ in the method of extracting the envelope of the analogue signal which resides in $L(t)$. The envelope extraction and subsequent power-law amplification used in the SCALE system is imperfect, but is much simpler to implement in hardware than it is in the s.c.d.s.m. codec.

The SCALE decoder, like the s.c.d.s.m. decoder, is the part of the encoder in the feedback loop plus a low-pass filter for removing out-of-band quantization noise.

7.5.1. Multiplier characteristic

A very convenient method of multiplying the binary output signal $L(t)$ by the sum of the extracted envelope signal $E(t)$ and the idling constant A is to use the arrangement shown in Figure 7.11. When the $L(t)$ signal is $+V$, switch S_2 is opened by $L(t)$ and switch S_1 remains closed. $E(t)+A$ is then connected to the amplifier having an amplification V via the path abc. Switch S_2 inhibits the path adc. The output signal $y(t)$ has a positive

Syllabically companded delta modulation

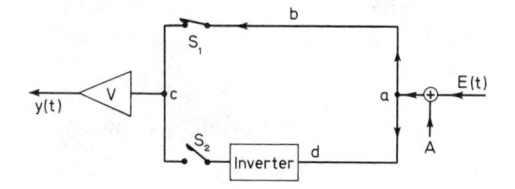

Fig. 7.11. The multiplier

magnitude

$$H = AV + E(t)V$$

When $L(t)$ has a negative binary value switch S_1 opens, thereby inhibiting path abc, and switch S_2 closes, enabling path adc to be used. The inverter in this path results in $y(t)$ having a negative magnitude $AV + E(t)V$.

A hardware version[10] of Figure 7.11 is achieved by one double output, a differential input, an operational amplifier and two transistors acting as switches S_1 and S_2.

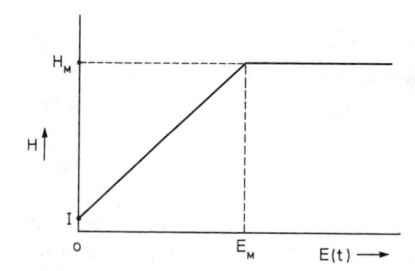

Fig. 7.12. Multiplier characteristic (S_1 closed, S_2 open)

The graph of H against $E(t)$ which gives the characteristic of the multiplier is shown in Figure 7.12. When $E(t)$ is zero only voltage A is fed to the multiplier to give $H = I$. When companding commences, i.e., $E(t) > 0$, the value of H increases linearly with $E(t)$ until $E(t) = E_M$ when $H = H_M$. The multiplier is now saturated and further increases in $E(t)$ do not affect H.

7.5.2. Threshold of companding

Consider the case when a sinusoidal input is present which causes the $L(t)$ waveform to develop sequences of ones and zeros < 3 bits. $E(t)$ consequently remains at zero volts and the fedback voltage remains at $\pm I$. This is in contrast to the analogue system where companding would have started to occur because the envelope can be extracted from $L(t)$ by analogue means. Increasing the input amplitude will eventually cause sequences of ones and zeros to occur which results in $E(t)$ becoming positive. Consequently the SCALE encoder has a 'threshold of companding' effect.

Observe that the constant A does more than establish an idling pattern, because it enables the SCALE codec to behave as a linear d.s.m. for low level input signals.

The simple encoding algorithm used in SCALE means that the threshold of companding is nearly independent of the frequency components in the input signal.

7.5.3. Overload condition

Consider applying a sinusoidal input signal to the encoder. If the amplitude of this sinusoid is continually increased a condition of severe overload will be reached where the binary output signal $L(t)$ is a square wave having the same periodicity as that of the sinusoid. This is illustrated in Figure 7.13 for a ratio $f_p/f_s = 16$ where f_s and f_p are the frequency of the sinusoidal input and the pulse repetition rate of the sampling, respectively. The waveform $g(t)$ at the output of the logic is also shown and it can be seen that following a change of binary values in $L(t)$ the logic does not have three identical type digits and $g(t)$ changes to zero. After three identical binary values in the $L(t)$ signal have occurred $g(t)$ returns to its positive level V_L volts. This results in $g(t)$ zero for two bits, every half cycle of the input.

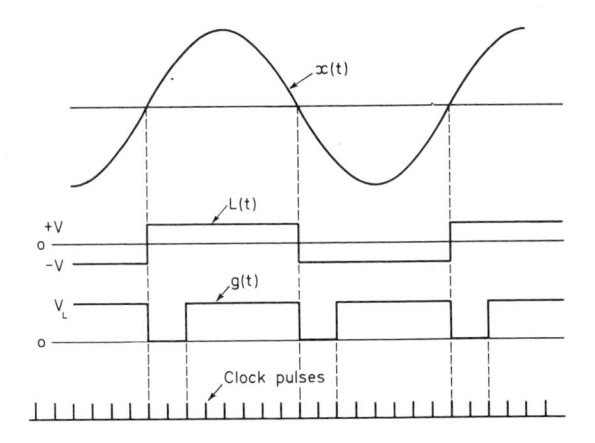

Fig. 7.13. $L(t)$ and $g(t)$ waveforms for a sinusoidal input $x(t)$ which grossly overloads the encoder, $f_p = 16 f_s$

Restated, the number of bits in one cycle of the input signal is f_p/f_s. In a half-cycle the number of sampling periods when the logic output is a one is

$$(f_p/2f_s) - 2$$

It can be seen in Figure 7.13 that the periodicity of $g(t)$ is twice that of the input sinusoid. The full wave rectifying effect can be seen here since $g(t)$ is positive when the input sinusoid has positive or negative polarity.

As the syllabic time constant is very much longer than that of the sinusoid input the waveform $E(t)$ at the output of syllabic filter is the average of $g(t)$, i.e.

$$E_0(t) \doteq V_L \left(1 - \frac{4f_s}{f_p}\right) \qquad (7.54)$$

$E_0(t)$ is the value of $E(t)$ when the encoder is totally overloaded. For a given encoder the register length and logic rails are fixed, and if the encoder is clocked at a rate f_p then $E_0(t)$ is clearly a function of the frequency of the input sinusoid, being larger for low frequencies. Note that

$$E_0(t) = V_L, \quad \text{when} \quad f_s = 0 \qquad (7.55)$$

$$E_0(t) = V_L/2, \quad \text{when} \quad f_s = f_p/8 \qquad (7.56)$$

$$E_0(t) = 0, \quad \text{when} \quad f_s = f_p/4 \qquad (7.57)$$

Equation 7.57 is of more general interest as at this frequency companding cannot occur irrespective of the amplitude of the input sinusoid, because the $L(t)$ waveform can never have the sequence 1, 1, 1 or 0, 0, 0.

It is also apparent from the above that when the frequency of the input sinusoid is high, for example as defined by Equation 7.56, then $E_0(t) = V_L/2$, but at much lower input frequencies the waveform at the output of the envelope extractor will exceed $V_L/2$ although the encoder is not overloaded.

The value of $E(t) = E_M$ (Figure 7.12) is selected so that the highest frequency sinusoid which the SCALE system is required to encode can cause this value of E_M to be produced before severe amplitude overloading of the encoder occurs. If, for example, $E_M = V_L/2$ then the frequency defined by Equation 7.56 gives $H = H_M$ only when severe overloading occurs; signals that can be decoded with high signal-to-noise ratio would have frequencies considerably less than $f_p/8$. The important point is that E_M should be determined by the highest frequency which the encoder is required to accommodate.

Elaborating, if E_M is chosen as a small percentage of V_L then at very low input frequencies, when the amplitude of the input is relatively low, $E(t) = E_M$. This means that H will 'catch' E_s before $H = H_M$, the signal-to-noise ratio will have a maximum at low values of E_s and the curve will have a shape similar to that shown for $B = 2$ in Figure 7.4. As the frequency increases the shape of the s.q.n.r. against E_s changes until for a particular frequency the value of E_s just equals H, when $H = H_M$, and the curve $B = 10$ in Figure 7.4 applies.

If E_M is a large percentage of V_L then only for sinusoids having low frequencies will $E(t) = E_M$. The higher frequencies cannot produce an $E(t)$ voltage as great as E_M and the maximum s.q.n.r. is unobtainable. This corresponds to the curve $B = 50$ in Figure 7.4. A further discrepancy between the analogue envelope method described in Section 7.2.2 and the digital extraction method is that $E(t)$ is independent of the frequency of the input sinusoid in the former and dependent on the frequency in the latter.

The selection of E_M requires that the highest frequency of consequence will cause $E(t)$ to equal E_M when E_s just catches H_M. In the hardware version $E_M = V_L/2$ and consequently an input sinusoid whose frequency $f_s = f_p/8$ can just produce the maximum value of H if its amplitude is sufficiently large to cause severe overloading. This means that an input sinusoid is required to have a frequency $\ll f_p/8$ to produce s.q.n.r. when $E_s = H_M$.

7.5.4. Performance

A SCALE system uses the following parameters. Integration time constant $T_1 = 0.2$ ms, syllabic time constant $T_2 = 10$ ms., $I = 25$ mV, $H_M = 2.4$ V, $f_p = 19.2$ kbits/s, $f_{c1} = 300$ Hz and $f_{c2} = 2,400$ Hz. Consider the performance of this system when a sinusoidal input is applied. At low frequencies, say 300 Hz, where the duration of the sinusoidal input is comparable with the syllabic time constant the value of H is not constant in spite of the constant envelope of the input signal. This is because the $E(t)$ signal (see Figure 7.10) is increased in both halves of the input cycle at a rate comparable with the changes in the input signal, causing H to be modulated at twice the input frequency which in turn produces distortion in the final filtered signal. However, this effect is only a second order one as the closed loop tends to minimise these variations in H.

When the highest frequency sinusoid, 2,400 Hz, is applied to the encoder in each cycle the 8 bits in $L(t)$ can only occur in 16 different patterns because with sinusoidal inputs the pattern is symmetrical about its centre. The logic only responds to sequences of three or more ones or zeros; as the $L(t)$ pattern tends to be cyclic for a large number of clock periods, only two basic $L(t)$ patterns will result from companding. They are (a) 1 1 1 1 0 0 0 0 and (b) 1 1 1 0 1 0 0 0. The result of this limited number of patterns is that the magnitude of $y(t)$ decreases at the syllabic rate when the $L(t)$ pattern is (b) until the $L(t)$ pattern changes to (a). The signal $y(t)$ is now increased and charges up C until H becomes sufficiently large to change the pattern back to (b), and the process repeats. The two possible $L(t)$ patterns cause $y(t)$, and hence the decoded output signal, to be amplitude modulated. These effects can be greatly reduced by increasing the clock rate thereby allowing many different $L(t)$ patterns having sequences of three or more ones and zeros. For

$$f_p = 40 f_{s\,max} = 96 \text{ kHz}$$

excellent encoding conditions are obtained and the SCALE system closely resembles companded d.s.m. rather than a form of d.m.

For frequencies in the middle of the band, say 1 kHz, the modulating effects experienced by H due to the time constant of the syllabic filter and the low clock rate largely disappear and H is constant except for a small ripple whose amplitude is 5% of H. This means that the encoder behaves as a linear d.s.m. encoder with the distortion which is inherent when the integrator has a short time constant and the clock rate is relatively low.

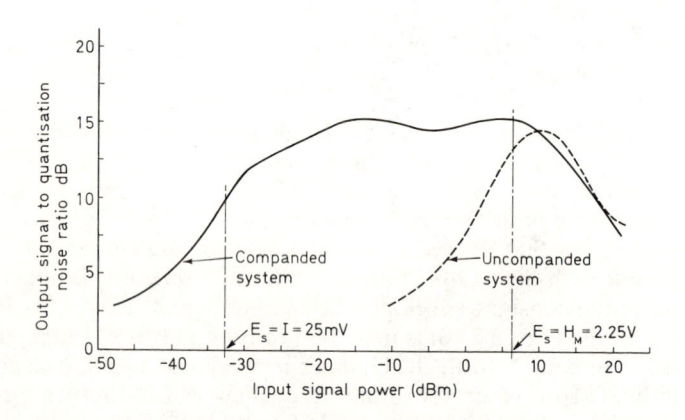

Fig. 7.14. SCALE System. Measured signal-to-noise ratio versus input power in dBm (i.e. in dB relative to 1 mW and 600 Ω). Band-limited white noise input signal, $f_p = 19 \cdot 2$ kbits/s

The behaviour of the pulse height H as a function of input amplitude E_s is approximately the same for all frequencies in the message band. Directly the threshold of companding is exceeded, i.e. at about 25 mV, H increases exponentially as a function of E_s until $E_s \doteq 1 \cdot 5$ V, when

Syllabically companded delta modulation 213

E_s catches H. Further increases in E_s result in a slower increase in H and in the SCALE encoder using the parameters stated at the beginning of this section H_M reaches 2·4 V when $E_s \doteq 3$ V. For further increases in E_s, H_M remains constant and the encoder functions as a linear deltasigma modulator.

Figure 7.14 shows the s.q.n.r. versus input signal power performance of the SCALE system when white noise band-limited between 450 and 550 Hz is applied to the encoder. The graph was obtained by measurements for a SCALE encoder having $H_M = 2\cdot25$ V, rather than 2·4 V, and $I = 24$ mV, i.e. the dynamic range of the multiplier is 39 dB. The values of sinusoidal power associated with I and H_M are also marked in this graph and indicate how the performance of the SCALE system is similar to that predicted for s.c.d.s.m. described in Section 7.2.9.

For register lengths of 3, 4 and 5 bits the effect on the dynamic range is small., i.e. about 2 dB at 19·2 kHz. A two-bit register tends to degrade the performance.

7.6. DIGITALLY CONTROLLED D.M.

Digitally controlled delta modulation described by Geefkes and Riemens[13] is very similar to the SCALE system. It is therefore more appropriate to mention some of their differences. The digitally controlled d.m. codec as its name suggests is a delta modulator system unlike SCALE which closely resembles delta sigma modulation. The digitally controlled d.m. encoder uses logical processes for extracting the companding signal from $L(t)$ which are similar to those used in the SCALE encoder, but inspects four rather than three $L(t)$ bits and uses an additional shift register to extend the $g(t)$ signal by three bit periods. It uses double rather than single integration, of the type mentioned in connection with Equation 2.30. These are probably the essential differences.

The performance of this system is very good. It has a signal-to-noise ratio for telephony which is well above the CCITT (Comite Consultatif International de Telegraphie et Telephonie specification) recommendation. At a bit rate of 56 kbits/s digitally companded d.m. performs better than logarithmic p.c.m.

The SCALE system has been specifically designed for low bit rate encoding of speech when the latter has come from a pre-emphasised microphone, hence the emphasis on delta-sigma rather than delta modulation. If the position and types of integrators are interchanged, SCALE and digitally companded d.m. systems are very similar. Although the author has found the performance of both these systems excellent, more space has been given to describing SCALE simply because the author has had more experience in using this system.

7.7. SPEECH-REITERATION D.M.

The d.m. systems described in this book are essentially of the waveform tracking kind. In some cases specific characteristics of the input signal have been accommodated by the codecs, for example, syllabically companded and Bosworth–Candy delta modulation have been conceived especially for binary transmission of speech and television, respectively. As the bit rate is reduced the design of the d.m. systems requires that greater emphasis be placed on the characteristics of the particular signal which is likely to be encoded.

In the case of digital transmission of speech at low bit rates, i.e. 10 kbits/s or less, analysis-synthesis methods have been seriously considered since Dudley[14] invented the vocoder in 1928. Basically these methods do not recreate the original signal at the output of the receiver, but produce a signal which is subjectively acceptable. Digital transmission systems have been designed[4] which neither track the input speech waveform nor utilise analysis-synthesis techniques. They involve removing some of the redundancy from the speech and then waveform tracking the residual signal. At the receiver the digital signal is decoded and the recovered residual signal is used to operate a type of feedback transversal filter to recover the speech, i.e. to replace the redundancy removed at the transmitter end. These systems use forms of differential p.c.m.

The subject of low-bit-rate transmission of speech is complex, and requires a knowledge of speech characteristics which is outside the scope of this book. Instead the reader who wishes to pursue this subject is referred to References 4 and 14–25.

However, Baskaran[26] has developed some low-bit-rate d.m. systems which although utilising more characteristics of speech compared to other d.m. systems described in this book, are still essentially waveform-tracking encoders. The characteristics in the speech signal which are exploited by these d.m. systems are

(1) When the larynx is excited the speech is said to be 'voiced speech'. Examination of voiced speech shows that it tends to be periodic. This is illustrated in Figure 7.15(a).

(2) The dynamic range of the voiced speech is much larger than unvoiced sounds, i.e. when the larynx is not excited. This is illustrated by Figure 7.15(b) which shows a speech waveform containing segments of voiced and unvoiced sounds.

(3) Generally, during every pitch period of voiced sound there is a rapid rise in the speech waveform followed by an exponential decay. This characteristic is apparent from speech waveforms.

The speech-reiteration d.m. system is shown in Figure 7.16. The adaptive encoder has its step-size suddenly increased to ten times its minimum value at the beginning of a pitch period when a pitch pulse is

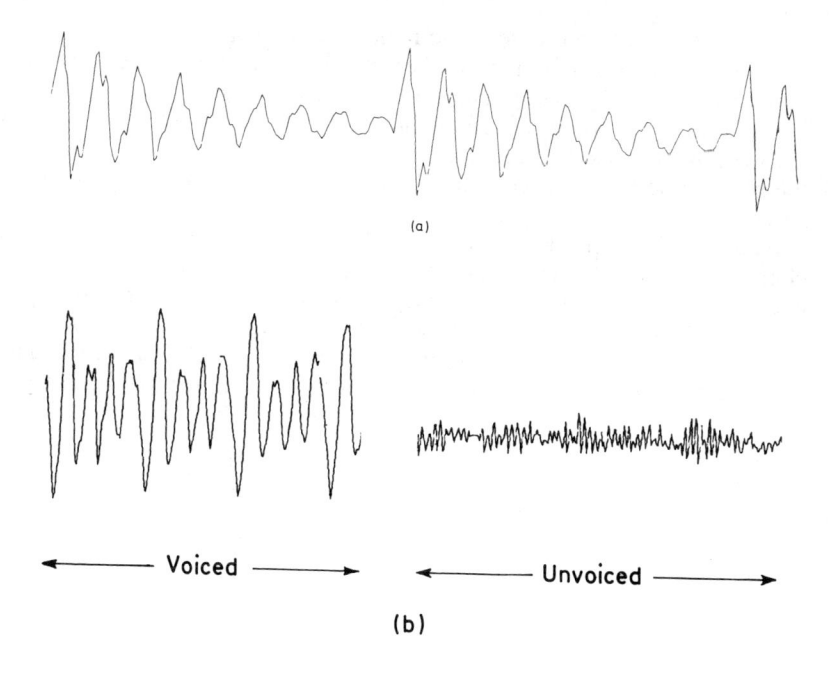

Fig. 7.15. Specimen of speech wave: (a) voiced-speech, (b) voiced and un-voiced speech

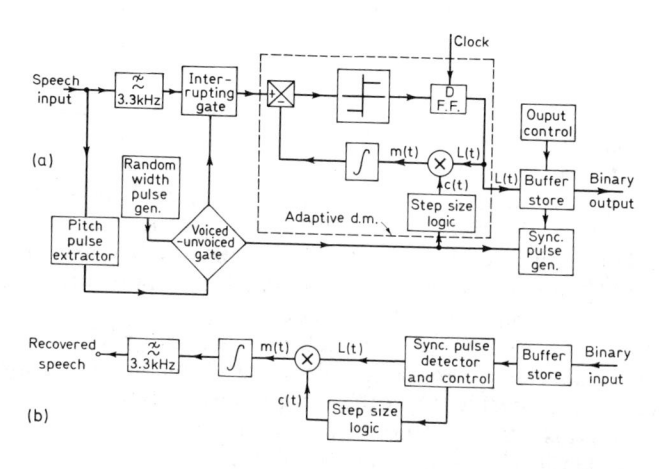

Fig. 7.16. Speech reiteration d.m. system: (a) encoder, (b) decoder

discovered; it is thus able to avoid being overloaded by the speech signal at its input. The step-size is now allowed to exponentially decay with the time constant of a typical pitch period, i.e. 10 ms. In this way the encoder is instantaneously companded at the beginning of the pitch period and then companded as if it were an s.c.d.s.m. encoder after the first pulse in the pitch period. The speech reiteration d.m. encoder thus acknowledges speech characteristic (3).

Because of the quasi-periodicity in the voiced speech, as described in characteristic (1), the d.m. encoder is arranged to track the speech signal every other pitch period. The binary data from the d.m. encoder is put into a buffer store together with the synchronising information. The data is removed from this store at a lower bit-rate than that used by the d.m. encoder to give continuous data transmission. The timing arrangement is illustrated in Figure 7.17.

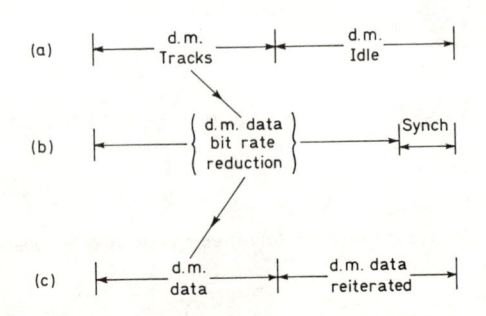

Fig. 7.17. Timing arrangement of speech reiteration d.m.: (a) d.m. functions for 50% of time, (b) transmitted data, (c) data fed to d.m. decoder

When unvoiced sounds are present the encoder operates as a linear delta modulator. This is satisfactory because although the frequency components in the speech may be high the magnitudes are relatively small, with the result that the moderate slopes can easily be accommodated by the encoder. The interruption of the d.m. encoder and the reiteration of the d.m. data is performed randomly, and the average interruption cycle is 50%. It is necessary to apply random interruption in order to avoid a line spectrum occurring in the decoded signal.

Returning to Figure 7.16, the pitch pulse extractor which can be very complex in analysis-synthesis encoders is a simple device in this system. It merely has to decide when the largest pulse in the voiced sound has arrived and this can be arranged by some simple circuit. If the pitch pulse extractor identifies a voiced sound it allows the speech to enter the delta modulator for encoding. The input to and the output from the delta modulator is inhibited during the next pitch period, and the encoded binary signal obtained during the first pitch period is removed from the buffer store at nearly half the d.m. clock rate. It would be exactly half the d.m. clock rate except that the synchronising information has to be transmitted to the receiver. This information may be a sequence of all ones to inform the receiver that voiced sounds are being transmitted as distinct from a sequence of all zeros which indicates the presence of unvoiced data.

The receiver stores the incoming binary signal then speeds up the data rate to that used by the delta modulator in the encoder and reiterates the data once. From the synchronising data the decoder senses that voiced sounds are present and consequently increases the height of the first binary bit by a factor of 10, and then allows the magnitude of the subsequent bits to decrease exponentially. In this way the pulses $m(t)$ fed to the integrator are identical, assuming no channel perturbations, to those at the encoder and the final output closely resembles the speech signal.

If the pitch extractor in the encoder does not produce an output a random width generator is employed which allows the speech to be applied to the delta modulator on a random basis. At the same time the step-size logic in the delta modulator is set to its smallest value and maintained there, and synchronising data is included in the output to inform the decoder that unvoiced speech is being transmitted.

When the random generator inhibits speech from entering the delta modulator, the previous sequence of d.m. digits is time scaled by the store for transmission in the manner previously described.

The behaviour of the receiver is similar to that for the voiced data, i.e. expansion of the bit rate and iteration, but the step-size generator is set to its minimum size to ensure that the unvoiced speech is recovered.

7.7.1. Pitch companded d.m.

A derivative of speech-reiteration d.m. called pitch-companded d.m.[26] modifies the previous encoding algorithm as follows.

The step size adaptation in the adaptive delta modulator is made to match speech characteristic (3). Accordingly when an abrupt increase in the amplitude of the speech signal is experienced by the delta modulator producing at the output a sequence of ones or zeros the step size in the feedback loop is increased exponentially in the ratio 1:2:4:8, etc. depending on the memory length used by the encoder. The step size remains at the maximum value until the output sequence changes sign when it exponentially decays with a time constant of about 3 ms. In this scheme, the speech is continuously encoded by the delta modulator and all the

218   Syllabically companded delta modulation

previous peripherals used in Figure 7.16, i.e. pitch extractor buffer, random gating etc. are removed. While pitch companded d.m. is much simpler to implement, when compared to speech reiteration d.m., the transmission rate required is higher. Typically, the bit rate would be at least 1·5 times that required by speech reiteration d.m. for approximately the same signal-to-noise ratio, if it is assumed that the synchronising pulses in the latter occupy one half of a reiteration cycle.

An improvement to both d.m. systems can be achieved by making use of the fact that an increase in the first formant frequency is generally associated with an increase in the energy and mean slope of the speech waveform. An appropriate estimate of the first formant can be made with a zero crossing detector operating on the decoded signal which has been suitably low-pass filtered to contain frequencies below 1,000 Hz. Although this estimate can be in error by as much as half the pitch frequency, it can

Input to encoder

(a)

Output from decoder at 8k bit/s

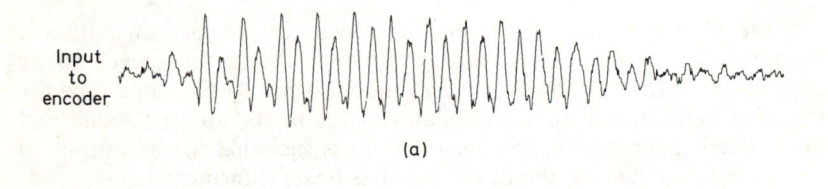

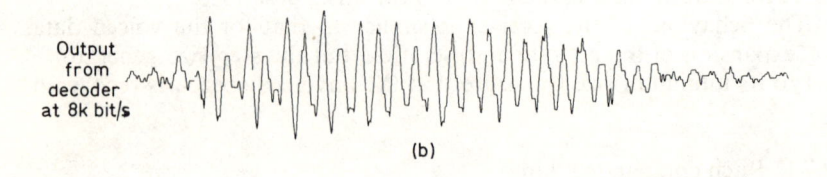

(b)

Fig. 7.18. Pitch companded d.m. system: (a) speech input signal, (b) waveform at the output of the decoder for a transmitted bit rate of 8 kbits/s

be used to change and improve the step-size adaptation even when the first format moves to the upper half of its range.

Another improvement which can be made to pitch companded d.m. is to delay the output from the decoder by say 6 bits. If there is a sequence of similar bits indicating an overload condition then the amplitude of the pulses fed into the integrator can be increased to give a sinusoidal prediction of the speech waveform. By this means the limited dynamic range of the encoder is enhanced by prediction in the decoder.

Figure 7.18 gives an idea of the performance of the pitch companded d.m. using the improvements mentioned above. Figure 7.18(a) shows a segment of a speech signal, band-limited to 3 kHz, which is applied to the encoder. The waveform (b) is the output from the decoder for a transmission bit rate of 8 kbits/s.
A typical speech signal having a dynamic range of 26 dB when applied to speech reiteration d.m. and pitch companded d.m. systems operating at an 8 kHz bit-rate, have decoded signal-to-noise ratios of 13 and 10 dB, respectively.

7.8. COMPRESSION AND EXPANSION CIRCUITS

An adaptive delta modulator[27] designed for encoding speech signals produces its 'syllabic signal' $V_s(t)$ with the aid of a JK flip-flop, a combination of NOR gates and an averaging circuit whose time constant is

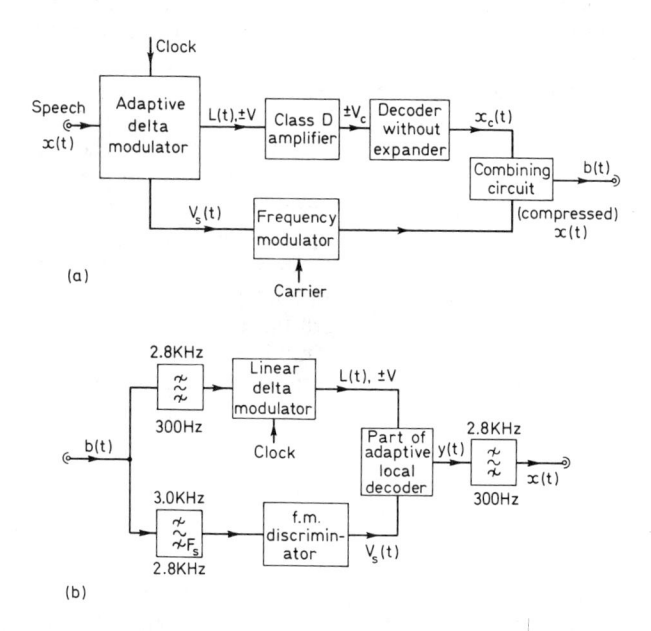

Fig. 7.19. Compression and expansion of speech using d.m. techniques: (a) syllabic compressor, (b) syllabic expandor. (After Betts[28])

20 ms. $V_s(t)$ is compared to a fixed level V_{so} and the difference signal is multiplied by the output binary signal $L(t)$ to produce a pulse height

modulated waveform which is integrated to give the feedback signal $y(t)$. The principles of this system are those of syllabically companded d.m. described in most of the preceding sections.

Betts[28,29] has used this adaptive d.m. system to produce speech compression and expansion circuits. The former, shown in Figure 7.19(a), encodes the speech signal $x(t)$ in the adaptive delta modulator. The binary signal $L(t)$ has its level suitably amplified by a Class-D amplifier to $\pm V_c$ and is then decoded by an RC circuit and band-pass filter whose cut-off frequencies are 300 to 2,800 Hz. This decoder is only suitable for a linear d.m. signal, not for an adaptive one. The result is that for $x(t)$ over a range of 30 dB the decoded signal $x_c(t)$ is compressed and has a nearly constant amplitude. The slowly varying signal $V_s(t)$ frequency modulates a carrier and the resulting f.m. waveform, occupying a frequency band from 2,800 to 3,000 Hz, is combined with $x_c(t)$ to form the baseband signal $b(t)$. This signal contains the compressed speech signal $x(t)$ and the compression information obtained from the $V_s(t)$ signal.

The expander is shown in Figure 7.19(b). The analogue signal $b(t)$ is filtered by F_s and demodulated by the f.m. discriminator to recover the syllabic signal $V_s(t)$; $b(t)$ is also encoded in a binary signal $L(t)$ having levels $\pm V$, by a linear delta modulator.

The signals $L(t)$ and $V_s(t)$ so produced are identical to those employed in the local decoder of the adaptive delta modulator in the compressor of Figure 7.19(a), to expand the binary levels of $L(t)$ and generate the feedback signal $y(t)$ which tracks the speech input signal. Consequently $y(t)$ is reproduced in the expander of Figure 7.19(b) from $L(t)$ and $V_s(t)$ with the aid of circuitry identical to the relevant part of the local decoder. On band-limiting $y(t)$ the original speech signal is recovered.

The main advantage of this technique compared to conventional methods of implementing syllabic companders and expanders is that the compression and expansion characteristics are exactly matched.

The system shown in Figure 7.19 has been designed to transmit the compressed signal $b(t)$ using single side-band modulation via an ionospheric transmission channel.

REFERENCES

1. Carter, R. O., 'Theory of syllabic compandors', *Proc. I.E.E.*, **111**, No. 3, 503–511, March (1964)
2. Smith, B., 'Instantaneous companding of quantized signals', *Bell Systems Tech. J.*, **36**, 654–709, May (1957)
3. Cartmale, A. A. and Steele, R., 'Calculating the performance of syllabically companded delta–sigma modulators', *Proc. I.E.E.*, 1915–1921, (1970)

4. Moye, L. S., 'Digital transmission of speech at low bit rates', *Electrical Com.*, **47**, No. 4, 212–223, (1972)
5. Tomosawa, A. and Kaneko, H., 'Companded delta modulation for telephone transmission', *I.E.E.E. Trans. Com. Tech.*, **COM-16**, 149–157, February (1968)
6. Schindler, H. R., 'Delta modulation', *I.E.E.E. Spectrum*, 68–77, October (1970)
7. Dalton, C. J., 'Delta modulation for sound-signal distribution. A general survey', *B.B.C. Engng.*, 4–14, July (1972)
8. Greefkes, J. A. and de Jager, F., 'Continuous delta modulation', *Philips Research Report*, No. 23, 233–246, (1968)
9. Brolin, S. J. and Brown, J. H., 'Companded delta modulation for telephony', *I.E.E.E. Trans. Com. Tech.*, **COM-16**, 157–162, (1968)
10. Wilkinson, R. M., 'A companded delta sigma modulation system', *Report No.* 68012, U.D.C. No. 621, 395. 6+621. 376.5, *Signals Research & Development Establishment*, Christchurch, Hants., July (1968)
11. Petford, B. and Clarke, C. M., 'A companded delta-sigma speech digitizer', *I.E.E. Conference Publication*, No. 64, 59–63, May (1970)
12. Wilkinson, R. M., 'Speech coding in digitally switched networks', *I.E.E. Conference Publication*, No. 64, 124–128, May (1970)
13. Greefkes, J. A. and Riemens, K., 'Code modulation with digitally controlled companding for speech transmission', *Philips Technical Review*, **31**, No. 11/12, 335–353, (1970)
14. Dudley, H., 'Remaking speech', *J. Acoust. Soc. Am.*, **11**, No. 2, 169–177, October (1939).
15. Coker, C. H., 'Real-time formant vocoder, using a filter bank, a general-purpose digital computer, and an analogue synthesizer', *J. Acoust. Soc. Am.*, **38**, 940, November (1965)
16. Flanagan, J. L., *Speech Analysis, Synthesis and Perception*, 2nd edn., Springer-Verlag, Berlin, (1965)
17. Schroeder, M. R., 'Vocoders, analysis and synthesis of speech', *I.E.E.E. Proc.*, **54**, No. 5, 720–734, May (1966)
18. Bially, T. and Anderson, W. M., 'A digital channel vocoder', *I.E.E.E. Trans. Com. Tech.*, **COM-18**, No. 4, 435–442, August (1970)
19. Moye, L. S., 'Self-adaptive filter', Brit. Pat. 1 184 653 (3.1.67)
20. Dunn, J. G., 'An experimental 9600-bit/s voice digitizer employing adaptive prediction', *I.E.E.E. Trans. Com. Tech.*, **COM-19**, No. 6, 1021–1032, December (1971)
21. Campanella, S. J., 'A survey of speech bandwidth with comparison techniques', *I.R.E. Trans. Audio*, 104–116, September–October (1958)
22. Atal, B. S. and Hanauer, S. L., 'Speech analysis and synthesis by linear prediction of the speech wave', *J. Acoust. Soc. Am.*, **50**, No. 2, Pt. 2, 637–655, August (1971)
23. Atal, B. S. and Schroeder, M. R., 'Adaptive predictive coding of speech signals', *Bell Systems Tech. J.*, **49**, No. 8, 1973–1986, October (1970)
24. Atal, B. S. and Schroeder, M. R., 'Predictive coding of speech signals', *Proceedings Speech Communications and Processing Conference*, Cambridge, Mass., 360–361, November 6–8 (1967). I.E.E.E.
25. Moye, L. S., 'Self-adaptive filter predictive-coding system', *Proceedings International Zurich Seminar on Integrated Systems for Speech, Video, and Data Communications*, Zurich, March 15–17 (1972). I.E.E.E.
26. Baskaran, P., 'Digital coding of speech', *Loughborough University of Technology, Reports for Joint Speech Research Unit*, London, October (1972) to April (1973)
27. Betts, J. A. and Ghani, N., 'Adaptive delta modulator for telephony', *Electronic Letters*, No. 6, 336–338, (1970)
28. Betts, J. A., 'Adaptifon system for telephony—an alternative implementation of the Lincompex concept', *Electronic Letters*, **6**, No. 17, 542–543, August 20 (1970)

29. Betts, J. A., 'Adaptive delta modulator for telephony and its application to the Adaptifon System—an alternative implementation of the Lincompex Concept', *I.E.E.E. Trans. Com. Tech.*, **19**, 547–551, August (1971)
30. Das. J., 'A critical review of delta modulation systems', *Electro Technology, Journal Society Electronic Engineers, Bangalore.* **XVI**, No. 2, 41–60, March–April (1972)

Chapter 8

INSTANTANEOUSLY COMPANDED DELTA MODULATION SYSTEMS

8.1. INTRODUCTION

In an instantaneously adaptive delta modulator the feedback signal to the error point makes significant changes during each sampling instant in response to the present and recent polarities of the output binary signal $L(t)$. The word 'recent' in the above context means approximately 2 to 8 sampling instants. The instantaneously adaptive d.m. therefore adapts its feedback signal $y(t)$ to the input signal over a relatively small number of clock periods when compared to the 200 or so which are required by the syllabically companded d.m. discussed in the previous chapter.

An interesting feature of instantaneously adaptive d.m. is their resistance to mathematical analysis, which means that they are frequently conceived rather than designed. The analysis is further complicated by the variety of performance criteria. These criteria depend upon the types of signals to be encoded, the properties of the transmission channel, the particular application etc. For example, if the communication is of the point-to-point type having zero error rate then a useful design criterion might be to minimise the mean square error in the encoder. On the other hand, if the encoder is used to broadcast information, an important design factor is for the local decoder to give a similar performance over a wide range of tolerance values. This is because each receiver has a local decoder plus filtering, and these local decoders may have considerable differences. An important situation arises when the instantaneously adaptive delta modulator has to operate in a military environment where the transmitted data may have to pass through noisy channels. The design pre-requisite here is the ability to function at a high error rate and care must be taken to reduce the hierarchical nature of the transmitted signal.

This chapter discusses a variety of instantaneously adaptive delta modulators. Because some of these d.m. systems have been given similar titles by their inventors, it has been necessary for the author to change the names of these systems in appropriate cases in order to avoid confusing the reader. As the chapter proceeds and different encoders are described it emerges that although the encoders have often completely different block diagrams they are more similar to each other than is apparent from cursory observations of these diagrams. The result is that some encoders are more general than others, but are not necessarily the easiest to construct. Consequently they have all been included in order to accommodate a wide variety of requirements unforeseen by the writer but known to an individual reader.

8.2. STATISTICAL D.M.

Statistical d.m. originated by Fine[1] and discussed in depth by Bello, Lincoln and Gish[2] is designed so that the mean square error of $y(t) - x(t)$ is minimised, where $x(t)$ is the signal applied to the input of the encoder and $y(t)$ is the signal at the output of the local decoder in the receiver. This minimisation is achieved assuming certain statistical properties of $x(t)$, and if these properties change the performance of the statistical d.m. system will be sub-optimum. This characteristic is not unusual. For example, a d.m. system optimised for speech signals will probably have a very sub-optimum performance if it is required to accommodate television signals.

The approach used[2] to establish the properties of statistical d.m. is to assign known stationary statistical properties to the input signal, assume that the encoder produces a binary signal and consider that this binary signal is perfectly recovered at the preceiver prior to decoding.

If the encoded binary sequence is $\{L_n\}$ where $L_n = \pm 1$ then the decoder operates on this sequence to produce the reconstructed analogue samples $\{y_n\}$ which are subsequently filtered to give the decoded analogue signal $o(t)$. The optimization procedure does not include the final filtering in the decoder. The value of $y(t)$ at the nth sampling interval is a function of the current value of L_n and the past m values

$$y_n = f_n(L_n, L_{n-1}, \ldots, L_{n-m}) \qquad (8.1)$$

This means that the local decoder has a memory of length m.

The optimisation procedure involves finding the optimum encoder given a decoder, and then finding the optimum decoder, given an encoder. Simultaneous optimisation of encoder and decoder is achieved by requiring the equations defining the conditionally optimum encoder, and the conditionally optimum decoder to be satisfied simultaneously. In principle, these two equations can be solved.

The main problem for the encoder is to decide at a sampling instant whether to make the binary output ± 1. The statistical delta modulator chooses L_n to minimise the error e_n, i.e.

$$e_n = |x_n - y_n| \qquad (8.2)$$

The final filter in the decoder is not considered, although it could be, and therefore this error is the error signal usually experienced in d.m.

It is the difference between the input signal and the feedback signal. If $L_n = +1$, then

$$y_{n+} = f_n(+1, L_{n-1}, \ldots, L_{n-m})$$

and if $L_n = -1$, then

$$y_{n-} = f_n(-1, L_{n-1}, \ldots, L_{n-m})$$

The arithmetic mean of these possible y_n values is

$$l_n = \tfrac{1}{2}(y_{n+} + y_{n-})$$

The decision on whether L_n is to be ± 1 is made by comparing the input sample x_n with a threshold level l_n

$$L_n = +1, \quad \text{when } x_n \geq l_n$$
$$L_n = -1, \quad \text{when } x_n < l_n$$

Notice that l_n depends on the previous m values of $L(t)$ and the time (n), and that by deciding the value of L_n in this way the error e_n is minimised.

To determine the optimum decoder for a given encoder the value of y_n is the conditional mean of x_n; given the set $L_n, L_{n-1}, \ldots, L_{n-m}$,

$$y_n = \langle [x_n | L_n, L_{n-1}, \ldots, L_{n-m}] \rangle \tag{8.3}$$

For simultaneous optimisation Equations 8.1 and 8.3 must both be satisfied. However, the solution of these equations is prohibitive in terms of computer time and therefore an iterative procedure is adopted. This procedure basically applies for an input signal having known stationary statistical properties and assumes an initial encoder and threshold level set as prescribed by the previous theory, enabling the $\{L_n\}$ sequence to be provided. Using Equation 8.3 the y_n values are determined giving minimum mean square error for the assumed encoder. The next step is to produce a slightly different encoder for the optimised decoder, and so on.

By this iterative procedure a situation is reached where a further iteration has a negligible effect and the encoder and decoder in this condition give the statistical d.m. system.

A general idea regarding the performance of statistical d.m. can be gained from the following conclusions when the input signal is a second-order Gaussian–Markov process. This process is formed by passing white noise through a filter having a transfer function $A/(1+T_m s)^2$ where A is selected so that the variance of $\{x_n\}$ is unity. T_m is the time constant of the process.

(1) The curve of signal-to-noise ratio as a function of memory length m for fixed $T_m f_p$, where f_p is the clock rate, initially rises rapidly with increasing m and then saturates. A compromise between the performance of the system and its complexity favours a value m of about 5.

(2) For fixed m the signal-to-noise ratio increases with $T_m f_p$ and eventually saturates. For $m = 5$ the value of f_p just prior to saturation is $40/T_m$.

(3) The effect of transmission errors degrades the signal-to-noise ratio to a greater extent for systems having large m, such that at high error rates there is little difference between systems having different memory lengths.

(4) Statistical d.m. while designed for optimum performance for a given input process and sampling rate, is not found to be markedly sensitive to changes in these parameters.

8.2.1. Mapping d.m.

Bertora[3] has proposed an adaptive delta modulation system which is conceptually similar to statistical d.m. The mapping d.m. encoder is shown in Figure 8.1. A short memory in the form of a shift register stores the last m bits of the binary output signal $L(t)$. These m bits, together with the current value of $L(t)$ are converted, i.e. mapped to a multilevel signal $c(t)$ which can have one of 2^{m+1} possible values. $c(t)$ is integrated to form the feedback signal $y(t)$.

The decoder consists of the local decoder, which is shown dotted in Figure 8.1, plus a band-pass filter whose critical frequencies are the same as the maximum and minimum frequencies in the $x(t)$ signal.

The encoder-decoder system is designed by iterative methods for an input signal having known stationary statistical properties using the criterion that the mean square error defined by Equation 8.2 is minimised.

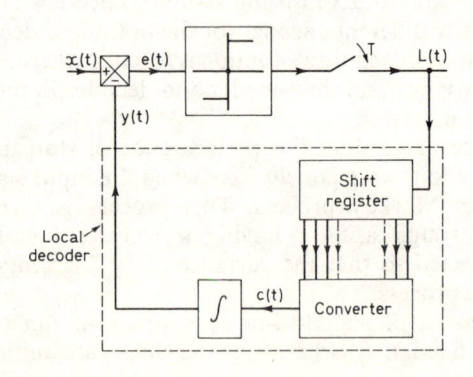

Fig. 8.1. Mapping d.m. encoder. (After F. Bertora[3])

Selecting $x(t)$ to closely resemble low-pass white noise, with a critical frequency of 3·6 kHz and mean square value $\sigma^2 = 1$, the iteration procedure commences with the initial encoder having a fixed step-size for near optimum s.q.n.r., a clock rate of 50 kbits/s and $m = 3$. After 13 iterations the $(m+1)$ possible binary words considered by the local decoder produce 16 different step-sizes in the waveform $c(t)$ which are close to optimum.
When compared to linear d.m. this system has a superior s.n.r. of 3 to 5 dB over the range of the input signal. This corresponds to an extension in the dynamic range of about 10 dB.

8.3. DISCRETE ADAPTIVE D.M.

This system was created by Abate.[4] The discrete d.m. encoder feeds back a range of step sizes depending on the polarity of the present and previous m bits of the transmitted binary signal $L(t)$. Suppose the minimum step-size is γ. Then in response to a series of all ones or zeros at the output of the encoder, the step height will change consecutively from γ say to

$$\gamma K_2, \gamma K_3, \gamma K_4, \ldots, \gamma K_m$$

where $K_2, K_3, K_4, \ldots, K_m$ are gain factors and

$$K_2 < K_3 < K_4 \cdots < K_m$$

The step size incrementally decreases when the error changes sign. The polarity of the feedback step is identical to the polarity of the $L(t)$ pulse. The block diagram of the discrete d.m. encoder is shown in Figure 8.2. The decoder is composed of the local decoder and a band-pass filter which removes the unwanted frequencies in $y(t)$. This system is similar to High Information d.m., which will be discussed in Section 8.4, when $K_2, K_3, K_4, \ldots$ becomes $1, 2, 4, \ldots$, respectively.

The evaluation of the performance[4] of discrete d.m. shows that like syllabically companded d.s.m. (described in Chapter 7) the maximum s.q.n.r. is almost the same as that obtained for linear d.m. The improvement in the dynamic range is determined by the final gain factor K_m which is closely analogous to the dynamic range of the multiplier in syllabically companded d.s.m. This statement is only applicable if a sufficient number m of K values are used and that they have suitable magnitudes; typically $m = 4$. The s.q.n.r. performance is the same irrespective of whether the input signal has a Gaussian or an exponential probability density function.

If the memory stores the last N bits in the waveform $L(t)$ and the sequence detector operates on those N bits and the current $L(t)$ pulse to select the appropriate gain factor, then $N + 1 = \log_2 m$, where m is the

228   Instantaneously companded d.m. systems

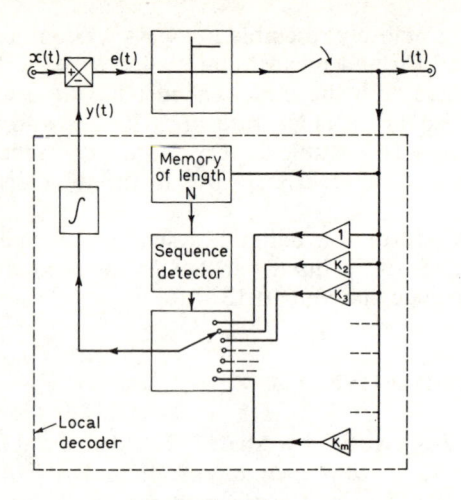

Fig. 8.2. Discrete adaptive d.m. encoder. (After Abate[4])

number of gain factors. The performance of the system is clearly dependent on the care taken in choosing the number and values of the gain factors. It seems that it should be possible to produce a higher maximum s.q.n.r. with this system than with linear d.m. This point is discussed in detail in Section 9.5.3.

8.4. HIGH INFORMATION D.M.

High information delta modulation, abbreviated h.i.d.m., was conceived by Winkler[5,6] for encoding video signals. These signals may make rapid transitions, e.g. from black to white, which cannot be satisfactorily encoded by a linear delta modulator due to its over-damped step response. The h.i.d.m. encoder overcomes this tracking deficiency and when presented with a step-input has a response, i.e. feedback signal, which is rather under-damped. Modulators to be described in subsequent sections will exhibit a nearly critical damped response. However, Winkler's system was probably the first significant delta modulation system for encoding television signals.

A possible arrangement for the h.i.d.m. encoder is shown in Figure 8.3. As usual, the forward path is the same as the one used in the linear delta modulator, i.e. after the error point there is a quantizer and a sampler. The adaptive local decoder in the feedback loop of the encoder inspects the binary output $L(t)$, and a simple logic system generates two binary signals, known as the 'down-pulses' and the 'up-pulses'. The 'down

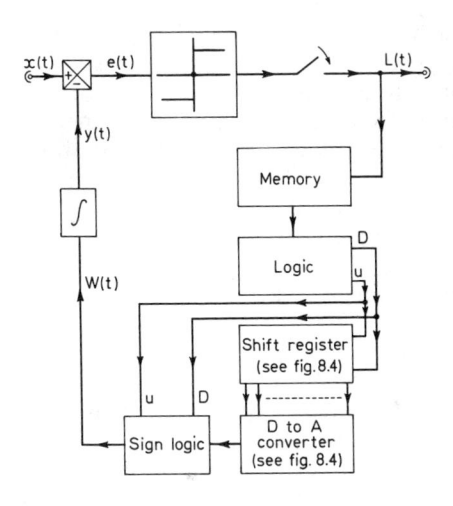

Fig. 8.3. *Possible arrangement of h.i.d.m. encoder.*

pulses' are generated every time the polarity of the $L(t)$ pulses changes. The 'up-pulses' are produced at the second and following $L(t)$ pulses after an $L(t)$ pulse has changed polarity. If the transmitted binary stream has a sequence 1 1 1 1 following a zero, the 'down-pulses' are 1 0 0 0 and the 'up-pulses' are 0 0 1 1.

A down pulse (D) is produced whenever the Boolean expression $\bar{L}_{r-1}L_r + L_{r-1}\bar{L}_r$ is satisfied, while an up-pulse is generated if $L_{r-2}L_{r-1}L_r + \bar{L}_{r-2}\bar{L}_{r-1}L_r$ occurs. However, neither a down nor an up-pulse is formed if the sequence of the output binary pulses satisfies $\bar{L}_{r-2}L_{r-1}L_r + L_{r-2}\bar{L}_{r-1}\bar{L}_r$; the bar above the symbol signifies a logical inversion of the symbol, and r refers to the rth sampling instant, $(r-1)$ to the previous sampling instant, and so on.

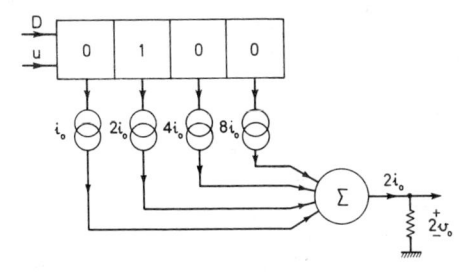

Fig. 8.4. *Shift register and digital to analogue converter in Fig. 8.3*

The adaptation of the feedback signal $y(t)$ to the error point can be formed by having a solitary 'one' in a shift register. This 'one' is moved forward, i.e. to the right in Figure 8.4, every time an up-pulse is generated, and backwards every time a zero is generated. If neither an up nor down pulse is generated the one does not change its location. Attached to the outputs of each stage in the shift register are binary weighted current generators which are turned-on by the presence of the solitary 'one'. If the 'one' is in the nth stage it is decoded into a current of magnitude $2^n i_0$, where i_0 is the basic unit of current, and thence to a voltage $2^n v_0$. Figure 8.4 shows that when the 'one' is in the second stage, $n = 1$ and the decoded voltage is $2v_0$.

This decoded binary weighted output is connected to a sign-logic unit which reverses the sign of this output *relative to its previous sign* only when a down pulse occurs. The output of the sign-logic unit is integrated to produce the signal $y(t)$ which is subtracted from the input signal $x(t)$ to produce the error signal $e(t)$. From the above remarks and Figures 8.3 and 8.4, it is apparent that successive 'up-pulses' increase $y(t)$ according to 2^n, i.e. 2, 4, 8, 16 etc. The condition[7] for the output to always decay to an oscillation of minimum step-size when the input is a constant, e.g. after the application of a step input, is that each weight be no greater than the sum of all the previous weights in the sequence. This is just fulfilled for this encoder.

8.4.1. Step input response

Figure 8.5 shows the step response $y(t)$ to a step $x(t)$ which starts at $+\frac{1}{2}$ unit and instantaneously increases to $19\frac{1}{2}$ units. When the step is at $+\frac{1}{2}$ unit the $y(t)$ signal oscillates between 0 and $+1$, as the smallest allowable incremental change in $y(t)$ is 1 unit. As $x(t)$ instantaneously increases the error signal equals $19\frac{1}{2}$ units causing the encoder to be totally overloaded; the resulting $L(t)$ waveform is a sequence of positive pulses.

These $L(t)$ pulses together with the up- and down-pulse sequences, and the decoded binary weighted signal $W(t)$ from the sign-logic unit, are also displayed in Figure 8.5. All these signals are shown for convenience as impulses. The weighting of $W(t)$ is proportional to ni_0, where n has a binary value 1 to 16.

Table 8.1 shows the position and movement of the 'one' in the shift register depending on whether an up, (U), down, (D), or no, (N), pulse was formed. The corresponding signals $W(t)$ and $y(t)$ for successive clock periods relating to Figure 8.5, are also shown.

The $y(t)$ signal in Figure 8.5 is observed to exceed the input step in 6 clock periods, and is tracking this input with its minimum step-size

Instantaneously companded d.m. systems

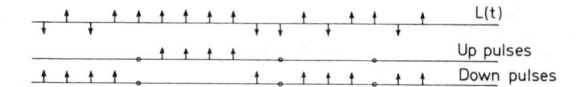

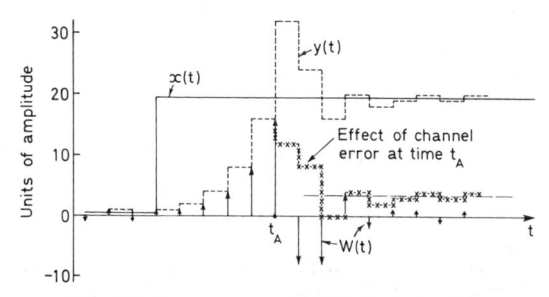

Fig. 8.5. *Step response of high information d.m. system.*

after 11 clock periods. By comparison, the linear delta modulator takes 20 sampling periods to exceed the input signal.

The maximum value which the $W(t)$ signal can assume is limited for circuitry reasons to W_M, say. When the encoder becomes overloaded it behaves as a linear delta modulator whose feedback impulses are W_M. Winkler[6] chose W_M to be 16 units.

Table 8.1

Position of 'one' in shift register					Logic output	$W(t)$	$y(t)$
1	0	0	0	0	D	−1	0
1	0	0	0	0	D	+1	1
1	0	0	0	0	D	−1	0
1	0	0	0	0	D	+1	1
1	0	0	0	0	N	+1	2
0	1	0	0	0	U	+2	4
0	0	1	0	0	U	+4	8
0	0	0	1	0	U	+8	16
0	0	0	0	1	U	+16	32
0	0	0	1	0	D	−8	24
0	0	0	1	0	N	−8	16
0	0	1	0	0	D	+4	20
0	1	0	0	0	D	−2	18
1	0	0	0	0	D	+1	19
1	0	0	0	0	N	+1	20
1	0	0	0	0	D	−1	19
1	0	0	0	0	D	+1	20

232   Instantaneously companded d.m. systems

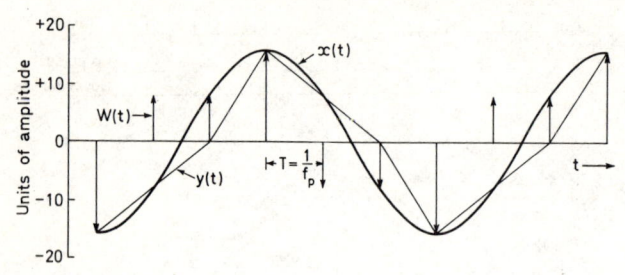

Fig. 8.6. *Response of high information d.m. system when on the verge of overload in the steady-state.*

If a sinusoid $E_s \sin 2\pi f_s t$ is applied to the encoder the latter will be on the verge of overload when

$$2\pi f_s E_S = W_M f_p \qquad (8.4)$$

Figure 8.6 shows the $y(t)$ responses when the input sinusoid has an amplitude equal to W_M and a frequency of $f_p/6$. The response has been drawn (as have those in Figure 8.7, 8.8 and 8.9) on the assumption that a sample and hold circuit has been used in the forward path of the encoder. This steady-state $y(t)$ waveform when the encoder is on the verge of overload can track a sinusoid whose amplitude is ten times greater than the amplitude which overloads a linear delta modulator.

Figure 8.7 is drawn for $E_s \doteq 2W_M = 30$ units and $f_p = 12f_s$, i.e. a near overload condition. The graphs start at the point 0 and are shown in cyclic form, i.e. Curve (1) is the first cycle of $y(t)$, Curve (2) the second cycle and so on. The curves show how the $y(t)$ signal rapidly increases until it

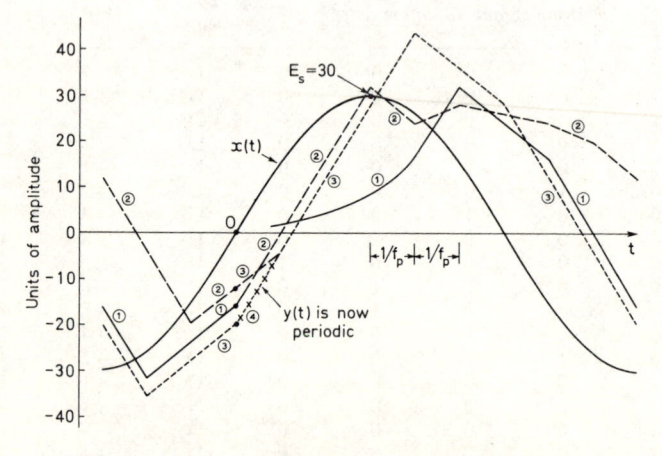

Fig. 8.7. *The first few cycles of $y(t)$ when the encoder is nearly overloaded, $W_M = 16$*

Instantaneously companded d.m. systems

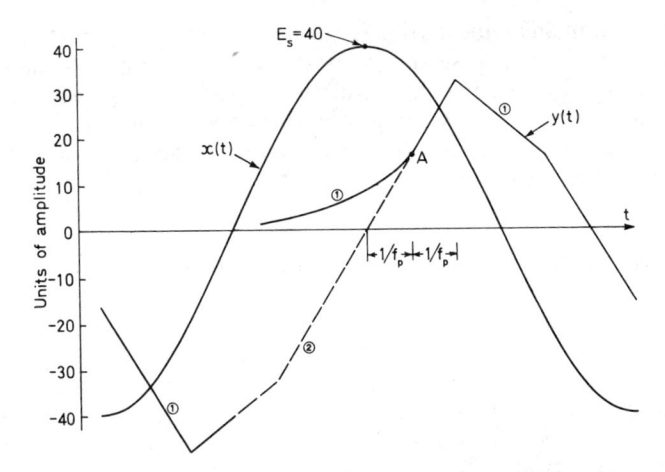

Fig. 8.8. *The y(t) waveform when grossly overloaded*, $W_M = 16$

exceeds $x(t)$, and after a few cycles the pattern

$$\tfrac{1}{2}W_M, \tfrac{1}{2}W_M, W_M, W_M, W_M, W_M, -\tfrac{1}{2}W_M, -\tfrac{1}{2}W_M, -W_M, -W_M, -W_M,$$
$$-W_M, \text{etc.}$$

emerges. Figure 8.8 shows a grossly overloaded condition where just after one cycle, at point A, the $y(t)$ waveform has become periodic with a constant phase difference relative to the sinusoid. In Section 4.2.1 we saw that a linear delta modulator when grossly overloaded develops a triangular $y(t)$ waveform.

8.4.2. Effect of transmission errors

With a linear delta modulator there is no hierarchy in the transmitted pulses. The effect of companding requires the decoder to operate on the received pulses in hierarchical manner (as in p.c.m.) such that some pulses are more important than others. Syllabically companded systems are less hierarchical than instantaneously companded ones. Making a political analogy, linear and instantaneously companded delta modulators can be thought of as transmitting pulses which are organised on a plebian and aristocratic basis, respectively.

If a digital error occurs in a linear delta modulator, $y(t)$ is in error by two units (see Section 5.2). If a digital error occurs in a h.i.d.m. system the error can be as little as two units or as great as $1.5\,W_M$. Figure 8.5 shows the step response, in the absence of errors, oscillating about $19\tfrac{1}{2}$ units. Should the decoder wrongly interpret the 9th $L(t)$ pulse from the left, then instead of the decoder generating $W(t) = W_M = 16$ units it produces

$W(t) = -4$ units. This means that the decoder produces a step of magnitude $3\frac{1}{2}$ units. This is of course the greatest error resulting from one transmission error. The error is accumulative if perfect integrators are used, but by employing a leaky integrator this effect disappears due to the zero and d.c. levels frequently encountered in speech and television encoding, respectively.

8.5. BOSWORTH–CANDY DELTA-SIGMA MODULATION

The so-called direct feedback encoder (Bosworth and Candy[8]) will be known here for convenience as the Bosworth–Candy delta-sigma modulator. The encoder and decoder, i.e. the codec, abbreviated B.C.d.s.m. is specifically designed for encoding picture-phone signals. It differs from h.i.d.m. in that its step response is more damped. Before describing its adaptation algorithm the block diagram shown in Figure 8.9 will be discussed.

The block $A(j2\pi f)$ can be produced by an RC integrator, where RC is much greater than the sampling period T. In the absence of $H_1(j2\pi f)$ and $H_2(j2\pi f)$ the system is a companded delta-*sigma* type. The local decoder is seen to consist of a sequence detector which inspects $L(t)$ and controls the amplitude weighting network, and therefore controls the amplitude of $y(t)$. The polarity of the multi-level signal $y(t)$ is identical to the polarity of the binary signal $L(t)$.

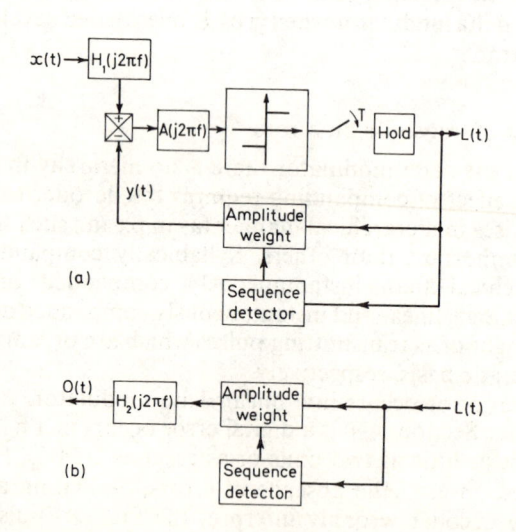

Fig. 8.9. Bosworth–Candy[8] delta-sigma modulation system: (a) encoder, (b) decoder

Instantaneously companded d.m. systems

When this encoder is used to encode picture-phone signals it is found subjectively that it is better to pre-emphasise the input $x(t)$ by the inclusion of the filter $H_1(j2\pi f)$. The high frequency components from the local decoder in the decoder are de-emphasised by the filter $H_2(j2\pi f)$. Filters $H_1(j2\pi f)$ and $H_2(j2\pi f)$ are nearly the inverse of each other and have short time constants compared to the filter $A(j2\pi f)$, i.e. the break frequency of $A(j2\pi f)$ is much lower than that for $H_1(j2\pi f)$ and $H_2(j2\pi f)$.

An advantage of having short time constants for $H_1(j2\pi f)$ and $H_2(j2\pi f)$ is that transmission errors have transient rather than accumulative effects on the recovered analogue signal $O(t)$.

Note that if $A(j2\pi f)$ is made the inverse of $H_1(j2\pi f)$ then the encoder becomes a companded delta-modulator, rather than a companded delta-sigma modulator with pre-emphasis.

8.5.1. Adaptation algorithm

When the output binary signal $L(t)$ has no more than two consecutive ones or zeros, the step size in the waveform $y(t)$ at the output of the amplitude weighting network is unchanged at its smallest value of I. Consequently the largest average values of $y(t)$ before the adaptation commences is $+I/3$ or $-I/3$ corresponding to $y(t)$ and $L(t)$ patterns of ...0 1 1 0 1 1 0 1 1... and ...0 0 1 0 0 1 0 0 1..., respectively.
When the $L(t)$ waveform has three or more consecutive ones or zeros the adaptation algorithm operates. The amplitude weighting network increases the step-size in $y(t)$ in response to a sequence of ones in the $L(t)$ waveform as follows

$$I, I, 2I, 3I, 5I, 5I \ldots 5I$$

Directly the first zero appears in $L(t)$ following the all 'one' sequence, the step in the $y(t)$ waveform changes from $5I$ to $-I$. When $L(t)$ has a sequence 0 0 0 0 ... 01 the corresponding $y(t)$ step-sizes are

$$-I, -I, -2I, -3I, -5I, -5I, \ldots, -5I, +I$$

The optimum weighting sequence was obtained from subjective tests on picture phone signals. It is $I, I, 2I, 3\cdot6I, 4\cdot7I, \ldots, 4\cdot7I$. However, for practical purposes the numbers presented above can be used.

8.5.2. Step and Gaussian inputs

Figure 8.10a(i) shows the response of the B.C.d.s.m. system to an arbitrary signal $x(t)$ which is chosen to have a step-like form. The curves have been drawn for convenience with $H_1(j2\pi f)$ approximating to the

236   Instantaneously companded d.m. systems

inverse of $A(j2\pi f)$. The $L(t)$ sequence is shown together with the generated weighting sequence and feedback waveform $y(t)$ and are displayed in Figure 8.10a(ii), a(iii) and a(i) respectively.

The response of h.i.d.m., Figure 8.10(b), and linear d.m., Figure 8.10(c) indicate that for step inputs the former is underdamped and the latter overdamped relative to B.C.d.s.m.

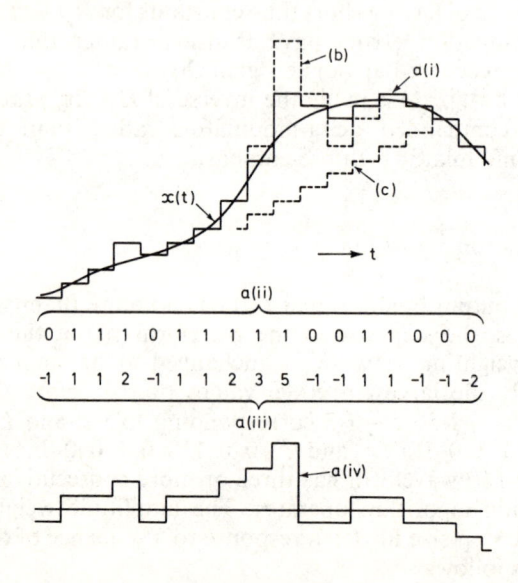

Fig. 8.10. Response of a B.C.d.s.m. system to an arbitrary input $x(t)$: (a) a (i) response, a (ii) $L(t)$, a (iii) weighting sequence, a (iv) $y(t)$; (b) response of h.i.d.m.; (c) response of linear d.m.

When Gaussian signals in the form of low-pass white noise are applied to the B.C.d.s.m. encoder the peak signal-to-noise ratio as a function of input signal power is approximately the same irrespective of the weighting sequence. When the maximum signal-to-noise ratio is 25 dB the dynamic range of the input signal (where the signal-to-noise ratio is equal to 20 dB) is approximately 12, 13 and 16 dB which corresponds to, an unweighted sequence, i.e. a linear d.m. codec, a weighting sequence of $I, I, 2I, 2I, \ldots, 2I$ and a weighting sequence of $I, I, 2I, 3I, 5I, \ldots, 5I$, respectively. The small improvement in the dynamic range for various weightings is because the encoder is designed for encoding video and not Gaussian signals.

However, when video signals are encoded the maximum s.q.n.r. is larger than the corresponding value in linear d.m.

8.5.3. Threshold control

Candy[7] modifies the local decoder to adjust the weighting of the steps in the waveform $y(t)$, to be

$$+I, +I, +8I, \ldots, +8I$$

in response to an output sequence of all logic ones. The appearance of the first logic zero at the output of the encoder causes the step-size in $y(t)$ to change to $-I$. When the output sequence is all zeros followed by a one the sequence of step-sizes in $y(t)$ is

$$-I, -I, -8I, \ldots, -8I, +I$$

This change in the encoding algorithm causes instability in the system as is demonstrated in Figure 8.11(a) where the response to an arbitrary input is displayed. This instability can be prevented by controlling the threshold level of the quantizer. The philosophy behind this control is best explained as follows. Consider a sampling instant nT where the error is $-I$. Without threshold control, this negative error will cause $L(nT)$ to be a logical zero and $y(t)$ will decrease by $8I$ as shown by Waveform (a). Now the weighting sequence related to Curve (a) corresponding to $(n-2)T$, $(n-1)T$ and nT is $-I$, $-I$ and $-8I$. After the first two consecutive $-I$ steps have occurred the encoder is restricted to choosing either $+I$ or $-8I$; the $-I, -I, -I$ sequence is not permitted because the decoder will act on the weighting sequence

$$\pm I, \pm I, \pm 8I, \ldots, \pm 8I, \mp I$$

However, out of the two possible magnitudes available at time nT, $+I$ is clearly going to produce a smaller error than $-8I$, and at the same time stabilise the response.

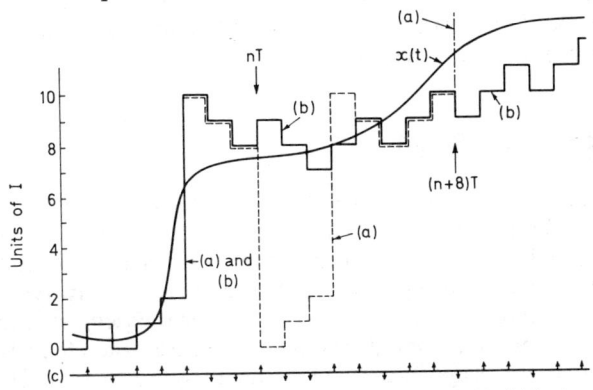

Fig. 8.11. Response of a B.C.d.s.m. system to an arbitrary input signal $x(t)$: (a) having a weighting sequence, 1, 1, 8; (b) having a weighting sequence, 1, 1, 8, and threshold control, and (c) with output pulses $L(t)$ corresponding to (b)

The selection of $+I$ rather than $-8I$ is achieved by changing the threshold level of the quantizer to the average of the possible weights, namely

$$(-8I+I)/2 = -(\tfrac{7}{2})I$$

The actual error $-I$ when compared to $-(\tfrac{7}{2})I$, rather than zero, is seen to be positive and as a result a positive pulse is generated at the output of the encoder. This causes the error to increase to $-2I$. If a zero had been generated the error would have increased to $-7I$. The solid Curve (b) in Figure 8.11 shows the effect of threshold control. Subsequent to nT the variations in Curve (b) is never greater than 1 unit per clock time until $(n+8)T$ when another critical decision must be made. During these 8 clock pulses the threshold level remains at zero as the choice is always $\pm I$, which has an average of zero. At $(n+8)T$ the encoder, without threshold control, rises by $8I$ although the error is $+(\tfrac{3}{2})I$. The encoder with threshold control averages the possible steps $8I$ and $-I$ to give $+(\tfrac{7}{2})I$, and sets the threshold to $+(\tfrac{7}{2})I$. As $e(t) = (\tfrac{3}{2})I$ ($<(\tfrac{7}{2})I$) the encoder generates a logical zero, i.e. negative pulse, and Curve (b) reduces by one unit.

Figure 8.11(c) shows the output pulses $L(t)$ for the encoder with threshold control. At the decoder these pulses will be passed to the local decoder which is identical to that for the $I, I, 8I$, weighting sequence encoder *without* threshold control. The pulse sequence of Figure 8.11(c) will clearly be interpreted by this logic to give Curve (b). Thus by introducing threshold control in the encoder without changing the decoder the response of the system can be significantly improved.

Threshold control can be implemented by continuously examining the last two bits in the $L(t)$ sequence and producing the next two possible steps Y_1 and Y_2 in the $y(t)$ waveform. The threshold voltage V_t in the quantizer is now made equal to $(Y_1 + Y_2)/2$; if $e(t) \geq V_t$ a logical one is produced and the more positive value of Y_1 and Y_2 is selected and fedback to the error point, and vice versa if $e(t) < V_t$.

8.6. PULSE GROUP D.M.

A delta modulator in its idle channel state has an output binary waveform $L(t)$ composed of alternate polarity pulses, i.e. a logical pattern of ... 1 0 1 0 1 0 .... When a positive ramp signal is applied to the encoder, this pattern changes such that occasionally two consecutive 'ones' occur in the ... 1 0 1 0 1 0 ... pattern. By continuing to increase the slope of the input signal $x(t)$, groups of three, then four, then five, etc. consecutive groups of 'ones' will occur until the encoder becomes overloaded and the output pattern is a series of all 'ones'. Generally for any particular slope of $x(t)$ the output pattern will contain groups of all 'ones' and all 'zeros' such that their average value is equal to the constant slope

Instantaneously companded d.m. systems

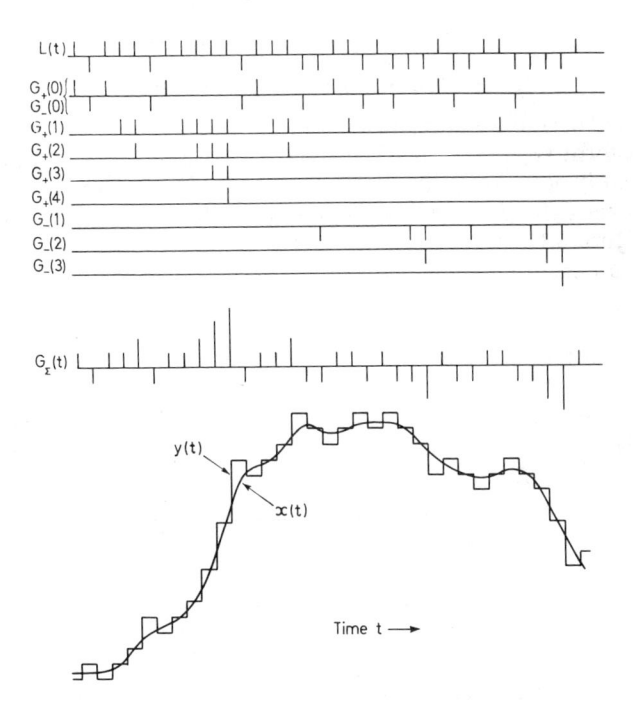

Fig. 8.12. Pulse group d.m. waveforms corresponding to an arbitrary input x(t)

of $x(t)$. Suppose that $L(t)$ has the pattern shown in Figure 8.12. The patterns corresponding to signals $G_+(1)$, $G_+(2)$, $G_+(3)$, etc., occur when 2, 3, 4, ... consecutive ones appear in $L(t)$, respectively. The $G_+(0)$ signal occurs when the $L(t)$ pulse changes polarity from negative to positive, and vice versa for $G_-(0)$. A negative subscript associated with the G signal implies a number of consecutive zeros, i.e. $G_-(2)$ means three consecutive zeros in $L(t)$.

Computer simulation results indicate that when the slope of $x(t)$ is increased to a value where a particular group occurs, then a further increase in the slope results in a linear increase in the frequency of occurrence of this group. This information has been used by Flood and Hawksford[9] to design an adaptive delta-sigma modulator. The delta modulator version is considered here and the modulation is called pulse group d.m.

The local decoder in the feedback loop of the pulse group d.m. operates on the output binary signal $L(t)$ and forms the group signals

$$G_+(0), G_+(1), G_+(2), \ldots, G_+(m)$$

and

$$G_-(0), G_-(1), G_-(2), \ldots, G_-(m)$$

with the aid of simple gates and an m-bit memory. In practice it is unnecessary[9] to form $G_+(0)$, $G_+(1)$, $G_-(0)$ and $G_-(1)$ as their sum equals $L(t)$; consequently only groups involving $m \geq 2$ need be produced and added to $L(t)$ to form $G_\Sigma(t)$. The feedback signal $y(t)$ is then formed by integrating $G_\Sigma(t)$. Figure 8.12 shows the waveforms produced in the encoder when tracking an arbitrary input signal $x(t)$.

8.6.1. Step input response

The step input response is shown in Figure 8.13 for a step which rises instantaneously from $\frac{1}{2}$ to $19\frac{1}{2}$ units. When compared to the response of the

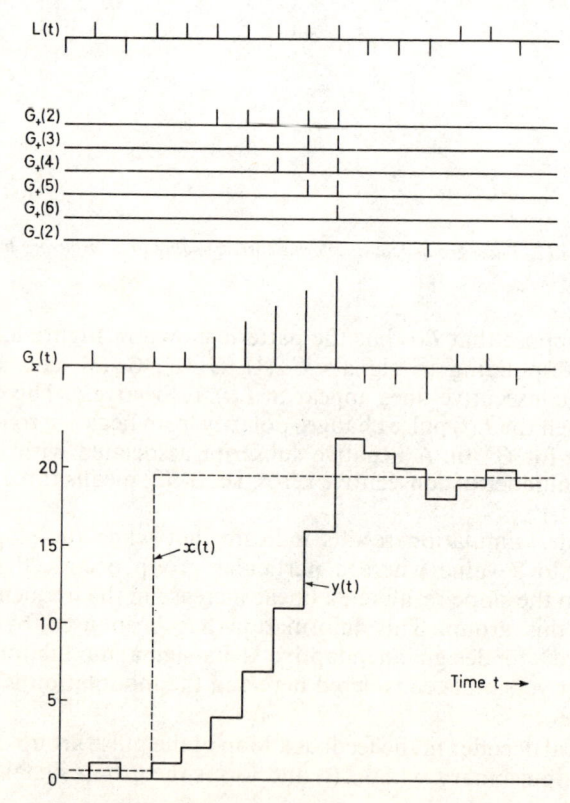

Fig. 8.13. Response of pulse group d.m. system to a step input of amplitude 0·5 to 19·5 units

h.i.d.m. encoder shown in Figure 8.5 the pulse group delta modulator has more damping, i.e. it does not respond so quickly but there is less overshoot. Pulse group d.m. was designed for television encoding.

8.7. COMPANDED DOUBLE INTEGRATION D.M.

Double integration d.m. has been discussed in detail in Section 2.2. It was found essential to use prediction if the oscillatory modes are to be damped. If the prediction is either too large or too small then the quantization noise may not be minimised. An obvious improvement in the performance of the encoder can be achieved by either making the prediction adapt to minimise the mean square error, or by using a non-linear integrator.

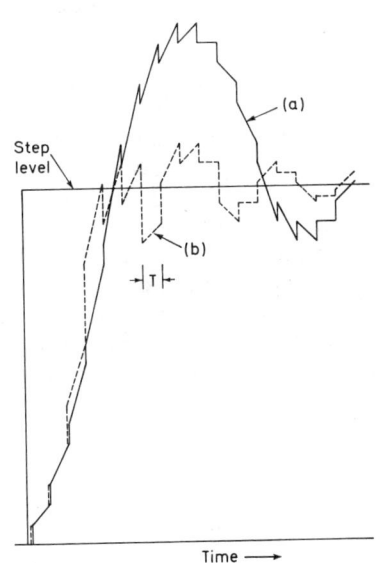

Fig. 8.14. Step response of double integration system: (a) fixed prediction; (b) adaptive prediction

Figure 8.14(a) shows the response of a double integration delta modulator to an arbitrary step input for a prediction time equal to the clock period. Figure 8.14(b) shows the response to the same step input when the prediction time is made into a variable. The algorithm controlling the vertical steps in this response correspond to those used in h.i.d.m., although any of the algorithms discussed in this chapter would probably improve the non-adaptive response. The adaptive response is faster and has less overshoot. However, for other step inputs the response of the

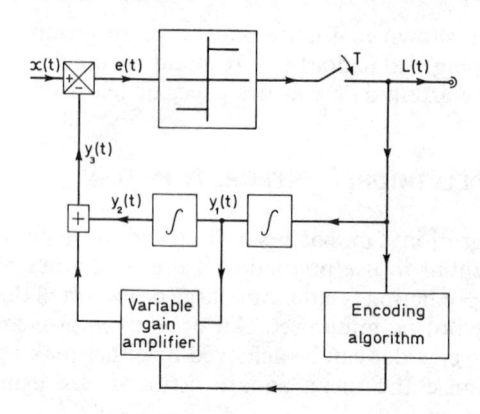

Fig. 8.15. Non-linear double integration d.m. encoder

adaptive encoder may exhibit large swings due to the nature of the h.i.d.m. algorithm and the effect of using two integrators in the feedback path. The block diagram of the encoder is shown in Figure 8.15. The amount of $y_1(t)$ added directly to $y_2(t)$ is determined by the gain of the amplifier which is in turn controlled by the encoding algorithm.

An alternative arrangement is to use the algorithm to control the magnitude of $y_1(t)$ which will alter both the step height and the slope of $y_3(t)$. This has been done by Childs[14] who replaced the single integrator in the first order c.f.d.m. encoder shown in Figure 8.18 by the double integration and prediction process illustrated in Figure 2.8. This action resulted in the encoder having a faster response to step inputs.

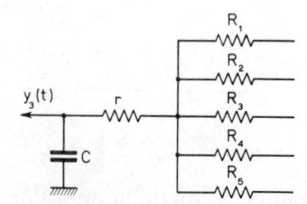

Fig. 8.16. A choice of integrators in the local decoder

Kikkert[10] has used an encoding algorithm to determine which resistor (Figure 8.16) should be connected to the $+E$ or $-E$ supply in order to to adapt $y_3(t)$ to the input signal. This arrangement has effectively five second integrators, one corresponding to each resistor R which controls the charging current. The resistor r achieves the prediction as described in Section 2.2.5.

Instantaneously companded d.m. systems

8.8. DIGITAL DELTA MODULATION

A low cost version of an adaptive encoder employs a linear delta modulator to produce a binary signal for a digital delta modulator. The arrangement is shown in Figure 8.17.

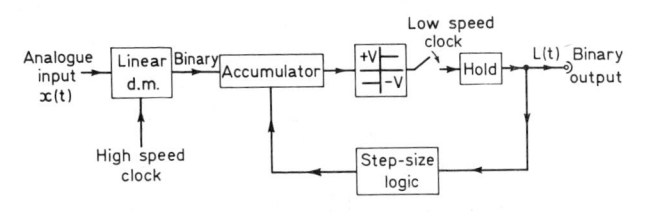

Fig. 8.17. Digital delta modulator (after Goodman[11])

The linear delta modulator operates at a high clock rate and low step-size thereby minimising the quantization noise and allowing a wide range of input slopes to be encoded without causing slope overload. The function of this linear delta modulator is to present to the digital delta modulator a binary input which is a faithful binary representation of the analogue input signal $x(t)$.

The digital delta modulator is completely digital and can therefore be produced in an economic form. The transmitted binary signal $L(t)$ is also fed back to the accumulator via the step-size logic. This logic can take a variety of forms but Goodman[11] uses the adaptation logic in the first order c.f. delta modulator, described in the next section. The accumulator is modified at each sampling instant by either ± 1 from the linear encoder and by the digital number at the output of the step-size logic. The advantage of this scheme is that only one step-size need be accurately controlled by analogue hardware, i.e. in the linear delta modulator. The step-size variation in the adaptive section is exact due to the digital implementation.

Typically for a line bit rate of 64 kbits/s the linear delta modulator is clocked at 8·192 Mbits/s, and the decoded signal-to-noise ratio for a speech signal is about 32 dB.

An all digital linear delta modulator is described in Section 11.8 in connection with digital filters.

8.9. FIRST ORDER CONSTANT FACTOR D.M.

This section describes an adaptive delta modulator having a one bit memory whose voltages fedback to the error point are discrete and

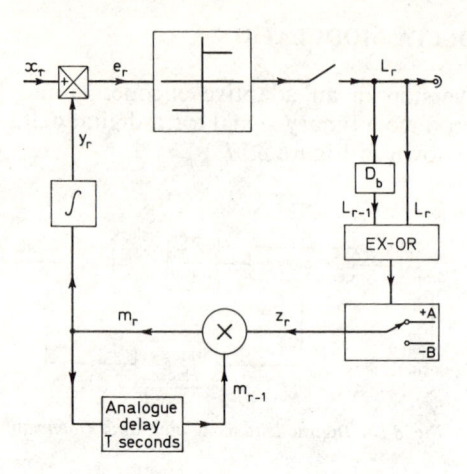

Fig. 8.18. 1st order constant factor d.m. encoder. (After Jayant[12])

multi-valued. The encoder responds to the instantaneous variations in the analogue signal and is suitable for encoding both speech and television signals.

This encoder, invented by Jayant[12,13] will be known here as a first order constant factor delta modulator, and its encoder-decoder system is abbreviated to first order c.f.d.m. The reason for this title will be apparent in due course. The encoder is shown in Figure 8.18. The output binary waveform $L(t)$ is of course transmitted via suitable processing (e.g. filtering, modulation etc.) to the receiver. It is also connected to the adaptation logic in the feedback network. This logic has a one bit digital delay D_b so that it can inspect the binary level of the $L(t)$ signal at say sampling interval r and also at the previous sampling interval $(r-1)$. Let the corresponding values of $L(t)$ be L_r and L_{r-1}, respectively.

If L_r and L_{r-1} are both ones or zeros it indicates that the error has not changed sign for two successive clock periods and some increase in the size of the step fed back to the subtractor is warranted. On the other hand if L_r and L_{r-1} are different the error has changed polarity between successive clock periods. This suggests that the step change in $y(t)$ should be reduced in magnitude.

The Exclusive-OR in Figure 8.18 has a logic one output when L_r and L_{r-1} are different, and a logic zero when they are the same, i.e. either both ones or both zeros. Suppose at a sampling instant the output of the Exclusive-OR is a logic 'one' and the switch is moved to the negative voltage source. If the output from the Exclusive-OR is a logic zero the switch connects the positive voltage to the input of the multiplier. This input $z(t)$ consists of impulses of strength $-B$ or $+A$ depending on whether the

Instantaneously companded d.m. systems

switch is connected to the negative or positive voltage source. Impulses are considered in order to simplify the description and analysis.

At a particular sampling instant r, $z(t) = z_r$ where

$$z_r \begin{cases} = \text{impulse } A, & \text{if } L_r = L_{r-1} \\ = \text{impulse } -B, & \text{if } L_r \neq L_{r-1} \end{cases}$$

The final waveform $y(t)$ fedback to the error point is the integral of $m(t)$ where $m(t)$ is the waveform at the output of the multiplier. At the rth sampling instant $m(t)$ is m_r and is formed by multiplying the previous value of m_r by z_r

$$m_r = z_r m_{r-1} \tag{8.4}$$

The production of m_{r-1} requires the use of an analogue delay of one bit duration. As m_{r-1} can occupy a wide range of values there are practical difficulties in producing an analogue delay having a wide voltage dynamic range.

z_r is an impulse and consequently m_r is also an impulse with the result that the waveform $y(t)$ contains steps of varying height. In particular, if y_r is the value of $y(t)$ at the rth sampling instant, then

$$y_r = y_{r-1} + m_r \tag{8.5}$$

This is because m_r is integrated to give the new value of $y(t)$, namely y_r. The integral of an impulse of strength m_r is a step of height m_r.

Substituting for m_r from Equation 8.4 in Equation 8.5 gives

$$\Delta y_r = y_r - y_{r-1} = z_r m_{r-1} \tag{8.6}$$

As z_r is either A or $-B$, the ratio between two successive steps in the feedback waveform differ by a constant factor z_r. It is because of this important property that this type of encoder is called here a constant factor delta modulator. The name distinguishes it from other delta modulator encoders which have short memories in the feedback loop and are of the instantaneous adaptive type.

The local decoder therefore operates on L_r and L_{r-1} to produce changes in y_r which attempt to follow x_r and minimise e_r. The forward path of the encoder undertakes the usual task of producing an output impulse L_r whose polarity is identical to the polarity of e_r at the sampling instant.

The decoder has the usual arrangement of a local decoder followed by a low pass filter to remove unwanted noise.

8.9.1. Overload condition

The impulses $m(t)$ applied to the integrator change strength by a constant factor of either A or $-B$ at every clock instant, hence the adaptation is instantaneous. This is illustrated by considering the situation when the encoder is suddenly overloaded by a positive step input which results in $L(t)$ having a sequence of all 'ones', i.e. $L_r = L_{r-1}$, over a number of clock periods. Suppose that at the instant when overload occurs the impulse at the output of the multiplier has a value of one unit. Because the logical output waveform is a sequence of all 'ones', a corresponding sequence of A adaptations occur and the impulse sequence applied to the integrator increases as $A, A^2, A^3, \ldots, A^n$ units. The feedback voltage at the output of the integrator sums the impulses applied to its input such that after n A-type adaptations, the feedback voltage has increased to

$$y_n = \sum_{r=0}^{n} A^r \qquad (8.7)$$

If the overload condition is caused by a negative step, the output binary sequence is all 'zeros', but as L_r still equals L_{r-1} in this sequence, the adaptation remains the A-type and y_n is

$$y_n = -\sum_{r=0}^{n} A^r$$

assuming that the initial impulse to the integrator is -1 unit.

Adaptation constants

In order for the sequence of A-adaptations to result in the increase of the feedback signal $y(t)$ the value of A must be greater than unity. In this way the impulses A^r applied to the integrator increase their strength with the increasing number r of clock periods. When $y(t)$ exceeds the level of the input signal $x(t)$ there will be a change in the sign of the error signal, $L_r \neq L_{r-1}$, and a B adaptation will be produced. The encoder is now required to generate a smaller impulse at the input to the integrator in order to reduce the magnitude of the error signal at the next clock instant. This reduction is achieved by making $B < 1$. It will be demonstrated in the next section that in order for $y(t)$ to converge to $x(t)$, where $y(t)$ step variations about $x(t)$ will decay to minimum amplitudes, the product AB must be less than unity. However, for a band-limited random input signal the optimum product AB will be close to unity. Observe that if $A = B = 1$, the encoder is a linear delta modulator.

8.9.2. Step input response

Consider the application of a step input to the 1st order c.f.d.m. encoder having an amplitude of 9 units. Suppose that the initial step in the $y(t)$

Instantaneously companded d.m. systems

waveform is one unit then the behaviour of the various signals in the encoder at successive sampling instants $r = 0, 1, 2, \ldots$ is shown in Table 8.2. The values of y_r and m_r are given by Equations 8.5 and 8.4 respectively.

The value of y_r is calculated for the adaptation constants having the same values as those used by Jayant,[12] namely $A = 1.5$ and $B = 0.66$. The response $y(t)$ is shown in Figure 8.19(a) where it can be seen that $y(t)$ hunts about the step input $x(t)$ with a ... 1 1 0 0 1 1 0 0 ... pattern. The amplitude of this hunting pattern is large compared to the small step sizes which $y(t)$ can have on different occasions. The $y(t)$ signal decays very slowly: if $A = 1/B$ then no decay occurs.

Consider Curve (a) in Figure 8.19. Applying the previous equations, and letting $m_r = AM$ when $y_{r-1} = H_1$, the values of m_r and y_r for various values of r are given in Table 8.3.

The hunting pattern shown in Figure 8.19 will be periodic and will not decrease with time if $H_5 = H_1$. As

$$H_5 = H_1 + AM + A^2M - BA^2M - BA^3M \tag{8.8}$$

putting $BA = 1$ yields $H_s = H_1$. Thus for unconditional instability with this particular step input, $BA = 1$.

The range of step magnitudes X which will cause hunting in this example are

$$H_2 < X < H_4$$

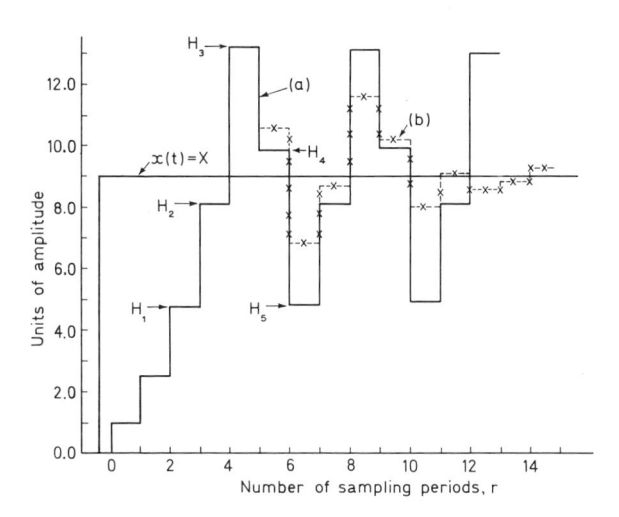

Fig. 8.19. *Step response of a 1st order c.f.d.m. codec: (a) $y(t)$ for $A = 1.5$, $B = 0.66$, $AB < 1$ (just); (b) $y(t)$ for $A = 1.5$, $B = 0.5$, $AB = 0.75$*

Table 8.2

r	L_r	L_{r-1}	z_r	$m_r = z_r m_{r-1}$	$y_r = m_r + y_{r-1}$	y_r $A=1.5$ $B=0.66$	$A=1.5$ $B=0.5$
0	1			1	1	1.0	1.0
1	1	1	A	$A \times 1 = A$	$1 + A$	2.5	2.5
2	1	1	A	$A \times A = A^2$	$1 + A + A^2$	4.74	4.75
3	1	1	A	$A \times A^2 = A^3$	$1 + A + A^2 + A^3$	8.12	8.12
4	1	1	A	$A \times A^3 = A^4$	$\sum_{n=0}^{4} A^n$	13.19	13.19
5	0	1	$-B$	$(-B) \times A^4 = -BA^4$	$\sum_{n=0}^{4} A^n - BA^4$	9.84	10.66
6	0	0	A	$A(-BA^4) = -BA^5$	$\sum_{n=0}^{4} A^n - BA^4(1+A)$	4.83	6.86
7	1	0	$-B$	$(-B)(-BA^5) = +B^2A^5$	$\sum_{n=0}^{4} A^n - BA^4 - BA^5 + B^2A^5$	8.14	8.76
8	1	1	A	$A \times B^2A^5 = B^2A^6$	$\sum_{n=0}^{4} A^n - BA^4 - BA^5 + B^2A^5 + B^2A^6$	13.10	11.61

Instantaneously companded d.m. systems

Table 8.3

r	$m_r = z_r m_{r-1}$	$y_r = y_{r-1} + m_r$
2	AM	$H_2 = H_1 + AM$
3	A^2M	$H_3 = H_1 + AM + A^2M$
4	$-BA^2M$	$H_4 = H_1 + AM + A^2M - BA^2M$
5	$-BA^3M$	$H_5 = H_1 + AM + A^2M - BA^2M - BA^3M$

From Table 8.3

$$H_1 + AM < X < H_1 + AM + A^2M - BA^2M$$

From Table 8.2, in this particular example, AM is the value of m_r when $r = 2$, i.e. $M = A^2$. Also

$$H_1 = \sum_{r=0}^{2} A^r$$

$$\sum_{r=0}^{3} A^r < X < \sum_{r=0}^{4} A^r - BA^4$$

Providing $BA = 1$

$$H_3 - H_2 = H_4 - H_5 = A^2M$$

Different input size step sizes still produce a ... 1 1 0 0 1 1 0 0 1 1 0 0 ... hunting pattern if $AB = 1$. The actual size of the hunting amplitudes change with the step level. In the example given above, following the first overshoot the error changed sign after 5 sampling periods. If following an overshoot the error changed sign at the next sampling instant the hunting amplitudes are comparatively reduced, but once the ... 1 1 0 0 1 1 0 0 ... pattern is established they do not decrease with time.

A requirement for the hunting amplitudes to decay is $BA < 1$. In the previous example it was $H_5 > H_1$, and from Equation 8.8 this gives

$$AM + A^2M - BA^2M - BA^3M > 0, \quad \text{or} \quad BA < 1$$

Curve (b) in Figure 8.19 shows the behaviour of the encoder to the same step when $A = 1.5$ and $B = 0.5$, i.e. $BA = 0.75$. The difference between $y(t)$ and $x(t)$ is seen to rapidly decay with time.

8.9.3. ENCODER WITH MINIMUM STEP SIZE

In order for the hunting waveform $y(t)$ shown in Figure 8.19(a) to decay in magnitude as a function of time in the manner shown in Figure 8.19(b) the product of the adaptation constants AB must be less than unity. As the magnitude of these oscillations about the step input level X become smaller a condition will occur where the minimum value of the impulse at the output of the multiplier produces the smallest magnitude change G

in $y(t)$. In other words the decay in the hunting amplitudes cannot proceed indefinitely and these amplitudes have a minimum value.

It will be shown in the next section that when a random input signal $x(t)$ is applied to the encoder, the optimum value of the product AB is unity, a condition which is sub-optimum when $x(t)$ is a step-input. If the encoder has to accommodate both of these types of input signals then a compromise choice for the product AB is a value which is less than but close to unity. For this value of AB the $y(t)$ waveform will decay relatively slowly as it 'homes-in' on an input step of magnitude X; the output binary pattern will be a sequence of pairs of two 'ones' followed by two 'zeros'. Given this encoding condition suppose that if the multiplier output had no minimum output level and a second consecutive zero is generated at the output of the encoder the resulting A adaptation decreases the magnitude of $y(t)$ to

$$y_{r+1} = y_r - AF$$

where $-F$ is the impulse applied to the integrator at the previous sampling instant r. However, because the signal level at the output of the multiplier cannot fall below a minimum value y_{r+1} is forced to take a smaller value, $y_r - G$, than that given above; $G > AF$ is the smallest practical step size in $y(t)$. As $y_{r+1} < X$ due to the generation of two consecutive logical ones at the output of the encoder a B adaptation occurs. However, y_{r+2} cannot increase to $y_{r+1} + BG$ because the magnitude of BG is less than the minimum step-size G, since $B < 1$. Consequently

$$y_{r+2} = y_{r+1} + G$$

But

$$y_{r+1} = y_r - G$$

therefore

$$y_{r+2} = y_r$$

If the difference between X and y_r, i.e. the error at the rth sampling instant is $-k$, then the error at the $(r-2)$th sampling instant is also $-k$ and the same is true at all even clock times. At odd sampling instants the error is positive and has a magnitude of $G - k$. Consequently the output binary waveform consists of alternate binary levels and the $y(t)$ signal has a square-wave oscillation of peak-to-peak value G asymmetrically located about the imput level X, except when $k = G/2$.

8.9.4. Adaptation constants

If the encoder is constrained to idle with a ... 1 0 1 0 1 0 ... output binary pattern all the adaptations are of the B type. Consequently, $P_B = 1$

Instantaneously companded d.m. systems   251

and $P_A = 0$, where P_B and P_A are the probabilities of B and A adaptations, respectively.
When the encoder is experiencing a partial overload then a sequence of all zeros will occur at the output of the encoder. During this sequence, $P_A = 1$ and $P_B = 0$.
Consider the case[12] when the band-limited input signal $x(t)$ has random properties, and a zero mean value. At sampling time r the step-size applied to the integrator is m_r. Suppose that during the following 5 sampling instants the logic progressively produces A, A, B, A and B adaptations. The corresponding values of $m(t)$ are Am_r, $A^2 m_r$, $-A^2 B m_r$, $-A^3 B m_r$, $A^3 B^2 m_r$, i.e.

$$m_{r+s} = A^3 B^2 m_r$$

During the N sampling instants, NP_A is the number of A adaptations and NP_B is the number of B adaptations made by the logic. Hence from the above example the magnitude of m_{r+n} is given by

$$|m_{r+N}| = A^{NP_A} \cdot B^{NP_B} \cdot |m_r| = (AB^{P_B/P_A})^{NP_A} \cdot |m_r|$$

For optimal adaptation

$$P_A = P_B = P = 0.5, \quad \text{as} \quad N \to \infty \qquad (8.9)$$

For this condition

$$\underset{N \to \infty}{\text{Limit}} \left| \frac{m_{r+N}}{m_r} \right| \qquad \underset{N \to \infty}{\text{Limit}} \{(AB)^{NP}\}$$

This limit approaches infinity if $AB > 1$ and zero if $AB < 1$. However if $AB = 1$

$$\underset{N \to \infty}{\text{Limit}} \left| \frac{m_{r+N}}{m_r} \right| = 1$$

The condition given by

$$AB = 1 \qquad (8.10)$$

is optimum assuming that P_A and P_B both approach 0.5 as N approaches infinity.
Observe that although $AB = 1$ is optimum for random signals and suitable when the codec is handling speech and television signals, it does produce the hunting patterns described in Section 8.9.2 when a step input is applied to the encoder.
Jayant[12] has shown that the bounds on the adaptation constants A and B in Equation 8.10 are

$$1 < A < 2$$
$$0.5 < B < 1$$

8.10. SECOND ORDER CONSTANT FACTOR D.M.

In the first order c.f.d.m. system just described the changes in the stepsizes of the waveform $m(t)$ fedback to the integrator were based on the present and previous bits at the output of the encoder.

Considerable improvement in the performance of this system can be obtained by controlling $m(t)$ according to an adaptation rule[15] based on the present and previous *two* bits in the output binary waveform, $L(t)$. Defining these bits as L_r, L_{r-1} and L_{r-2} respectively, where r indicates the rth sampling interval, then the 8 possible binary patterns obtained are displayed in Table 8.4.

Table 8.4

L_r	L_{r-1}	L_{r-2}	Group number	Name of group	Adaptation constant
0	1	0	1	Alternate polarity	A_1
1	0	1			
0	1	1	2	Sign reversal	A_2
1	0	0			
0	0	1	3	Semi-overload	A_3
1	1	0			
0	0	0	4	Overload	A_4
1	1	1			

Group 1 is called the alternate polarity group and occurs when the encoder is idling. Group 2, the sign reversal group, indicates that the encoder is coming out of an overload condition, whereas Group 3, the semi-overload group, indicates that overloading has commenced. The final group consisting of three consecutive 'ones' or 'zeros' occurs when the encoder is experiencing slope overload.

The block diagram is the same as shown in Figure 8.18, except the z_r is formed with the aid of a two-bit shift register and some combinational logic according to the equation

$$z_r = A_r \, \text{sgn}\,(L_r)$$

where A_r is an adaptation constant shown in Table 8.4, depending on the logical values of L_r, L_{r-1} and L_{r-2}.

Instantaneously companded d.m. systems

8.10.1. Selection of adaptation constants

Rather than apply complicated techniques to find the constants A_1, A_2, A_3 and A_4 a simple approach gives satisfactory results. This involves selecting an arbitrary value $A_4 = 2$. As A_4 is the adaptation constant when overloading occurs then clearly it must be greater than 1 so that the step-sizes fed to the integrator increase with each sampling interval until the overloading condition ceases. By selecting $A_4 = 2$ the encoder when suddenly overloaded responds in a similar way to the h.i.d.m. encoder until the error changes sign.

The values of the remaining three adaptation constants are calculated using computer simulation. The signal-to-noise ratios versus A_1 are computed for sinusoidal inputs with A_2, A_3 and A_4 constant. The computations are repeated with A_2 and A_3 as variables; for each computation the other three A values are held constant.

The final choice of parameters is $A_1 = 0.9$, $A_2 = 0.4$, $A_3 = 1.5$ and $A_4 = 2.0$. These parameters apply to $f_p/f_{c2} = 13$, and although chosen for sinusoidal input signals they do enable a satisfactory encoding performance to be achieved when Gaussian signals are applied to the encoder. For higher ratios the adaptation values selected are closer to unity than the values given above.

8.10.2. Step input response

Suppose that in response to a d.c. level of $+0.5$ units the encoder generates a $y(t)$ waveform which oscillates about this level with values $+1$, 0, $+1$, 0, etc. so that A_1 adaptations are continually generated. When this level suddenly increases to $+39.5$ units, as shown in Figure 8.20(a), the encoder responds by generating an A_3 adaptation; $m(t) = IA_3$ where I is the value of m_{r-1}, i.e. the value of $m(t)$ just prior to the rapid rise in the input signal. At the next clock instant all three $L(t)$ digits inspected are logical ones; accordingly adaptation constant $+A_4$ is generated and $m(t)$ is increased to $A_3 A_4 I$. The A_4 adaptation constant is reproduced for the next three clock periods, increasing $m(t)$ each time by a factor A_4 and raising $y(t)$, which is the integral of $m(t)$, to

$$y(t) = I\{1 + A_3 + A_3 \sum_{j=1}^{4} A_4^j\} = y_5$$

In Figure 8.20(a), I is unity and this value of $y(t)$ is in excess of 39.5. At the next clock instant the output binary pattern is a 0 1 1 and an A_2 adaptation is generated reducing $y(t)$ to

$$y_5 - A_3 A_4^4 A_2$$

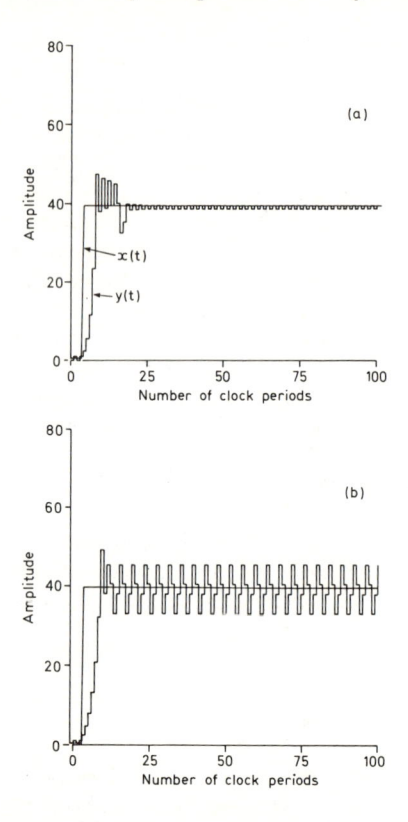

Fig. 8.20. Step input response of c.f.d.m. codecs: (a) 2nd order; (b) 1st order. The initial step-size of y(t) is unity

The output pattern changes to 1 0 1, A_1, is generated and $y(t)$ becomes

$$y_5 - A_3 A_4^4 A_2 + A_3 A_4^4 A_2 A_1$$

and so on.

8.10.3. Comparison between first order and second order constant factor systems[16]

Step inputs

Figure 8.20 shows the behaviour of the $y(t)$ signal for first and second order c.f.d.m. systems when a step input of 0·5 to 39·5 units is applied. The $y(t)$ signal in the second order c.f.d.m. encoder overtakes the step input before the first order one does. The response of first order c.f.d.m. system is shown for adaptation constants $A = 1.5$, $B = 1/1.5$–the hunting oscillations do not decay. Damping can be achieved with this system if

$$AB < 1 : A > 1, B < 1$$

However, if AB is considerably reduced, below unity, the signal-to-noise ratio is degraded when tracking Gaussian or speech signals. Consequently there is a conflict in selecting this product. No such serious conflict exists with second order c.f.d.m. codec. The response following the catching of

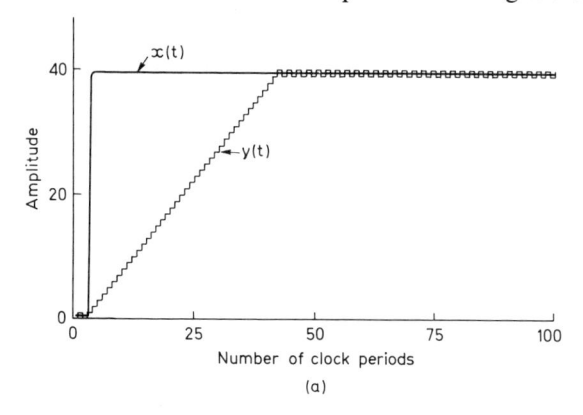

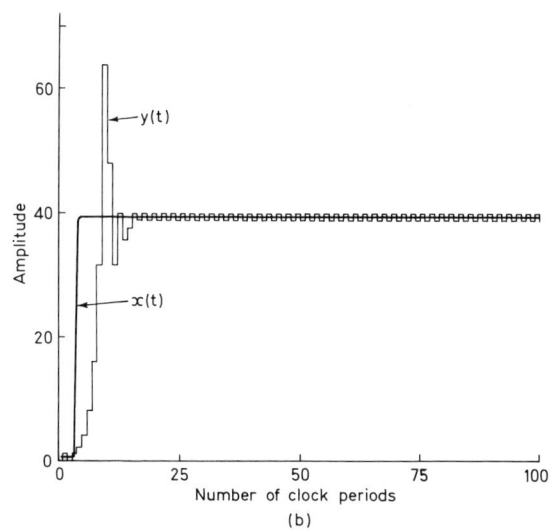

Fig. 8.21. Step response when the same input step as used in Fig. 8.20 is applied: (a) linear d.m. codec and (b) h.i.d.m. codec. The initial step-size of $y(t)$ is unity

the step is almost critically damped, yet the adaptation constants are not selected for step inputs, but sinewaves. The values of the adaptations used in Figure 8.20(a) have the values given in Section 8.10.1.

The response of linear d.m. and h.i.d.m. systems for the same step input as used in Figure 8.20 are given in Figures 8.21(a) and (b) respectively.

The linear d.m. and h.i.d.m. codecs have an overdamped and an underdamped response, respectively. The second order c.f.d.m. system has a response which increases quickly due to $A_4 = 2$ and in this respect is similar to h.i.d.m. However, it does not have the large overshoot of the h.i.d.m. system, but it does take a few more clock periods to 'home in' on $x(t)$. The response to a step input for B.C.d.s.m. shows a less rapid rise than those of c.f.d.m. and h.i.d.m. codecs but generally it 'homes in' with lower amplitude oscillations.

Pulse response

The response of 1st and 2nd order c.f.d.m., h.i.d.m. and linear d.m. systems to an arbitrary pulse rising from zero volts to 2·5 V in one clock period having a flat top for 6 clock periods and returns to zero level in a further clock period, is shown in Figure 8.22. The minimum step size in the waveforms for each encoder is 0·03 units. The second order c.f.d.m. system has the best response to this particular input pulse. The response of the linear d.m. and h.i.d.m. codecs resemble the pulse response of an *RC* integrator and differentiator, respectively. The $y(t)$ waveform of the first order c.f.d.m. system is unable to reach the peak amplitude of the pulse before this input signal returns to zero volts, whence it exhibits a large hunting oscillation. If for the second order c.f.d.m. system the value of its largest adaptation constant $A_4 = 2$, as in this case, then when

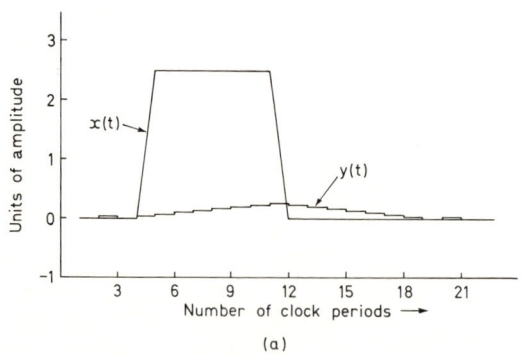

Fig. 8.22. Pulse response of (a) linear d.m.; (b) h.i.d.m.; (c) 1st order c.f.d.m. and (d) 2nd order c.f.d.m. systems

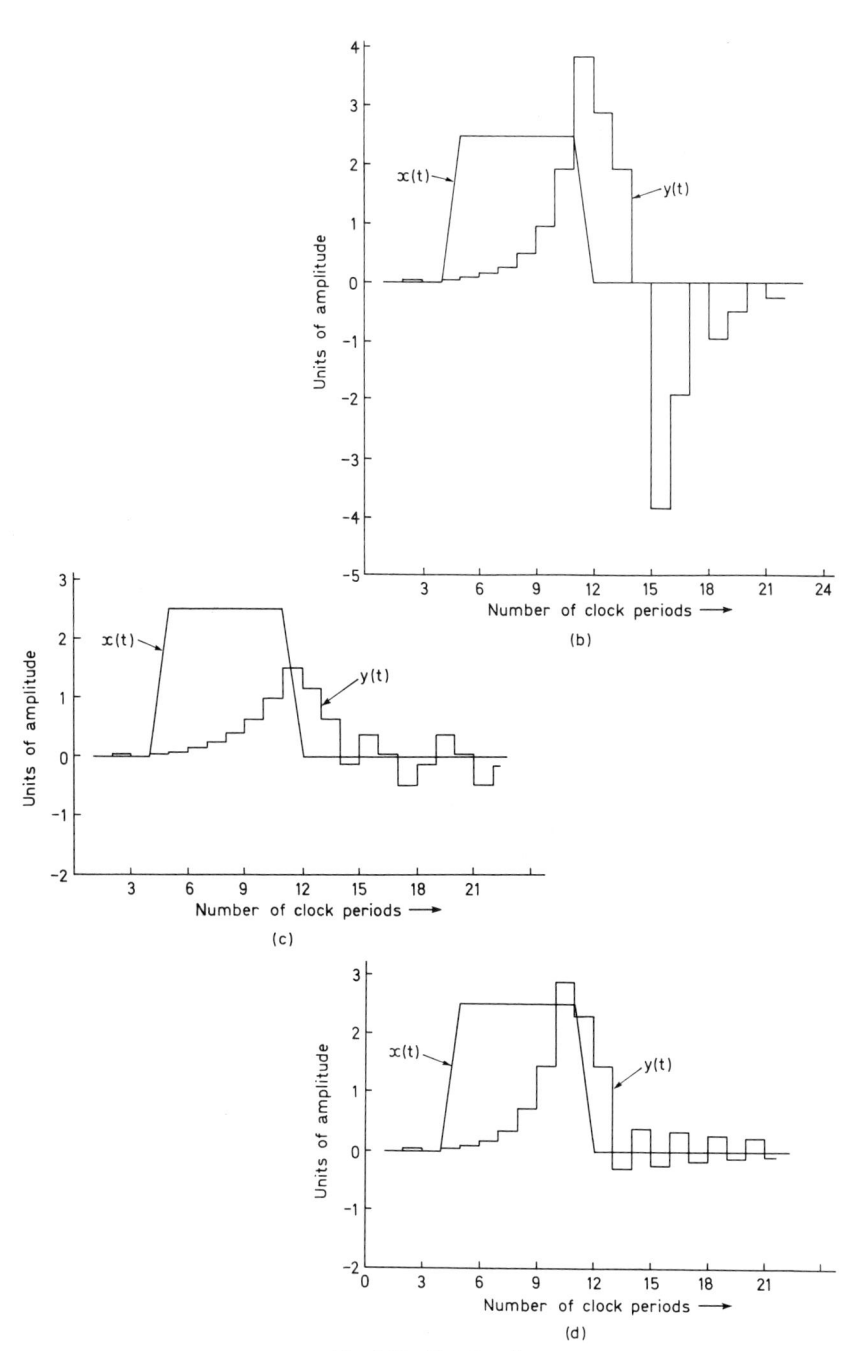

Fig. 8.22. (Continued)

overloaded it responds by doubling the size of the steps in the feedback signal at subsequent sampling intervals. It will be recalled from Section 8.4.1 that the h.i.d.m. codec also behaves in this manner when overloaded. Consequently it is found that the shape of the codec pulse response is dependent on the amplitude and duration of the input pulse; for some input pulses the h.i.d.m. system will be best, exhibiting negligible overshoot on the negative slope of the pulse, whereas the second order c.f.d.m. system may have a response similar to the one shown in Figure 8.22(b).

Tracking an arbitrary signal

Although the step-response and impulse response of second order c.f.d.m. systems are superior to that of the first order version, both systems behave similarly when tracking random type signals. Figure 8.23 shows the $y(t)$ signals for linear d.m. and first order and second order c.f.d.m. when an arbitrary input is applied to the encoders. Although linear d.m. experiences slope overload, c.f.d.m. systems easily track the signal. Observe that no violent oscillations occur in the $y(t)$ waveforms in the first order c.f.d.m. codec.

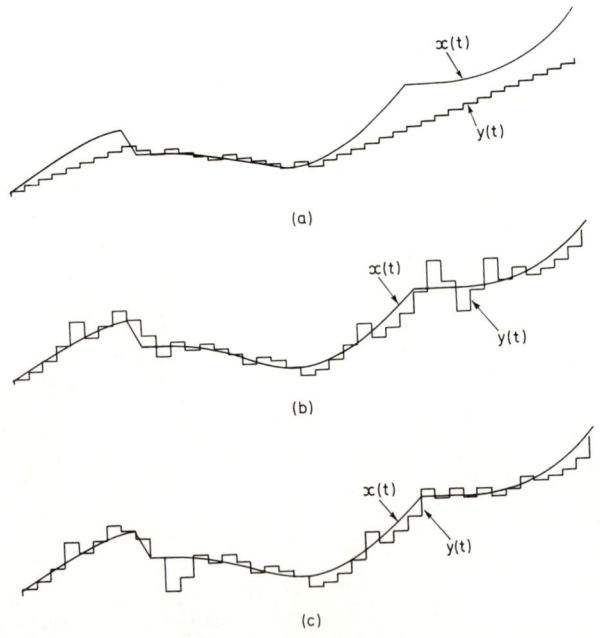

Fig. 8.23. *Response to an arbitrary input signal $x(t)$: (a) linear d.m.; (b) 1st order c.f.d.m. and (c) 2nd order c.f.d.m. systems*

Signal-to-noise ratio

The input signal $x(t)$ considered here to determine the signal-to-noise ratio of the c.f.d.m. and the linear d.m. systems is band-limited white noise. It is produced by passing wide-band white noise through a filter F_i having

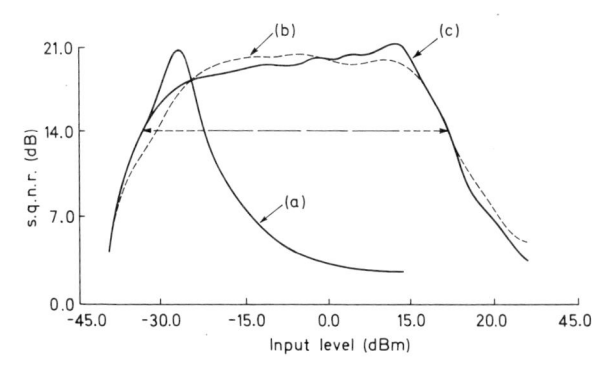

Fig. 8.24. Signal-to-noise ratio as a function of input signal power in dBm (i.e. in dB relative to 1 mW and 600 Ω): (a) linear d.m.; (b) 1st order c.f.d.m. and (c) 2nd order c.f.d.m. systems

the gain characteristic shown in Figure 8.25 where the cut-off frequency is 3·1 kHz.

The step-size of the linear delta modulator is ± 30mV which coincides with the minimum allowable step-size in the c.f.d.m. encoders. These encoders have a maximum step-size of ±5·0 V. All systems operate at

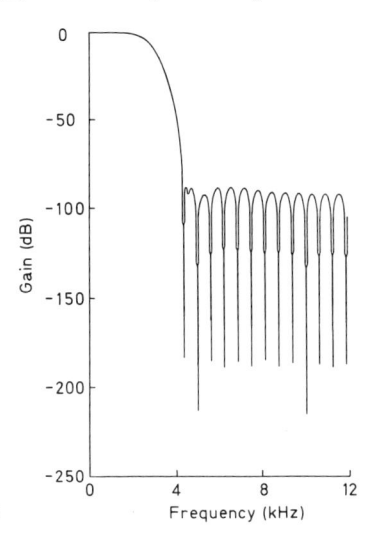

Fig. 8.25. Filter characteristic

40 kbits/s. The message signal $x(t)$ is recovered by filtering the ouput of the local decoder with a filter identical to F_i. The mean square value of $x(t)$ is varied and for each value the signal-to-noise ratio is determined. The results are displayed in Figure 8.24 and were obtained by computer simulation techniques.

The following general comments can be made.

(1) The peak signal-to-noise ratio is approximately the same for all systems.

(2) The curves for the first and second order c.f.d.m. show only marginal differences.

(3) For signal-to-noise ratios 6 dB below the peak value the improvement in the dynamic range of the c.f.d.m. compared to the linear d.m. system is approximately 43 dB.

8.11. ROBUST D.M.

An adaptive d.m. system due to Song[17] optimises the encoder and decoder separately to minimise the overall mean square error for a Markov–Gaussian source. The encoder has an optimum predictor in the feedback loop whose function is to ensure that the optimum step-size is obtained in the feedback signal. The decoder contains an optimum estimator. Like the first order c.f.d.m. system, this one has a short memory of only 2 bits in the encoder and decoder. The approach used is analytical and the resulting performance equations are relatively complex. By making some approximations to the theoretical findings, excellent computer simulation results are obtained, i.e. the signal-to-noise ratio although not greater than s.q.n.r. for linear d.m., is independent of the mean square value of the input signal. This means a large dynamic range as the quantization noise is highly dependent on the level of the input signal–almost completely the opposite to linear d.m. where the quantization noise is nearly independent of the mean square value of the input signal (Chapter 3). However, the Song system has been modified to enable the realisation of an experimental encoder called a Robust delta modulator which can be arranged to behave similarly to a discrete adaptive d.m. encoder, or a first order c.f.d.m. encoder (see Sections 8.3 and 8.9).

An encoder derived from the Robust delta modulator is shown in Figure 8.26. It is similar to the c.f.d.m. encoder but has two multipliers instead of one. The integrator is increased every sampling period by m_r, where

$$m_r = (\gamma + \alpha |m_{r-1}|) \operatorname{sgn}(L_r) + (\delta + \beta |m_{r-1}|) \operatorname{sgn}(L_{r-1})$$

and $\alpha > 1, \beta < 1$.

Instantaneously companded d.m. systems

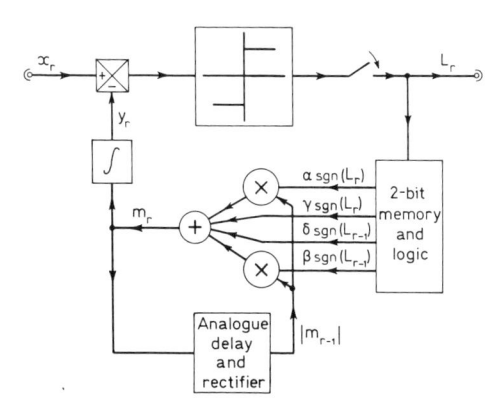

Fig. 8.26. Modified robust d.m. encoder

The values of γ and δ are small, i.e. $\gamma \ll \alpha$ and $\delta \ll \beta$ and ensure a minimum step size when $|m_{r-1}|$ is close to zero. For

$$\beta|m_{r-1}| \gg \delta, \quad \beta = 0{\cdot}5, \quad \alpha|m_{r-1}| \gg \gamma \quad \text{and } \alpha = 1$$

the performance is exactly as for a first order c.f.d.m. encoder, having $A = 1{\cdot}5$ and $B = -0{\cdot}5$. This is easily verified by drawing the response to a step input of nine units magnitude and observing that it is identical to the response shown in Figure 8.19(b).

8.12. SIMILARITY BETWEEN THE INSTANTANEOUSLY ADAPTIVE D.M. SYSTEMS

The modified Robust delta modulator shown in Figure 8.26 having $\alpha = 1$ and $\beta = 0{\cdot}5$ behaves identically to a first order c.f.d.m. encoder whose adaptation parameters are $A = 1{\cdot}5$ and $B = 0{\cdot}5$. The behaviour of the first order c.f.d.m. system having $A = 2$ and $B = 0{\cdot}5$ is almost the same as the h.i.d.m. codec.

The B.C.d.s.m. and pulse group delta modulators are closely related and have some resemblance to the statistical, mapping and discrete delta modulators in that the step-heights in their feedback waveforms are directly related to the size of the memory.

Encoders with a short memory using a recirculating loop and a multiplier, typified by a c.f.d.m. encoder, have effectively a very long memory. As a consequence the step-sizes in the feedback waveform can have a large range of values, i.e. orders of tens or hundreds.

8.13. TELEVISION PERFORMANCE

This section comments on the subjective performances of the h.i.d.m., the B.C.d.s.m., and the first order c.f.d.m. encoders, when encoding monochrome television pictures. Comparative performances cannot be ascertained because the evaluations of the different encoding systems were not made for the same criteria. The comments to be made are not comprehensive, but are presented with the object of providing an insight into the subjective effects produced by some of the different encoding algorithms discussed in this chapter.

8.13.1. h.i.d.m.

When the video signal makes a rapid excursion the error signal increases. This is no real disadvantage as the eye can tolerate large errors in picture areas containing detailed edges. The effect of the response of the encoder overshooting the video signal is not serious as three clock periods are used per picture element, and the response is often beginning to closely track the signal after this number of clock periods. Further the transients are averaged by the eye and their effect is mitigated by the dimensions of the spot size.

In regions when the variations in the signal amplitude is small, i.e. slowly changing grey areas, the encoder responses are step-sizes of unit magnitude similar to those obtained with linear delta modulation. The contouring effects however, tend to be absent and the performance in these regions is similar to that obtained with Pseudo Random Noise p.c.m.,[18] having three bits per picture element. Thus unlike p.c.m., h.i.d.m. does not usually produce contour effects.

The eye requires accurate values of luminance in regions of slowly changing grey scale. This is achieved with h.i.d.m., but it is apparent that while a burst of channel errors in a p.c.m. system are not too serious a similar burst in the h.i.d.m. system produces both transient and accumulative effects.

8.13.2. B.C.d.s.m.

For flat areas of the picture when the television signal changes slowly the adaptation logic keeps the step-size in the $y(t)$ waveform to its smallest value of one unit until more than two consecutive $L(t)$ pulses of the same polarity are produced, i.e. the encoder behaves identically to the linear and h.i.d.m. encoders.

Unlike Winkler, Bosworth and Candy consider that it is desirable to encode sharp changes in the television signal within three rather than one picture element if the distortion is not to be objectionable. For a picturephone signal having a bandwidth of 1 MHz, and the B.C.d.s.m.

Instantaneously companded d.m. systems

system operating at a bit rate of 6·3 Mbits/s, the sharp changes in the signal are encoded in less than 1·5 μs, i.e. in less than nine bits.

The maximum weighting sequence which will preserve stability is that chosen by Winkler. The rule[7] is that with steady input each weight should not exceed the sum of all preceding weights. B.C.d.s.m. has a slower weighting sequence than h.i.d.m. producing a more damped response. By making the weight return to its unit value after a sequence of similar polarity bits ceases the subsequent weight values are independent of the previous code. The effect is to reduce streaking in the picture due to transmission errors.

Subjective experiments

A television signal band limited to 1 MHz and displayed on a 271 line interlaced picture at 30 frames a second was viewed on a 5·5 × 5 in display. The peak luminance was set to 70 foot-lambert†, and the room illumination was approximately 100 foot-candle.[8]

The subjective tests were made by observers having a knowledge of the encoding technique and picture evaluation. They viewed the display at a distance of 3·5 ft. The observers displayed with equal contrast a selected television picture, and the same picture which had been encoded and decoded by the B.C.d.s.m. system, alternately. The picture which had *not* passed through the encoding process had band-limited white noise added, with a flat spectrum from 100 Hz to 0·6 MHz. The power in this noise signal was adjusted until both pictures had an equal overall quality. The ratio of television signal power to noise power produced the subjective signal-to-noise ratio. The television signal power was varied for the same picture and a graph of signal-to-noise ratio against the r.m.s. value of the signal was obtained. For the subjective experiments the televised picture was a still face and four observers were used. When the video level was low the effects of quantization were found to be objectionable by all the observers. When the encoder was frequently overloaded the observers showed considerable differences in estimating the signal-to-noise ratio. The highest signal-to-noise ratios were recorded where only a limited amount of slope overloading occurred. Similar results were obtained for a live scene.

The B.C.d.s.m. codec gives improved subjective maximum signal-to-noise ratio, by some 4 dB, and an increase in the dynamic range when compared to the linear delta modulator.

Using 3 bits per picture element, and for a 10 dB range of input levels the B.C.d.s.m. system results in a subjective degradation equivalent to −50 dB of added noise.

† 1 foot-lambert = 3·426 candela/m^2
1 foot-candle = 10·76 lux

Weight values

From the description of the adaptation algorithm given in Section 8.5.1 the feedback signal $y(t)$ has its amplitude 'weighted', i.e. varied in accordance with the sequence of binary levels of the waveform $L(t)$ at the output of the encoder. If the encoder is idling then the change in $y(t)$ at a sampling instant is $\pm I$; if $y(t)$ changes by $\pm 3I$, say when encoding an input signal then the encoder is said to generate 'weight three'. The weights used in this encoder are one, two, three and five. It is found that the choice of weights is not critical in the subjective signal-to-noise ratio, provided they do not cause the encoder to become unstable.

The system exhibits considerable tolerance to differences of the weight values in the encoder and decoder. For example, if all the weights in the encoder and decoder were mismatched by up to 30% a subjective signal-to-noise ratio of 50 dB, when the input video signal is 70 mV, is degraded by less than 2 dB. This feature of the codec indicates its possible broadcasting application, where the weights in the many decoders at each receiver would not be required to be closely matched to the weights in the encoder at the transmitter. Mismatching the weights at transmitter and receiver tends to distort the picture in areas where the signal is changing rapidly.

Transmission errors

If the received binary signal is erroneously detected on one occasion in every 10^5 bits transmitted then this error rate is troublesome when the signal being transmitted is of a live scene. The effect of each error is to produce a streak in the picture which is usually less than 20 elements long having a random amplitude. The length of these streaks is largely determined by the time constant (3 μs) of the filter at the receiver.

8.13.3. First order c.f.d.m.

This codec was simulated on a computer and used to process a single picture of 1 MHz bandwidth. The picture frame consisted of 250 lines, with approximately 275 picture elements per line. The scan rate was 30 frames/s.[12]

A study was made of signal-to-noise ratio averaged over the 'active' part of the picture as a function of the adaptation constants A and B. This signal-to-noise ratio was within 1 dB of its maximum value for $0.8 < AB < 1$; the value of 1.5 for A produces near optimum results. These conditions for A and B were virtually independent of the sampling rate.

A p.c.m. processed picture operating at 20×10^6 bits/s was compared with a first order c.f.d.m., operating at 10×10^6 bit/s. The first order c.f.d.m. codec operating at this lower clock rate was not as effective as the p.c.m. system when the television signal changed rapidly from black to white and vice versa as these sudden changes produced a hunting effect which appeared as a twinkle in the picture. Although no subjective tests have been made with the second order c.f.d.m. this objectionable twinkle effect should be removed.

REFERENCES

1. Fine, T., 'Properties of an optimum digital system and applications', *I.E.E.E. Trans. Information Theory*, **IT-10**, 287–296, October (1964)
2. Bello, P. A., Lincoln, R. N. and Gish, H., 'Statistical delta modulation', *Proc. I.E.E.E.*, **55**, No. 3, 308–320, March (1967)
3. Bertora, F., 'Short memory adaptive delta modulation', *Alta Frequenza*, **XXXIX-NII**, 974–979, November (1970)
4. Abate, J. E., 'Linear and adaptive delta modulation', *Proc. I.E.E.E.*, **55**, No. 3, 298–308, March (1967)
5. Winkler, M. R., 'High information delta modulation', I.E.E.E. International Conv. Rec., Pt. 8, 260–265 (1963)
6. Winkler, M. R., 'Pictorial transmission with h.i.d.m.', I.E.E.E. International Con. Rec., Pt. 1, 285–291 (1965)
7. Candy, J. C., 'Refinement of a delta modulator', *Bandwidth Reduction Symposium*, M.I.T. (1970). Reproduced in *Picture Bandwidth Compression*, Ed. Huang, T. S. and Tretiak, O. J., Gordon and Breach, 325–399, (1972)
8. Bosworth, R. H. and Candy, J. C., 'A companded one-bit coder for television transmission', *Bell Systems Tech. J.*, **48**, 1459–1479, July (1969)
9. Flood, J. E. and Hawksford, M. J., 'Adaptive delta–sigma modulation using pulse grouping techniques', Joint Conference on Digital Processing of Signals in Communications, University of Technology, Loughborough, 445–462, April (1972)
10. Kikkert, C. J., 'Digital techniques in delta modulation', *I.E.E.E. Trans. Com. Tech.*, **19**, 570–573, August (1971)
11. Goodman, D. J., 'A digital approach to adaptive delta modulation', *Bell Systems Tech. J.*, **50**, No. 4, 1421–1427, April (1971)
12. Jayant, N. S., 'Adaptive delta modulation with a one-bit memory', *Bell Systems Tech. J.*, **49**, No. 3, 321–342, March (1970)
13. Jayant, N. S. and Rosenberg, A. E., 'The preference of slope overload to granularity in the delta modulation of speech', *Bell Systems Tech. J.*, **50**, No. 10, 3117–3125, December (1971)
14. Childs, I., 'Adaptive 2nd-Order delta modulation system with 1-bit memory', *Electronic Letters*, **7**, No. 11, 295–296, June (1971).
15. Kyaw, A. T. and Steele, R., 'Constant-factor delta modulator', *Electronic Letters*, **9**, No. 4, 96–97, February (1973)
16. Kyaw, A. T., 'Constant factor delta modulation', Ph.D. Thesis, Electronic and Electrical Eng. Dept., University of Loughborough (1973).
17. Song, C. L., Garodnick, J. and Schilling, D. L., 'Available-step-size robust delta modulator', *I.E.E.E. Trans. Com. Tech.* **COM-19**, No. 6, 1033–1044, December 71.
18. Roberts, L. G., 'Picture coding using pseudo-random noise', *I.R.E. Trans. Information Theory*, **IT-8**, No. 2, 145–154, February (1962).

Chapter 9

DELTA AND PULSE CODE MODULATION

9.1. INTRODUCTION

In this chapter methods of producing uniform, i.e. linear pulse code modulation and A-law pulse code modulation by d.m. techniques are discussed in detail and comparisons are made between the performance of pulse code modulation and d.m. For readers who are not well acquainted with the subject of p.c.m. a brief description is given. For a sound understanding of pulse code modulation the reader is referred to *Principles of Pulse Code Modulation* by Cattermole.[1]

9.2. PRINCIPLE OF LINEAR PULSE CODE MODULATION

Pulse code modulation, abbreviated p.c.m., was invented by Reeves[2] in 1937. Its inclusion in commercial telecommunication networks was delayed until the advent of transistor technology. Like linear d.m., p.c.m. is conceptually simple. The block diagram of a linear p.c.m. codec is shown in Figure 9.1. The analogue input signal is band-limited to frequencies $f_{c1} \leq f \leq f_{c2}$ to give the signal $x(t)$. This signal is quantized by a linear staircase function to one of N levels spaced δ apart. This system

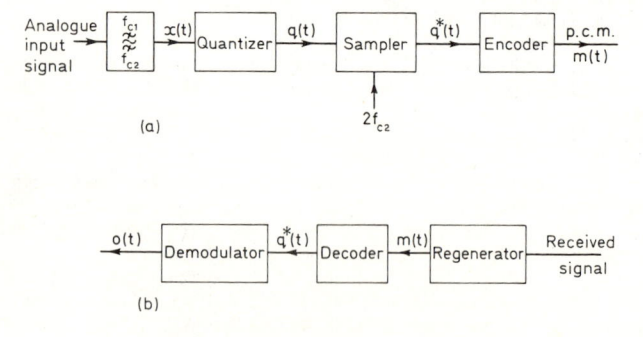

Fig. 9.1. Linear p.c.m. system: (a) encoder; (b) decoder

Delta and pulse code modulation

is called linear p.c.m. here because of the presence of this linear quantizer, and to distinguish it from the logarithmic p.c.m. discussed in Section 9.4. The quantized waveform $q(t)$ is identical to the waveform $y(t)$ in Figure 6.2. $q(t)$ is sampled at the Nyquist rate $2f_{c2}$ of the signal $x(t)$, and then encoded to give the p.c.m. signal $m(t)$. Each sample is encoded into a p.c.m. word of n bits. If each of these n bits can have any one of l levels, then $N = l^n$. The simplest case is when $l = 2$, i.e. $m(t)$ is a binary signal and $N = 2^n$. For example, if the $q(t)$ sample is encoded into a 7 bit word, $q(t)$ can have one of 2^7 ($= 128$) different values at any sampling time, provided of course that $x(t)$ is capable of having amplitudes which extend over the complete range of quantization levels. For simplicity it will be assumed that $m(t)$ is a binary signal.

The p.c.m. signal is transmitted via suitable terminal equipment, through a transmission channel. Assuming that there are no channel errors, the signal is reformed at the decoder, and decoded to produce the sampled waveform $q^*(t)$. This waveform is commonly referred to as a pulse amplitude modulated (p.a.m.) waveform. The final stage of the recovery process is to pass the p.a.m. waveform through a demodulator, essentially a band-pass filter, which extracts the message frequencies to produce a continuous signal, $o(t)$. $x(t)$ differs from $o(t)$ due to the quantization process in the encoder.

The calculation of the spectral density function $S_e(f)$ of the error signal in asynchronous d.m. is directly applicable to the p.c.m. encoder shown in Figure 9.1(a). $S_e(f)$ is given by Equation 6.15 and has a shape similar to the spectral density function at the output of an RC integrator when the input is white noise.

The 'break frequency' associated with $S_e(f)$ is f_n where $f_n \gg f_{c2}$. Consequently the Nyquist rate for the quantized signal $q(t)$ is very high as it has frequency components approaching infinity. However, the quantized signal is generally sampled at 10 to 20% above $2f_{c2}$ to allow for imperfect demodulation filtering. The result is that although the recovery requirement for the baseband signal $x(t)$ is satisfied, the quantization distortion is enhanced due to the sampling of the distortion signal

$$e(t) = x(t) - q(t)$$

well below its Nyquist rate, i.e. the degradation in the recovery of $x(t)$ is exacerbated by $S_e(f)$ being sampled at $2f_{c2}$.

The sampled function $S_e^*(f)$ has $S_e(f)$ located at $\pm 2f_{c2}$, $\pm 4f_{c2}$, $\pm 6f_{c2}$, etc. The effect is that over the message band, $-f_{c2}$ to $+f_{c2}$, $S_e^*(f) \doteq S_e(f)$ over the range $-\infty$ to $+\infty$. This is because multiple aliasing of $S_e(f)$, due to sampling, compresses almost all the $S_e(f)$ function into the message band. The quantization noise is

$$N_q^2 = \int_{-f_{c1}}^{f_{c2}} S_e^*(f)\,df \doteq \int_{-\infty}^{\infty} S_e(f)\,df$$

The mean square value of $e(t)$ is approximately determined by noting that $e(t)$ is generally sawtooth in shape and confined to peak values of $\pm \delta/2$. Representing $e(t)$ as mt for

$$-\{\delta/(2m)\} < t < \{\delta/(2m)\}$$

the mean square error is

$$N_q^2 = \frac{m}{\delta} \int_{-\delta/2m}^{\delta/2m} e^2(t)\,dt = \frac{\delta^2}{12}$$

If $x(t)$ is a sinusoid $E_s \sin 2\pi f_s t$ having a mean square value of $E_s^2/2$, then provided that N is large, s.q.n.r. $\gg 1$ and no peak overload occurs, the s.q.n.r. can be approximately represented by

$$\text{s.q.n.r.} = 6\left(\frac{E_s}{\delta}\right)^2 \qquad (9.1)$$

The peak s.q.n.r. occurs when E_s has a value $\delta N/2$ or $\delta 2^{n-1}$, which is the largest amplitude which does not overload the encoder.

$$\widehat{\text{s.q.n.r.}} = 1{\cdot}5 \times 2^{2n}$$

or the peak s.q.n.r. in dB is

$$\widehat{\text{s.q.n.r.}} = 1{\cdot}8 + 6n \qquad (9.2)$$

Replacing 1·8 by 3·0 gives the upper bound for this equation.[1]

9.3. LINEAR P.C.M. BY A D.M. TECHNIQUE

Consider the linear d.m. system shown in Figure 1.1. The binary output signal $L(t)$ consists of narrow pulses of amplitude V and duration τ spaced T seconds apart. If these pulses are represented by impulses of magnitude $V\tau$ volt seconds, then the feedback signal $y(t)$ consists of steps which change by $\pm \gamma$ volts every clock instant when encoding an input signal $x(t)$, as shown in Figure 1.2.

Suppose for convenience that γ is adjusted to have a value of unity. If the $L(t)$ impulses are applied to an up-down, i.e. reversible binary counter then the number X_r residing in this counter at any clock instant r is equal to the binary number of the instantaneous input amplitude x_r, to an accuracy of $\pm \gamma$, i.e. ± 1, provided that the clock rate of the d.m. is sufficiently large to prevent slope overload occurring. The number in the reversible counter increases or decreases by unity every sampling instant, and this is also true of the feedback analogue signal $y(t)$.

If the data in the up-down counter is inspected at the Nyquist rate it constitutes a p.c.m. word of N bits, where N is the number of stages in the binary counter. The signal $m(t)$ so produced is linear p.c.m., i.e. there is no compression.

In producing a p.c.m. signal by this technique the length of the p.c.m. word is fixed by signal-to-noise and bandwidth considerations. To guarantee tracking $x(t)$ the clock rate of the linear delta modulator must be many times greater than the Nyquist rate; a further practical difficulty is due to the inaccuracy of the integrator used in the feedback loop of the delta modulator. If the number in the counter is increasing by one unit per clock period and the $y(t)$ waveform is increasing at a slightly lower rate then the number in the counter is not representative of $y(t)$ and hence $x(t)$. Acknowledging that a perfect analogue integrator cannot be produced then an alternative is to make a 'leaky counter', such that the number in the counter is always a binary representation of the amplitude of $y(t)$.

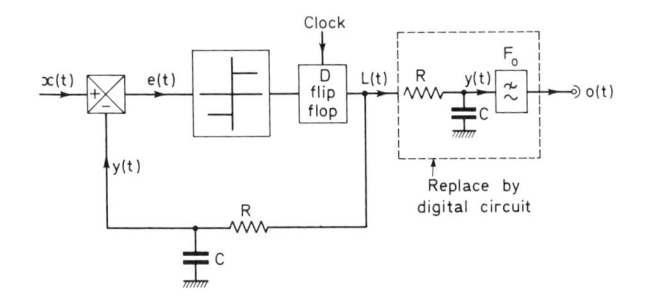

Fig. 9.2. Linear d.m. codec with RC integrators

Figure 9.2 shows a d.m. codec having 'leaky integrators', i.e. RC circuits. The object is to make a binary representation of $o(t)$ by replacing the analogue decoder by a digital circuit, e.g. a leaky counter. The change in the feedback waveform $y(t)$ during a sampling period depends on the amplitude of this waveform, as described in Section 2.3. Consequently in order that the binary representation of $o(t)$ is always a close approximation to $y(t)$ the binary number in the leaky counter, although primarily dependent on the $L(t)$ waveform, is also determined by the amplitude of $y(t)$.

Arrangements for producing linear p.c.m. have been proposed[3,4] where the input signal is encoded by a linear delta modulator and the binary signal $L(t)$ has been applied to an up-down counter and b.t.f. in cascade feeding a p.c.m. word generator. The up-down counter provides linear p.c.m., as previously described, the b.t.f. offers a digital interpolation and the p.c.m. word generator rounds off the least significant bits of the filter output.

9.4. A-LAW P.C.M. BY D.M. TECHNIQUE

The p.c.m. encoder to be described produces an A-law code by a delta modulation technique.[5] Before describing the action of this encoder a brief description will be given of the A-law, followed by a discussion on the segmented A-law p.c.m. code.

9.4.1. A-law

This law was first proposed by Cattermole[6] for the p.c.m. transmission of speech. Consider a non-linear element having a speech input signal $x(t)$ and an output signal $c(t)$, which is a monotonic function of $x(t)$ defined by

$$c(t) = \frac{Ax(t)}{1+\log_e A}, \qquad 0 \le x(t) \le V/A \tag{9.3}$$

and

$$c(t) = \frac{V\{1+\log_e \{Ax(t)/V\}\}}{1+\log_e A}, \qquad V/A \le x(t) \le V \tag{9.4}$$

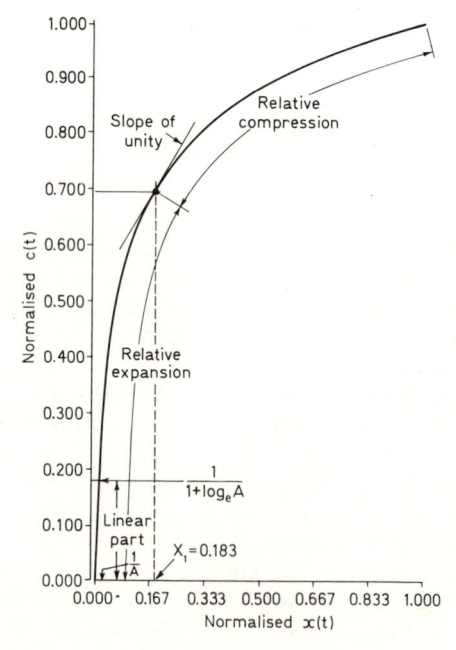

Fig. 9.3. A law characteristic for positive input levels

Delta and pulse code modulation 271

where V is the maximum value of both $x(t)$ and $c(t)$ and A is a constant after which the law is named.

The above equations for $c(t)$ relate to positive $x(t)$. However, $c(t)$ is a symmetrical odd function of $x(t)$. Consequently $x(t)$ is confined to the amplitude range $\pm V$, and when $x(t) = -V$, $c(t)$ also equals $-V$. Figure 9.3 shows the A-law for positive values of $x(t)$. $x(t)$ defined by Equation 9.3 has a linear characteristic of slope

$$A/(1+\log_e A)$$

The internationally recommended value of A is 87·6, which makes the initial slope 16; consequently the A-law expands small signal amplitudes by a factor of 16. The expansion of $x(t)$ decreases with increasing amplitudes. When the slope of the characteristic is unity there is neither expansion nor compression. From Equation 9.4

$$\frac{dc(t)}{dx(t)} = 1, \quad \text{when} \quad x(t) = \frac{V}{1+\log_e A} = X_1$$

For $x(t) > X_1$, $\{dc(t)/dx(t)\} < 1$ the amplitudes of $x(t)$ are relatively compressed, there being gain throughout the range of $x(t)$.

The variation of $c(t)$ with $x(t)$ over the signal range defined by Equation 9.4 is logarithmic.

Figure 9.4 shows a sinewave $x(t)$ having maximum amplitude V and the corresponding signal $c(t)$ emanating from the A-law non-linearity. Observe both $x(t)$ and $c(t)$ have the same peak values, but that $c(t)$ is greatly enhanced compared to $x(t)$ in the vicinity of the zero-crossings.

Consider a sinusoidal $x(t)$ which is uniformly quantized to N levels. If the amplitude E_s of $x(t)$ is just below the lowest quantization level a

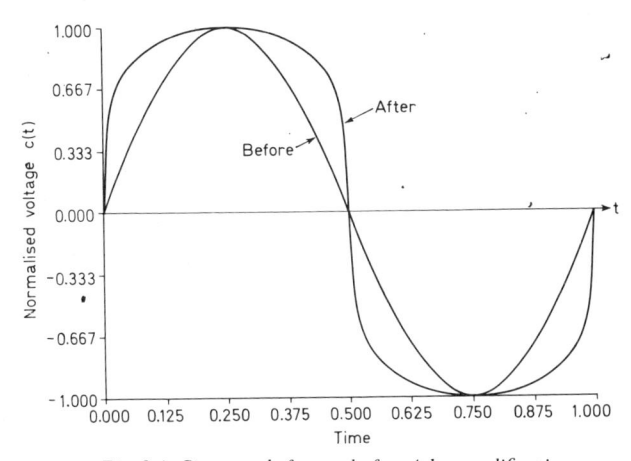

Fig. 9.4. Sinewave before and after A-law modification

linear p.c.m. signal $m(t)$ will be produced having binary words whose bits representing the magnitude of the quantized $x(t)$ will be zero for all samples of $x(t)$. This means that the input signal never has an amplitude which is above the threshold of encoding. However, if $x(t)$ is passed through the A-law non-linearity the signal

$$c(t) = 16x(t)$$

is produced which can be encoded over a substantial part of the $x(t)$ cycle. The A-law non-linearity therefore extends the range of $x(t)$ levels where encoding can occur.

The A-law non-linearity is often referred to as a compressor although the terminology is somewhat unfortunate. It is followed by a uniform quantizer which is in turn followed by an encoder. At the decoder, the demodulated analogue signal is 'expanded' to give the recovered signal. The transfer function of the expander is such that if a signal is passed through the compressor and expander in cascade the signal that emerges from the expander is identical to the signal that entered the compressor. It must be emphasised that the so-called expander actually compresses the signal when it has small amplitudes.

9.4.2. Segmented A-law

In international communications it is necessary to standardise transmission methods, and this function is provided by the International Telecommunication Union (ITU), working through the CCIR (Comité Consultatif International de Radiocommunications–International Consultative Radio Committee) and C.C.I.T.T. (Comité Consultatif de Telephone et de Télégraphie—International Consultative Committee for Telephony and Telegraphy) committees. The C.C.I.T.T. have made recommendations for the 30/32 channel p.c.m. system which uses an A-law characteristic composed of linear segments. By using a segmented A-law, the physical realisation of this law is simplified. Table 9.1 shows parts of the encoding law relevant to this discussion. There are 13 segments symmetrically arranged about the origin, with part of the Segment 1 located in both the negative and positive parts of the characteristic. The symbol ϕ is either unity or zero depending on whether the polarity of the signal $x(t)$ at the encoding instant is positive or negative, respectively. Figure 9.5 shows the curve $c(t)/V$ against $x(t)/V$ for the A-laws given by Equations 9.3 and 9.4 and the segmented curve from Table 9.1.

Over the range 0 to V, $x(t)$ is divided into units of $V/2048$. The numbers in column C3 are the number of $x(t)$ units at the end of each linear segment. The quantized signal $q(t)$ obtained by uniformly quantizing $c(t)$ has 128 equally spaced levels in the range $0 \leq q(t) \leq V$. This is because the 30/32

Delta and pulse code modulation 273

Table 9.1

C1	C2	C3	C4	C5		C6	C7
Linear segment number	(Number of ranges) × (Range size)	Amplitude at segment end Units of $V/2048$†	Decision amplitude number	\multicolumn{2}{c	}{8 Digit binary code}	Amplitude at decoder Output	Decoder output amplitude number
				Sign	Segment		
		2048	128			2016	128
				ϕ 1 1 1	1 1 1 1		
			127				
7	16 × 64						
			113				
				ϕ 1 1 1	0 0 0 0	1056	113
		1024	112				
6	16 × 32						
			97				
				ϕ 1 1 0	0 0 0 0	528	97
		512	96				
5	16 × 16						
			81				
				ϕ 1 0 1	0 0 0 0	264	81
		256	80				
4	16 × 8						
			65				
				ϕ 1 0 0	0 0 0 0	132	65
		128	64				
3	16 × 4						
			49				
				ϕ 0 1 1	0 0 0 0	66	49
		64	48				
2	16 × 2						
			33				
				ϕ 0 1 0	0 0 0 0	33	33
		32	32				
1	32 × 1						
			1				
				ϕ 0 0 0	0 0 0 0	0.5	1
		0	0				

† The smallest range size has been normalised to one unit of $V/2408$

channel system has 8 bit binary words which means that $x(t)$ is quantized to any of 2^8 or 256 levels.

In Segment 1 the quantizer has a uniform characteristic and over the range of $x(t)$ from 0 to $32 \times V/2048$, $q(t)$ changes from 0 to $32 \times V/128$. Hence $x(t)$ is expanded in this range as can be seen from Figure 9.5. Observe that the initial slope of the segmented A-law curve is $2048/128 = 16$ which is identical to the required law, $A = 87.6$. In Segment 2 each step in $q(t)$ corresponds to two steps in $x(t)$. Consequently the 16 steps

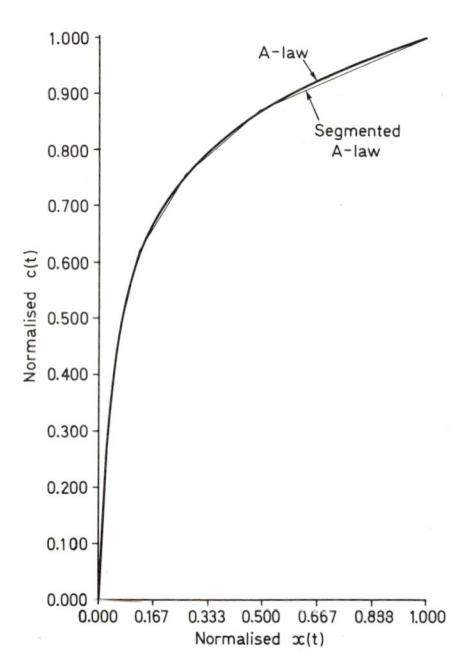

Fig. 9.5. A-law and segmented A-law characteristics

in $q(t)$ from 32 to 48 correspond to 32 steps in $x(t)$, and the slope of the curve is halved to 8. Progressing from Segment 2 to Segment 7 results in the range of $x(t)$ doubling relative to its value in the previous segment, while $q(t)$ always has 16 steps in each segment. This causes the slope of the segmented A-curve to be reduced by half relative to the previous segment. In the last segment the slope has reduced to 0·25. Observe that in Segment 5 the slope is unity. This means that in Segments 1, 2, 3 and 4 there is some expansion of $x(t)$, but in the remaining Segments 6 and 7, $x(t)$ is compressed. Note that over most of the range of $x(t)$ the latter is compressed.

Suppose that the input signal $x(t)$ has a value between x_i and x_{i+1}, where x_i and x_{i+1} are the values of $x(t)$ which exactly produce quantized signals $q(t)$ equal to the ith and $(i+1)$ levels, respectively. Now the A-law p.c.m. system is organised such that $x(t)$ will be quantized to the lower level q_i, even though $x(t)$ may be greater than $(x_{i+1}+x_i)/2$.

For example, from columns C3, C4 and C5, if $x(t)$ lays between 256 and 272 units, then the signal is quantized to 80 units and encoded ϕ 1 0 1 0 0 0 0. At the decoder the quantized signal is not q_i but q_{i+1}, and the recovered output amplitude is $x_i + \{(x_{i+1} - x_i)/2\}$. In the above

Delta and pulse code modulation 275

example, $q_{i+1} = 81$ units, and the output amplitude is $256+(16/2) = 264$ units. This is shown in columns C6 and C7.

Thus at the encoder the signal $x(t)$ is always quantized to the lower quantization level and at the decoder the expanded amplitude is increased by $(x_{i+1} - x_i)/2$. The A-law d.m. encoder to be considered must make use of the former requirement.

9.4.3. A-law delta modulator

The A-law delta modulator,[5] abbreviated A-law ΔM, is a p.c.m. encoder having the segmented A-law companding characteristic described in the previous section and implemented by a digital technique. Its basic components are an adaptive delta modulator and an adaptive binary counter.

The A-law ΔM has been specifically conceived to operate on one channel, rather than time shared with a number of channels. The reasons for using one p.c.m. encoder per channel are discussed in the next section.

The speech input signal $x(t)$ is sampled at the Nyquist rate to produce a p.a.m. signal. Each p.a.m. sample is held for the Nyquist interval $1/(2f_{c2})$ while it is encoded into a segmented A-law p.c.m. code by the A-law ΔM. As an example, if at the rth sampling instant $x(t) = X_r$ the latter is held in a zero order hold circuit and the A-law ΔM receives a d.c. level of X_r for the Nyquist interval, at the end of which the next d.c. level of X_{r+1} is applied.

The block diagram of the A-law ΔM is shown in Figure 9.6. The voltage at the output of the integrator is always set to zero just prior to the beginning of a Nyquist interval. This means that because of the sample and hold circuit the delta modulator section of the encoder is required to encode a step input. As a general rule at the beginning of the Nyquist interval the delta modulator experiences slope overload.

The feedback signal $y(t)$ in the delta modulator section can have step-sizes of either $\pm 8E$ or $\pm E$ where E is one unit in the segmented A-law,

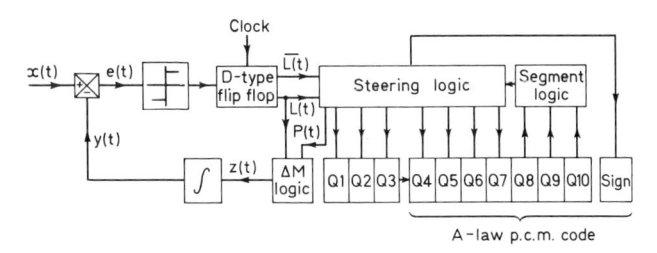

Fig. 9.6. A-law ΔM

i.e. $E = V/2048$, where $|V|$ is the maximum permitted magnitude of the input samples. The reasons for selecting these levels of $8E$ and E will be appreciated when the encoding behaviour is described.

The A-law ΔM has two main sections. The delta modulator, which performs the initial analogue to digital conversion, and the 10 stage synchronous binary counter together with its steering logic. The segmented A-law code is located in the 7 most significant digits of the counter at the end of the Nyquist interval.

The ΔM logic accepts signal $L(t)$ and $P(t)$ and produces a signal $Z(t)$. The sign of $Z(t)$ is the same as the sign of $L(t)$ and its magnitude is $8E$ if $P(t)$ is a logical zero and E if $P(t)$ is a logical one.

At the beginning of a Nyquist interval $y(t)$ is set to zero, $P(t)$ is a logical zero and the counter contains only zeros. Directly the held sample X_r is applied an error will be produced. If $e(t) \geq 0$, i.e. $X_r \geq 0$ the sign bit of the A-law p.c.m. signal is deemed positive, and vice versa if $e(t) < 0$. Therefore in the first d.m. clock period, the polarity of the sign bit is established and placed in the sign bit store shown in Figure 9.6. This means that the magnitude of the A-law code does not begin to be evaluated until the second d.m. clock period and consequently X_r will be quantized to the lower quantization level, a required segmented A-law condition described at the end of the last section. The sign bit when produced enables the steering logic to steer subsequent $L(t)$ or $\overline{L(t)}$ pulses to the synchronous binary counter according to the polarity of the sample. This is because the counter can only count upwards, and if $X_r < 0$ then a sequence of $\overline{L(t)}$ pulses will be produced.

$Z(t)$ has a magnitude of $8E$ and $y(t)$ increases by $8E$ following each $Z(t)$ impulse. The result is a sequence of all 'ones' or 'zeros' depending on the sign of X_r. For convenience of explanation assume X_r is positive. From Table 9.1 step-sizes of $8E$ occur only in Segment 4. The steering logic accepts the $L(t)$ pulses and inserts them into the binary counter in Stage Q7. After one $8E$ pulse the counter has 'zeros' in all of its stages with the exception of Q7, and this corresponds to an A-law code of 8. The second $8E$ pulse inserted in Q7 moves the 'one' to Q8 to give an A-law code of 16. The fourth $8E$ pulse applied to Q7 produces an A-law code of 32, i.e. ϕ 0 1 0,0 0 0 0. The A-law code has now moved from segment S1 to S2. From Table 9.1 it is seen that the range size in $x(t)$ doubles each time the segment number increases, i.e. it becomes coarser. Because the delta modulator is still operating with a step-size of $8E$, when the fifth $8E$ pulse is generated, the counter is updated via location Q6 rather than Q7 to increase the A-law code by 4. The situation is illustrated in Table 9.2. For example, in the 8th d.m. clock period $y(t)$ has been incremented 7 times to reach 56 E units, but the A-law code is 44. Provided $y(t) < X_r$ the $L(t)$ pulses are steered into the counter at locations Q5, Q4, Q3, Q2 and Q1, when the A-law code is in segments 3, 4, 5, 6 and 7,

Table 9.2

d.m. period	$y(t)$	A-law code in decimals	Segment number	Q_{10}	Q_9	Q_8	Q_7	Q_6	Q_5	Q_4	Q_3	Q_2	Q_1
1	0	0	1	0	0	0	0	0	0	0	0	0	0
2	8	8	1	0	0	0	1	0	0	0	0	0	0
3	16	16	1	0	0	1	0	0	0	0	0	0	0
4	24	24	1	0	0	1	1	0	0	0	0	0	0
5	32	32	2	0	1	0	0	0	0	0	0	0	0
6	40	36	2	0	1	0	0	1	0	0	0	0	0
		etc.						etc.					
9	64	48	3	0	1	1	0	0	0	0	0	0	0
10	72	50	3	0	1	1	0	0	1	0	0	0	0
		etc.						etc.					
16	120	62	3	0	1	1	1	1	1	0	0	0	0
17	128	64	4	1	0	0	0	0	0	0	0	0	0
18	136	65	4	1	0	0	0	0	0	1	0	0	0
		etc.						etc.					
32	248	79	4	1	0	0	1	1	1	1	0	0	0
33	256	80	5	1	0	1	0	0	0	0	0	0	0
34	264	80	5	1	0	1	0	0	0	0	1	0	0
35	272	81	5	1	0	1	0	0	0	1	0	0	0
		etc.						etc.					
65	512	96	6	1	1	0	0	0	0	0	0	0	0
66	520	96	6	1	1	0	0	0	0	0	0	1	0
		etc.						etc.					
69	544	97	6	1	1	0	0	0	0	1	0	0	0
		etc.						etc.					
129	1024	112	7	1	1	1	0	0	0	0	0	0	0
130	1032	112	7	1	1	1	0	0	0	0	0	0	1
		etc.						etc.					
137	1088	113	7	1	1	1	0	0	0	1	0	0	0
		etc.						etc.					
256	2040	127	7	1	1	1	1	1	1	1	1	1	1

Table 9.3

Segment number	Q_{10}	Q_9	Q_8	Boolean statement	Counter entry stage
1	0 0	0 0	1 1	$S_1 \overline{p(t)}$	$Q7$
2	0	1	0	$S_2 \overline{p(t)}$	$Q6$
3	0	1	1	$S_3 \overline{p(t)}$	$Q5$
4	1	0	0	$S_4 \overline{p(t)}$	$Q4$
5	1	0	1	$S_5 \overline{p(t)}$	$Q3$
6	1	1	0	$S_6 \overline{p(t)}$	$Q2$
7	1	1	1	$S_7 \overline{p(t)}$	$Q1$

278   Delta and pulse code modulation

respectively. The $y(t)$ signal has the required magnitudes of 32, 64, 128, 256, 512, 1024 and 2048 at the end of segments 1, 2, 3, 4, 5, 6 and 7, respectively. The rules for entering the counter depending on the segment number are given in Table 9.3.

Eventually the feedback signal $y(t)$ becomes greater in magnitude than the input sample X_r, and the first zero is generated. The behaviour of the encoder when this occurs depends upon which segment the code is in, i.e. the state of Q8, Q9 and Q10. If $X_r \geq 128E$ when the first zero is generated one of the signals S4, S5, S6 or S7 will be in existence at the output of the segment logic corresponding to Segments 4, 5, 6 or 7, respectively in which the code is in. From Table 9.1 it can be seen that quantum steps of $8E$ are sufficient to specify A-law in all of these segments and the counting is stopped, the code being one below because of the exclusion of the first pulse. If the first zero occurs with a segment signal S1, S2 or S3 then the code is in error because the quantum steps in these three segments are E, $2E$ and $4E$, respectively, i.e. less than $8E$. Thus for small signals, i.e. $X_r < 128E$, after the first zero has occurred the counter is emptied, the output of the integrator is set to zero, $P(t)$ is changed to a logical 1 and encoding starts anew. On the $P(t)$ change occurring $Z(t) = E$. The encoding therefore restarts with $y(t)$ being incremented by E each d.m. clock period according to the rules given in Table 9.4. The behaviour of the encoder is readily appreciated by referring to Table 9.5. When the error changes sign the counting stops.

Table 9.4

Segment number	Counter stages Q_{10}	Q_9	Q_8	Boolean statement	Counter entry stage
1	0	0	0	$S_1 p(t)$	Q4
2	0	0	1	$S_2 p(t)$	Q3
3	0	1	0	$S_3 p(t)$	Q2
	0	1	1		

The input sample will be encoded in less than the Nyquist interval, except when the maximum input sample V occurs. Just before the acceptance of the next input sample the 7 most significant digits in the counter which forms the A-law code are read-out into the multiplexing unit.

The highest count occurs when encoding a sample of $2048E$. The delta modulator step-size is maintained at $8E$ and the count lasts for 256 d.m. clock periods (Table 9.2). The choice of $8E$ is now apparent; on the maximum count it means that the d.m. clock rate must be 256 times the Nyquist rate, or $256 \times 8,000$ bits/s. As this is the channel rate for

Delta and pulse code modulation

Table 9.5

$y(t)$	A-law code in decimals	Segment number	Q_{10}	Q_9	Q_8	Q_7	Q_6	Q_5	Q_4	Q_3	Q_2	Q_1
0	0	1	0	0	0	0	0	0	0	0	0	0
1	1	1	0	0	0	0	0	0	1	0	0	0
	etc.							etc.				
15	15	1	0	0	0	1	1	1	1	0	0	0
16	16	1	0	0	1	0	0	0	0	0	0	0
	etc.							etc.				
31	31	1	0	0	1	1	1	1	1	0	0	0
32	32	2	0	1	0	0	0	0	0	0	0	0
33	32	2	0	1	0	0	0	0	0	1	0	0
34	33	2	0	1	0	0	0	0	1	0	0	0
35	33	2	0	1	0	0	0	0	1	1	0	0
36	34	2	0	1	0	0	0	1	0	0	0	0
	etc.							etc.				
64	48	3	0	1	1	0	0	0	0	0	0	0
65	48	3	0	1	1	0	0	0	0	0	1	0
	etc.							etc.				
68	49	3	0	1	1	0	0	0	1	0	0	0
69	49	3	0	1	1	0	0	0	1	0	1	0
70	49	3	0	1	1	0	0	0	1	1	0	0
71	49	3	0	1	1	0	0	0	1	1	1	0
72	50	3	0	1	1	0	0	1	0	0	0	0
73	50	3	0	1	1	0	0	1	0	0	1	0
	etc.							etc.				
127	63	3	0	1	1	1	1	1	1	1	1	0
128	64	4	1	0	0	0	0	0	0	0	0	0

the multiplexed signal it follows that only one clock need be used for both the A-law ΔM and the channel bit stream. The data is removed from the counter as a parallel word at the Nyquist rate. Another reason for selecting the step-size $8E$ is that it is a binary number in the range size of Segment 4 and enables the signals in the lower segments to be encoded in fewer than 256 d.m. clock periods. This can be demonstrated by considering the maximum input in Segment 3, i.e. an input whose value is fractionally less than $128E$. Initially the d.m. section of the A-law ΔM provides steps of $8E$, and after 16 d.m. clock periods $y(t)$ has increased to $16 \times 8E = 128E$, when the error changes sign. Because $y(t)$ has not moved into Segment 4, the A-law ΔM is reset, i.e. $y(t) = 0$ volts for one d.m. clock period. The encoding now recommences with $y(t)$ having step increments of E instead of $8E$, and after 128 clock periods $y(t)$ equals $128E$. As $y(t)$ has exceeded the input level the error again changes sign and this time the d.m. clock is inhibited and the A-code corresponding to the input sample now resides in the counter. The total number of d.m. clock periods

280 Delta and pulse code modulation

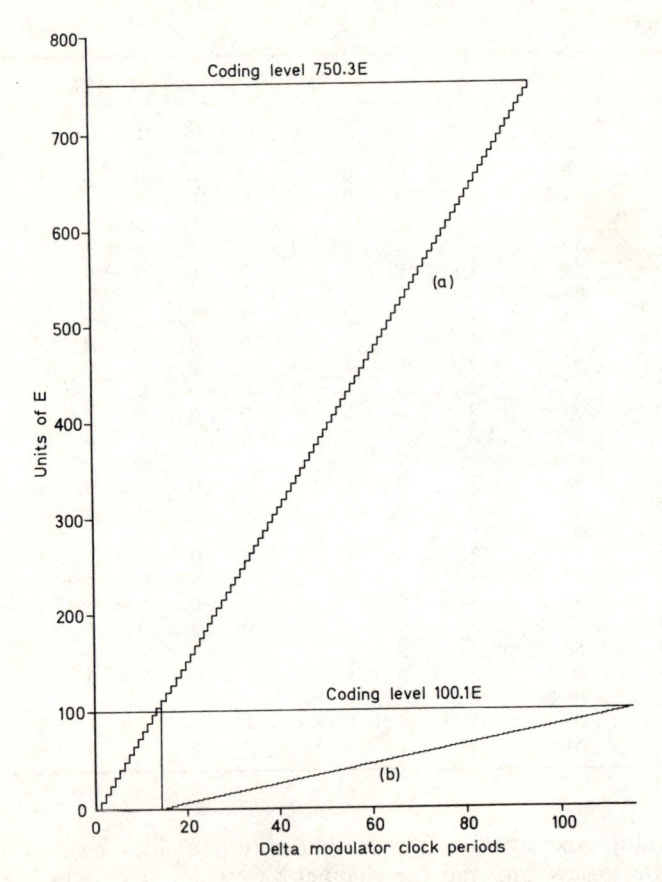

Fig. 9.7. $y(t)$ *waveforms for two arbitrary inputs:* (a) *input sample* $= 750 \cdot 3$ E; (b) *input sample* $= 101 \cdot 1$ E

used in the encoding after the production of the polarity bit is

$$16 + 1 + 128 = 145$$

9.4.4. Multiplexing A-law delta modulations

The 30/32 channel p.c.m. system has two channels per frame carrying synchronising, supervisory and signalling information, and the remaining 30 carrying speech signals. When the A-law ΔM is used on the 30/32 channel p.c.m. system, one A-law ΔM is provided for each of the 30 subscriber channels. Consider first of all Channel 1. This channel is sampled and the magnitude of the sample is held for the Nyquist interval of 256 d.m. clock periods during which the A-law ΔM encodes this sample into an A-law code. This A-law code resides in the counter in the A-law ΔM and is removed in a parallel form into a buffer store directly the Nyquist interval is dended. The A-law ΔM is then instantly reset and the encoding of the next input speech sample recommences. The removal of the A-law code from the A-law ΔM and the resetting of this p.c.m. encoder takes place within a single d.m. clock period; this means that an input sample is applied to the A-law ΔM every Nyquist interval. The A-law code which has been read into the buffer store in a parallel form is serially removed from the store at the d.m. clock rate and presented to the terminal equipment for transmission. In order to keep a constant transmission rate, equal to the d.m. clock rate, directly the last bit in the A-law code of Channel 1 has been transmitted from the buffer store the A-law code in the A-law ΔM connected to Channel 2 is inserted into the buffer store in a parallel form, and serial transmission of this code immediately starts. Notice that the Nyquist interval for Channel 2 is the same as that for Channel 1, but delayed in time by the length of the A-law code, i.e. by 8 d.m. clock periods. In fact every one of the 30 speech channels has the same Nyquist interval of 125 μs, but every adjacent channel is displaced in time by 8 d.m. clock periods. For example, the speech samples are applied to the A-law ΔM connected to Channels 2, 3, 4, 5, etc. after 8, 16, 32, 40, etc. d.m. clock periods have elapsed following the presentation of a sample to Channel 1 for encoding. After the A-law code for Channel 2 has been inserted into the buffer store it is serially transmitted, whence the buffer store is reset to zero. The A-law code for Channel 3 is then placed into the store and the serial transmission for this 8-bit code commences. The procedure is repeated for Channel 4 onwards until the bits from Channel 15 have been dispatched. The following sequence of bits are then serially transmitted; 8 bits of signalling information; 8×15 bits representing the A-code of the speech samples in Channels 17 to 31 inclusive; 8 bits of frame alignment information. The next 8 bits transmitted correspond to the A-law code of the sample in Channel I and so on.

By arranging for the input samples applied to adjacent channels to be spaced by 8 d.m. clock periods the transmitted rate is equal to the d.m. clock rate used by each A-law ΔM. Only one clock rate is therefore required for this time division multiplex system and it is equal to the Nyquist rate of 8,000 samples per second multiplied by the 256 possible steps in the $y(t)$ waveform of the A-law ΔM, i.e. 2·048 Mbits/s.

When only one p.c.m. encoder is time shared between many subscriber channels the signals in these channels are sampled in rotation at a rate of 8,000 per second. This method requires that each subscriber channel must be electronically switched into the encoder and this results in signals from these channels interfering with each other due to the imperfections of the electronic switch. The effect is referred to as 'cross-talk'. The time division multiplexing arrangement using A-law ΔM has one of these encoders per channel, and consequently the absence of electronic switching means that cross-talk is negligible. There are other advantages to be gained from using one encoder per channel. There is the ability to partially equip terminals and add channel encoder cards as required. If forms of traffic other than speech, for example computer data, are required to be multiplexed then this arrangement of one encoder per channel is an advantage. A failure of one encoder results in only one channel failing rather than the complete t.d.m. signal which would occur if only one encoder is time shared with all the input channels.

There are of course disadvantages; the power consumption and cost of each encoder must be of the order of one-thirtieth of that required by the time shared encoder. L.S.I. circuit techniques can help to make this figure realisable.

9.4.5. A-law p.c.m. using analogue compression

It has been argued in Section 9.3 that linear p.c.m. can now be produced by conveying the binary signal at the output of the delta modulator into an up-down counter and strobing the latter at the Nyquist rate to give the p.c.m. signal. An obvious extension of this idea is to precede the delta modulator with a non-uniform element having an A-law characteristic as shown in Figure 9.8. The number in the counter at the

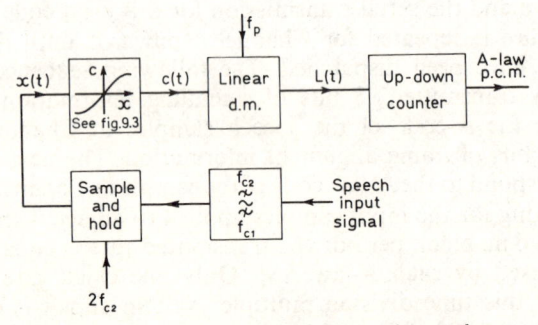

Fig. 9.8. A-law p.c.m. using a linear delta modulator and an analogue compressor

end of each Nyquist interval is now an A-law p.c.m. word. The speech signal $x(t)$ is sampled and held at the Nyquist rate and passed to the A-law non-linearity. The delta modulator therefore encodes a step equal to the input sample modified by the A-law characteristic. As $x(t)$ is required to be quantized over ± 128 levels in the d.m. clock rate

$$f_p = 128 \times 8{,}000 = 1{\cdot}024 \text{ Mbits/s}$$

This encoder is considered by the writer to be inferior to the previous A-law ΔM which achieves A-law compression by digital rather than analogue techniques.

9.5. COMPARISONS BETWEEN THE PERFORMANCE OF P.C.M. AND D.M.

This section makes some comparisons between p.c.m. and d.m. This comparison[15,18] cannot be comprehensive because of the many different types of system which exist. For example, there are the numerous companded d.m. systems and the derivatives, described in Chapters 7 and 8, all operating with slightly different encoding algorithms. Then for p.c.m.,[1] there is a range of different logarithmic companding laws although only the A-law has been described here. There are also the syllabically companded p.c.m. systems, one of which is described in Section 10.5, and differential p.c.m., etc.

The objectives of this section then are very modest. They are to offer some comparisons between linear d.m. and linear p.c.m., and to make some generalised comments about companded d.m. and logarithmic p.c.m.

9.5.1. Linear d.m. and linear p.c.m.

When a sinusoidal input $E_s \sin 2\pi f_s t$ is applied to linear, i.e. uniform p.c.m. the decoded s.q.n.r. can be estimated as a function of the normalised input signal level E_s/δ with the aid of Equations 9.1 and 9.2. Curve (a) in Figure 9.9 shows this characteristic for $n = 5, 6$ and 7 bits.

The corresponding characteristic for linear d.m. systems is obtained from Equations 1.9 and 1.10, where the normalised input signal level is E_s/γ, plotted as Curve (b) in Figure 9.9. This curve applies for $f_p = 56$ kbits/s, $f_c = f_{c2} = 3400$ Hz, $f_s = 800$ Hz, $K_q = 0{\cdot}3$.

Curve (c) displays the characteristic of the double integration d.m. codec. The normalised input used is $E_s/y_3(0)$, where $y_3(0)$ is the step increase in the feedback waveform due to the presence of prediction in the encoder as described in Sections 2.2.4 and 2.2.6. The frequency parameters f_p, f_c, and f_s are as for the linear d.m. system, and the empirical constant in Equation 2.26 is $C_q = 19$. The peak value of s.q.n.r. shown for Curve (c) is calculated from Equation 2.28.

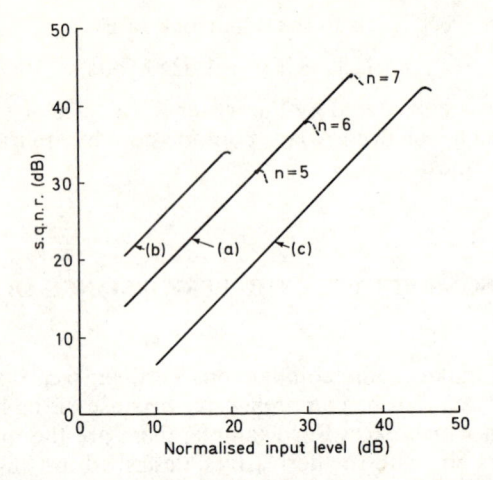

Fig. 9.9. s.q.n.r. versus normalised input level: (a) linear p.c.m.; (b) linear d.m. and (c) double integration d.m. codecs. Curves (b) and (c) are drawn for $f_p = 56$ kbits/s, $f_{c2} = 3{,}400$ Hz, $f_s = 800$ Hz

All three curves in Figure 9.9 have the same slope. When the p.c.m. system has a 7 bit code, $n = 7$, the transmitted bit rate is $2f_{c2}n = 47 \cdot 6$ kbits/s, but allowing for imperfect filtering it is usual to use a bit rate of 56 kbits/s for $f_{c2} = 3400$ Hz,. Thus the curves are comparable when

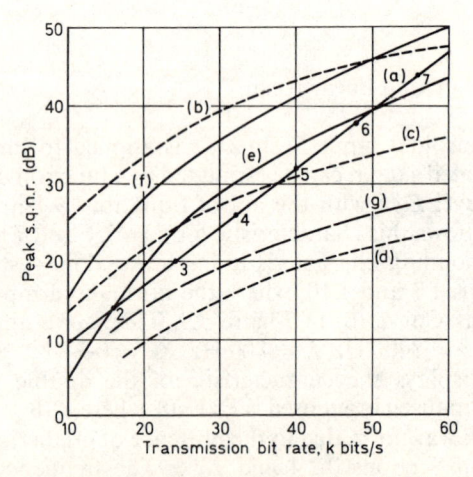

Fig. 9.10. Peak s.q.n.r. versus transmission bit rate for sinusoidal inputs. (a) Linear p.c.m., (b)(c)(d) are for linear d.m. where $f_s = 200, 800$ and $3{,}200$ Hz respectively, (e) Double integration d.m., $f_s = 800$ Hz, (f) Modified double integration d.m., $f_s = 800$ Hz, (g) Linear d.s.m., f_{c2} for all these systems is $3{,}400$ Hz

Delta and pulse code modulation

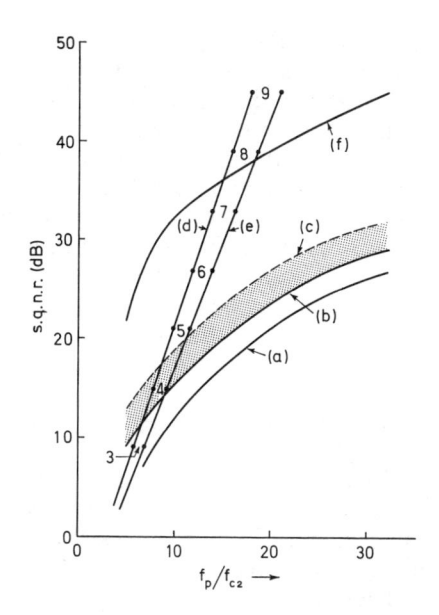

Fig. 9.11. s.q.n.r. as a function of normalised bit rate. Curves (a) to (e) apply to a band-limited white noise input signal. (a) Linear d.m. when no overloading occurs; (b) and (c) are the bounds of the peak s.q.n.r. for linear d.m. and (d) and (e) apply to linear p.c.m. sampled at and slightly above the Nyquist rate, respectively. Curve (f) applies to linear d.m. when the input signal is RC filtered band-limited white noise

Curve (a) has $n = 7$, and it is apparent that the performance of the double integration d.m. system is similar to the uniform p.c.m. and superior to the d.m. system.

For Gaussian input signals the curves of peak s.q.n.r. against normalised input power for linear d.m. are shown in Figure 3.18 for various f_p/f_c ratios. The curves have the same general shape as those shown in Figure 9.9. The s.q.n.r. against input power curves are similar for the linear p.c.m. and linear d.m. systems. The main effect of a Gaussian or speech input signal compared to a sinusoidal one is to reduce the peak s.q.n.r. for a given set of conditions in both linear d.m. and p.c.m. systems. This is illustrated in Figure 9.11.

In a linear p.c.m. system, the bit rate is $2f_{c2}n$, i.e. the Nyquist rate multiplied by the number of bits in the code.

Consider a sinusoidal input signal $x(t)$ which is fed through a low-pass filter having a critical frequency $f_{c2} = 3400$ Hz, and sampled slightly in excess of the Nyquist rate to produce a bit rate of $8000 n$. Figure 9.10(a) shows the variation of s.q.n.r. against bit rate, where the numbers on the curve represent the values of n. The curve is obtained using Equation 9.2.

For the linear d.m. codec the s.q.n.r. is given by Equation 1.10. If the filter in the decoder is of the low-pass type (having critical frequency $f_{c2} = 3400$ Hz) then putting $f_{c2} = f_c$ and $K_q = 0.3$ in Equation 1.10, the variation of s.q.n.r. as a function of f_p for three different sinusoidal input signals f_s having frequencies 200 Hz, 800 Hz and 3200 Hz are displayed by Curves (b), (c) and (d), respectively in Figure 9.10. These three curves highlight the characteristic of linear d.m.–that the s.q.n.r. decreases by 6 dB for every octave increase in f_s. The s.q.n.r. increases by 9 dB for every octave increase in clock rate f_p. At low values of f_p the slope of the curves for linear p.c.m. and linear d.m. systems are nearly the same, but at high clock rates the slope of Curve (a) exceeds those for the linear d.m. codec, indicating that for these conditions the p.c.m. system is more efficient at exchanging bit rate for higher peak signal-to-noise ratio.

Curve (e) applies for double integration d.m. and is plotted, using Equation 2.28, for the same f_{c2} and $f_s = 800$ Hz. By modifying the local decoder in the double integration d.m. codec such that Equation 2.30 is relevant, Curve (f) is produced and also applies for $f_s = 800$ Hz. The supremacy of double integration d.m. over linear d.m. is apparent except at very low clock rates.

Curve (g) relates to delta-sigma modulation and was plotted using Equation 2.50. The s.q.n.r. for d.s.m. depends critically on f_{c2} and is independent of f_s. If f_{c2} is restricted to 1700 Hz the s.q.n.r. rises by 9 dB, a slight improvement on Curve (c).

These curves relating to the delta modulators must not however be taken too literally. For example, the value of K_q in linear d.m. is not a constant, and Equation 1.10 is only valid over a restricted range of frequency parameters. For example the s.q.n.r. does not continue to increase, ad infinitum, by 9 dB for every octave increase in f_p.

When the bit-rate is low the d.m. system has a higher s.q.n.r. than the linear p.c.m. system, but this advantage diminishes as the bit rate is increased until eventually the p.c.m. system becomes superior.

When a Gaussian signal having a spectrum of low-pass white noise is applied to a linear delta modulator such that the encoder virtually never overloads, the expression for s.q.n.r. is obtainable from Equation 1.8 and from Equation 3.1 when the equality condition prevails and is

$$\text{s.q.n.r.} = \frac{1}{K_q} \cdot \frac{3}{(8\pi)^2} \left(\frac{f_p}{f_{c2}}\right)^3 \tag{9.5}$$

This formula is approximate and depends on the variation of the so-called constant K_q, which can have values ranging from 0·3 to 1·0 for Gaussian signals resulting in s.q.n.r. differences of up to 5·3 dB. In Section 3.10 the exact formula for s.q.n.r. was established and graphical results presented in Figures 3.18 and 3.19. However, because of the complexity of the exact formula for s.q.n.r. it is difficult to calculate it for a wide variety of conditions without the aid of a computer. Consequently, the simplified formula of Equation 9.5 will be pursued. Setting $K_q = 0.3$, the s.q.n.r. is plotted as a function of f_p/f_{c2} and displayed in Curve (a) of Figure 9.11. This curve does not give the maximum s.q.n.r. because no overloading occurs. Using the computer simulation results[7] in Figure 3.18 which give peak s.q.n.r.'s for different f_p/f_{c2} ratios, Curve (b) in Figure 9.11 is produced. Curve (a) and (b) are nearly parallel, and lead to the formulation of a simple empirical equation based on Equation 9.5

$$\text{s.q.\hat{n}.r. (dB)} \doteq -15 + 30 \log_{10}(f_p/f_{c2}) \qquad (9.6)$$

Eliminating f_p/f_{c2} from Equations 1.8 and 3.1, and applying the equality condition in Equation 3.1, then

$$\text{s.q.n.r.} = \frac{1}{K_q} \cdot \frac{8\pi}{\sqrt{3}} \left(\frac{\sigma}{\gamma}\right)^3 \qquad (9.7)$$

Using this equation where $K_q = 0.3$, an empirical formula can be deduced for the peak s.q.n.r. From Figures 3.18 and 3.19 where $(\sigma/\gamma)_m$ is the value of σ/γ to give s.q.$\hat{n}$.r.

$$\text{s.q.\hat{n}.r. (in dB)} = 14 + 30 \log_{10}(\sigma/\gamma) \qquad (9.8)$$

The (σ/γ) and f_p/f_{c2} ratios used in Equations 9.8 and 9.6 for s.q.$\hat{n}$.r., and the results shown in Figure 3.19, enables an approximate relationship to be found which gives the peak s.q.n.r. This relationship is

$$\frac{f_p}{f_{c2}} = 9 \left(\frac{\sigma}{\gamma}\right)_m \qquad (9.9)$$

Performance results of SCALE and second order c.f.d.m. system shown in Figures 7.14 and 8.24, respectively, give s.q.$\hat{n}$.r.'s which lie in the shaded zone in Figure 9.11. They have s.q.$\hat{n}$.r. values which are greater by 2 to 3 dB than Curve (b), which accounts for the imperfect Gaussian probability density functions and different filters, F_0, in the decoder. Curve (c) gives an indication of the upper bound of the s.q.$\hat{n}$.r.

In the case of linear p.c.m. the criterion suggested by Cattermole[1] is used, namely that the number of quantization levels multiplied by the spacing between these levels should be 10 times the r.m.s. value of the imput signal, i.e.

$$2^n \delta = 10\sigma$$

This criterion ensures that the amount of peak clipping of the imput signal is negligible. The variation of s.q.n.r. as a function of f_p/f_{c2} is shown by Curve (d) in Figure 9.11. For this curve the s.q.n.r. is the ratio of the signal power σ^2 to the noise power

$$N_q^2 = \delta^2/12$$

and the ratio

$$f_p/f_{c2} = 2n$$

where n is the number of bits in the p.c.m. code. Curve (e) in Figure 9.11 shows the performance of linear p.c.m. when the sampling is slightly above the Nyquist rate, to allow for imperfect filtering; it corresponds to the sampling rate used in establishing Curve (a) in Figure 9.10.

Curve (f) in Figure 9.11 relates to the performance of a linear d.m. system when encoding RC filtered low-pass white noise.[16] The RC time constant is $15/f_{c2}$. This type of input signal is well suited to the delta modulator because its spectrum is approximately shaped to the overload characteristic of the encoder, resulting in a better tracking performance. Curve (f) is some 16 dB above Curve (b) which applies to a 'flat' Gaussian input signal.

The curve relating to the case of double integration d.m. is not presented in Figure 9.11 as this system has an overload characteristic which falls by 12 dB per octave; its performance is likely to be poor when presented with an input signal whose spectrum is flat. However, if the input signal has a spectrum which matches that of the overload characteristic of the encoder then the performance curve will lie above Curve (f). It is found that when encoding speech signals this d.m. system has a subjective performance equivalent to 9-bit p.c.m. when $f_p/f_{c2} = 18$.

9.5.2. Syllabically companded d.m. versus logarithmic p.c.m.

When syllabic companding is applied to a delta modulator the effect on the signal-to-noise versus input level curve is to prise apart the curve at its peak, and insert a nearly horizontal portion. The effect is illustrated in Figures 7.5 and 7.14.

The important point is that when a d.m. system is syllabically companded its peak s.q.n.r. is unaffected. This result is hardly surprising as the variation of the step-size H in a syllabically companded system is very slow compared to the clock rate and therefore it behaves as a linear d.m. codec over many clock periods. However, in order to maintain the same s.q.n.r. over a wide range of signal level the delta modulator has to behave in an optimum fashion. To achieve this H must be automatically adjusted by the local decoder in the feedback loop to ensure that the slope of the signal $y(t)$ feedback to the error-point has the appropriate value. When sinusoidal inputs are applied Equation 1.5 must be satisfied, whereas with Gaussian signals Equation 9.9 is relevant when γ is replaced by H. A syllabically companded d.m. system is therefore similar to having a number of linear d.m. systems which are selected according to the slope of the input signal. Each linear system selected has the correct step-size $\hat{\gamma}$ to produce s.q.n.r.

The accurate calculation of the s.q.n.r. in linear d.m. is relatively complex, see Section 3.10, and becomes more difficult in companded systems. It was because of this difficulty that the theory of the syllabically companded delta-sigma modulation system, abbreviated s.c.d.s.m., presented in Section 7.2 was based on the simplified expression for quantization noise N_q^2 discussed in Sections 1.2.7 and 2.4. By adopting this approach the behaviour of the s.c.d.s.m. codec was investigated and useful results established. However, when the magnitude of the peak s.q.n.r. is required the inaccuracies latent in the simplified expression for N_q^2 become evident because the equation contains the term K_q whose value depends upon the nature of the input signal and encoder parameters. When sinusoidal signals are used $K_q \doteqdot 0.3$ is a reasonable value. With Gaussian input signals the value of K_q can vary from 0.3 to 1.0, but there are guide lines in Section 1.2.7 which gives values of K_q for different encoding parameters.

In the case of the SCALE codec, which is an approximate system due to the presence of its digital compander, the above values of K_q are not applicable. However, some values can be found which agree with experimental results. Using Equation 7.3 the s.q.n.r., or R, can be calculated. For Gaussian input signals and $K_q = 0.48$, it gives a calculated peak s.q.n.r. of 15 dB which agrees with the measured value displayed in Figure 7.14. At double the clock rate, i.e. 38.4 kbits/s, the peak s.q.n.r. increases by 9 dB as predicted by the equation. It is probably safe to extrapolate to 56 kbits/s where the s.q.n.r. = 29 dB. If the dynamic range of the pulse height modulator in the local decoder is 40 dB the dynamic range of the codec (as defined in Section 7.2.7) at a s.q.n.r. of 20 dB is approximately 48 dB. In the case of sinusoidal inputs it appears that Equation 7.3 can be applied if K_q is approximately 0.2, or Equation 2.50 can be used if $K_q = 0.3$ and $f_1 < f_{c2}/10$.

When the system is syllabically companded delta modulated, rather than delta-sigma modulated, the s.q.n.r. can be estimated from Figures 9.10 and 9.11. When double integration d.m. is used to encode speech it seems[14,8] that Curve (f) in Figure 9.10 is approximately representative of the peak s.q.n.r. If the multiplier has a dynamic range of 40 dB then for a clock rate of 56 kbits/s the peak s.q.n.r. = 48 dB; the dynamic range is close to 66 dB for a s.q.n.r. = 20 dB.

P.C.M. systems for commercial quality speech transmission use instantaneous companders of the A or μ types.[1] Consequently it is these types of p.c.m. systems which will be compared to syllabically companded d.m., although instantaneously companded d.m. can be used for speech transmission. Syllabically companded p.c.m. tends to be used for low bit rate speech transmission and such a system is discussed in Section 10.5.

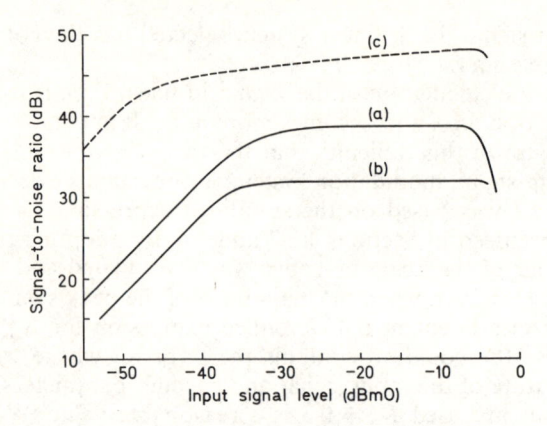

Fig. 9.12. Signal-to-noise ratio versus input level in dBm0, i.e. in dBm where the noise power is measured at the zero reference point of the system.[17] *Curves (a) and (b) are for 8- and 7-bit A-law p.c.m. codecs with Gaussian inputs respectively, and curve (c) is for the modified double integration d.m. system when a 800 Hz sinusoid is applied*

The instantaneously companded p.c.m. system has the same general shape of signal-to-noise ratio against input signal level curve as those of the delta modulation systems just discussed.

Curve (a) in Figure 9.12 shows the C.C.I.T.T. signal-to-distortion ratios R_M for the 30/32 channel p.c.m. system. The curve relates to a Gaussian input signal, variance σ^2, and the segmented A-law described in Section 9.4.2. The formula for R_M is[19]

$$R_M = 10 \log_{10} (m^2 \sigma^2 / \sigma_e^2)$$

where σ_e^2 is the variance of the distortion components and m is the slope of the regression line of output on input. Curve (a) in Figure 9.12 is for 8 bit p.c.m. and applies to the 30/32 channel system, whereas Curve (b) is for $n = 7$, a 24 channel p.c.m. system. The horizontal portion of these curves may exhibit wavelike ripples of about 1 dB if the input signal has a rectangular probability density function[1] or be smooth if this function is exponential. Curve (c) in Figure 9.12 applies to a hypothetical syllabically companded d.m. system using a double integrator with predictor. The multiplier is assumed to have a dynamic range of 40 dB, the clock rate is 56 kbits/s and the input signal is a sinusoid of 800 Hz, rather than a Gaussian signal. This curve is therefore drawn with the aid of Equation 2.30 and is considered to have similarities with the curve obtained with speech inputs, as mentioned in Section 2.2.6. Greefkes and Riemens[14] show that their digitally companded d.m. system, discussed in Section 7.6, does have a curve above Curve (a).

Curve (c) in Figure 9.12 highlights an important characteristic of companded d.m. Because it is a tracking encoder it can in principle continue to adjust its step-size as required for all input levels. There is no need to pay particular attention to low level signals at the expense of high level ones or vice versa. Companded d.m. can therefore be constructed relatively easily with a large dynamic range, but the peak s.q.n.r. is approximately the same as an uncompanded d.m. system. Consequently it is advantageous to compand the linear codec which has the highest s.q.n.r., namely the double integration d.m. system using prediction which is described in Section 2.2.

In Section 10.2.3 the performance of differential p.c.m. is discussed, and when encoding speech this system has a 6 dB advantage over logarithmic p.c.m.

9.5.3. Potentialities of instantaneously companded d.m.

Instantaneously companded d.m. codecs are discussed in Chapter 8, where it is mentioned on a number of occasions that their peak s.q.n.r. is approximately the same as that obtained in a linear d.m. system. There are exceptions, e.g. the mapping d.m. system described in Section 8.2.1 which has an improvement in the peak s.q.n.r. of 3 to 5 dB. The equity of the peak s.q.n.r. in syllabically companded d.m. systems with linear d.m. codecs is predictable, as described in the previous section. However, there is no obvious reason why this equity should occur in instantaneously adaptive d.m. systems which can in principle operate with a small error signal because they can adjust their step-size very rapidly to meet the variations in the input signal. The question which therefore arises is how far are instantaneous systems from being optimum? A profitable line of investigation is to examine the instantaneous error in a d.m. system.

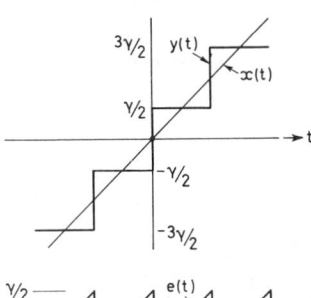

Fig. 9.13. Feedback and error waveforms in a linear delta modulator when the input sinusoid passes through a zero crossing

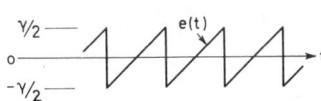

Consider first of all linear d.m. When a sinusoidal input $x(t)$ is applied the peak s.q.n.r. can occur when the maximum slope of the feedback waveform $y(t)$ is equal to the maximum slope of $x(t)$, which of course occurs at the zero crossings. Figure 9.13 illustrates this situation together with the error signal. When the sinusoid is in the vicinity of its maxima the error signal $e(t)$ increases, as shown in Figure 9.14. If these error signals were periodic, which of course they are not, the mean square value of the error signal in Figure 9.13 is $\gamma^2/12$, whereas in Figure 9.14 it is between $\gamma^2/4$ and $\gamma^2/2$ depending where $x(t)$ is located relative to $y(t)$.

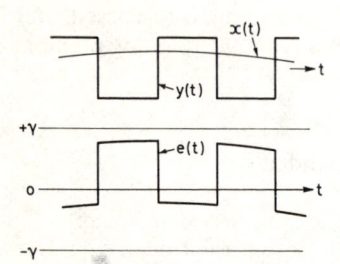

Fig. 9.14. Feedback and error waveforms in a linear delta modulator when the input sinusoid is near a peak value

It is apparent that for peak s.q.n.r. the major contributions to the error occur at the peaks of the sinusoid. This can be seen in Figure 2.23. From these observations if the step-size γ in the feedback signal $y(t)$ is reduced as the sinewave approaches its extremities the value of the peak s.q.n.r. should rise, due to $e(t)$ decreasing and having two zero-crossings per clock interval. Although the instantaneous encoder adjusts its step-size as a function of the slope of $x(t)$ it has been observed that the peak s.q.n.r. is not improved when compared to linear d.m. The reason may be that the damping of the system is not optimum—as the slope of $x(t)$ decreases $y(t)$ overshoots the input taking time to settle and in so doing generates a considerable amount of noise.

Suppose $y(t)$ is maintained as a step-like waveform and its step height is adjusted such that for every clock period the mean square error is

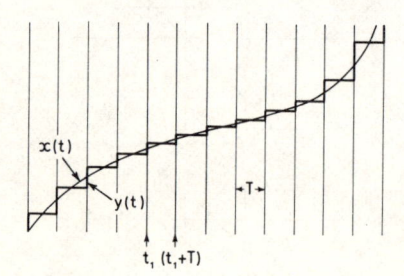

Fig. 9.15. Linear d.m. Optimum tracking of an arbitrary input signal $x(t)$, when the feedback signal $y(t)$ is kept constant during each sampling period

minimised. Figure 9.15 shows this $y(t)$ waveform for an arbitrary input signal $x(t)$.

Consider one time interval from t_1 to $t_1 + T$. The integral squared error between $x(t)$ and $y(t)$ during this period is

$$\mathscr{I} = \int_{t_1}^{t_1+T} |x(t) - y(t)|^2 \, dt \qquad (9.10)$$

and $y(t)$ has a constant amplitude h in the range $t_1 \leq t \leq t_1 + T$

$$\mathscr{I} = \int_{t_1}^{t_1+T} x(t)^2 \, dt - 2h \int_{t_1}^{t_1+T} x(t) \, dt + h^2 \int_{t_1}^{t_1+T} dt$$

$$\frac{\partial \mathscr{I}}{\partial h} = -2 \int_{t_1}^{t_1+T} x(t) \, dt + 2hT \qquad (9.11)$$

Setting Equations 9.11 to zero gives

$$h = \frac{1}{T} \int_{t_1}^{t_1+T} x(t) \, dt \qquad (9.12)$$

and this can be shown to be the value of h to minimise $\mathscr{I}$. This result is an obvious one, i.e. the value of h to give minimum $\mathscr{I}$ is the average value of $x(t)$ over the interval prescribed.

Extending the time interval to $T_s = kT$, where k is an integer, the integral squared error is

$$\mathscr{I} = \int_0^{T_s} x^2(t) \, dt - 2 \sum_{n=1}^{k} h_n \int_{(n-1)T}^{nT} x(t) \, dt + T \sum_{n=1}^{k} h_n^2 \qquad (9.13)$$

From Equations 9.12 and 9.13

$$\mathscr{I} = \int_0^{T_s} x^2(t) \, dt - \frac{T_s}{k} \sum_{n=1}^{k} h_n^2 \qquad (9.14)$$

Example
Let

$$x(t) = \sin 2\pi f_s t$$

The period T_s is divided into k zones where $h_1, h_2, \ldots, h_k$ are the values of the 1st, 2nd, ..., kth zones respectively. From Equations (9.12)

$$h_n = \frac{k}{T_s} \int_{T_s(n-1/k)}^{T_s(n/k)} \sin\left(\frac{2\pi t}{T_s}\right) dt = -\frac{k}{2\pi}\left[\cos 2\pi\left(\frac{n}{k}\right) - \cos 2\pi\left(\frac{n-1}{k}\right)\right]$$

If $k = 32$, i.e. n goes from 1 to 32

$$h_n = -5 \cdot 096 \left[\cos \frac{2\pi}{32} \cdot n - \cos \frac{2\pi}{32}(n-1)\right]$$

Delta and pulse code modulation

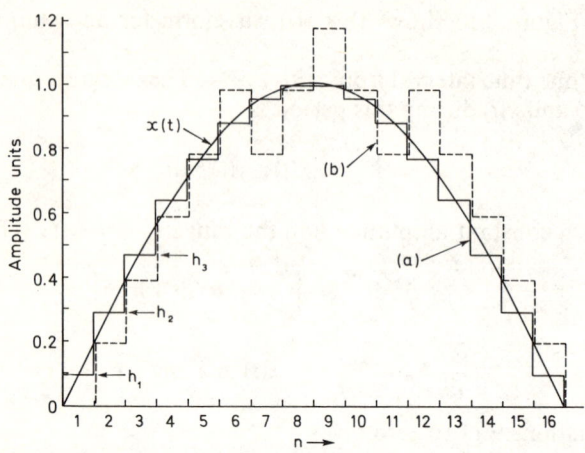

Fig. 9.16. *The feedback waveform y(t) for (a) optimum delta modulator, (b) linear delta modulator*

and

$$h_1 = 0.0978$$
$$h_2 = 0.291$$
$$h_3 = 0.469 \text{ etc.}$$

The results are plotted in Figure 9.16. During the first quarter of the sinewave

$$\int_0^{T_s/4} y^2(t)\,dt = T \sum_{n=1}^{8} h_n^2 = 3.97T$$

The area under $y^2(t)$ in one cycle of the sinusoid is $4 \times 3.97T$ V²s, and the average value is $4 \times 3.97T/T_s = 3.97/8$ V².

The mean square value of the sinusoid over one cycle is 0·5 and the error is 0·00161 V². The mean square value of the error is different from the quantization noise N_q^2 which is the mean square value of the in-band noise.

In the case of linear d.m., the value of $y(t)$ changes by a fixed amount $\pm\gamma$ per sampling instant. Optimum encoding is nearly obtained when the maximum slope of $x(t)$ equals the maximum slope of $y(t)$, and therefore for the sinusoid in Figure 9.16,

$$\gamma = (2\pi/k) = 0.196$$

The behaviour of the $y(t)$ waveform with this step-size is also shown in Figure 9.16, and it is obvious that the mean square error of this linear encoding is in excess of that corresponding to the optimum adaptive encoder. The $y(t)$ signal on the adaptive encoder is always adjusted each clock instant to minimise the mean square error, and as a consequence the polarity of the output binary pulses is the same as the polarity of the slope of $x(t)$. In the linear encoder $y(t)$ not only has a fixed step-size but the polarity of γ at each clock instant is not always the same as the polarity of the slope of $x(t)$. In the Author's opinion an instantaneously adaptive d.m. system could have a peak s.q.n.r. in excess of 4 dB of the value for the linear and syllabically companded systems, the actual improvement being greater for higher f_p/f_{c2} ratios. For example, if the adaptation parameters of the second order c.f.d.m. codec are not dispersed overshooting will be significant and slope overloading will be substantially avoided due to the high clock rate. The performance can probably be further enhanced by allowing $y(t)$ to vary during a sampling period according to some optimising law.

The instantaneously adaptive d.m. systems offer a wide dynamic range which is ultimately limited by the maximum and minimum step-size in the feedback signal $y(t)$. In the case of the second order c.f.d.m. system discussed in Section 8.10.3 the ratio of the maximum to minimum step-sizes is 44 dB which provides a good measure of the encoders dynamic range (see Figure 8.24).

The general conclusion is that having specified the bandwidth of the message and the channel, then for a given type of input signal the peak s.q.n.r. is at least equal to that obtained by a linear delta modulator. The selection of the encoding algorithm will have a profound effect on the behaviour of the codec, as discussed in detail in Chapter 8, and may increase the s.q.n.r. The dynamic range is ultimately determined by the ratio of maximum to minimum step-size in $y(t)$.

From the above discussion instantaneously companded d.m. codecs should have a greater signal-to-noise ratio than the syllabically companded d.m. codecs, and this factor taken in conjunction with the relative ease of achieving companding means that they can be an alternative to companded p.c.m. codecs in many applications.

9.5.4. Channel errors

Linear d.m. has a relatively good performance in the presence of errors due to the absence of any hierarchy in the transmitted signal. A digital error at the decoder causes the signal $y(t)$ at the output of the integrator to be in error by two steps (2γ). The effect of errors on $y(t)$ is accumulative and only becomes a serious problem when error burst occurs. When encoded speech is being transmitted the natural pauses in the speech effectively overcomes this accumulative effect. Section 5.2.1 showed that linear d.s.m. has better performance than linear d.m. in the presence of random errors.

The performance of linear p.c.m. is more degraded by channel errors compared with linear d.m. At worst the word-synchronisation may be lost. If synchronisation is maintained then the amount of error in the decoded signal depends upon which bit in the word is in error. The worst case is when an error occurs in the most significant digit when the error can be as great as 2^n quantization levels. The decoded signal-to-noise ratio is exponentially dependent on the signal-to-noise ratio at the input to the decoder.[9] In contrast to linear d.m., the errors in linear p.c.m. are generally larger, but non-accumulative.

S.C.D.S.M. performs well in the presence of errors because the step height in the feedback loop changes very slowly compared to the bit rate. When encoding speech, error-rates of 1% or more can be tolerated. For these high error rates s.c.d.s.m. is spectacularly better than logarithmic p.c.m.

Instantaneously companded d.m. is far more prone to channel errors than s.c.d.s.m., although the effect of these errors is related to the encoding algorithm used and the type of signals being transmitted. For example, in the case of television transmission utilising the B.C.d.s.m. system described in Section 8.13.2, one error bit in every 10^5 causes the decoded picture to be degraded, while for high quality colour reception using differential p.c.m. (described in Section 10.2) an acceptable error rate is of the order of 1 in 10^9 bits transmitted.

Generally, the more hierarchy in the transmitted binary signal the more sensitive is the system to channel errors; the effect of an isolated error is highly dependent on the immediate history of the encoding process, as discussed in Section 8.4.2.

For example, if the second order c.f.d.m. codec using perfect integration is encoding a sinusoid then the effect of an isolated error after the transient effects have subsided is to produce a sinusoid which has a phase change and a d.c. component. The magnitude of the d.c. component may be negligible or it may be many times the amplitude of the sinusoid, depending critically on which bit was erroneously detected.

If the integrators in the codec are 'perfect' the effect of an error is present at the output of the decoder for all time. It is therefore essential that the integrators are made deliberately leaky so that the effect of an error on the decoded signal will not persist longer than a few integrator time constants.

9.5.5. Synchronisation

D.M. systems encode on a bit-by-bit basis as distinct from p.c.m. systems which transmit n-bit code words. This means that p.c.m. requires word and bit synchronisation. The advantages of not having word synchronisation means that the d.m. decoder can operate with a clock, if one is used, which is not precisely synchronised to that at the transmitter. The other advantage is a more simpler decoder.

9.5.6. Time sharing

Each subscriber in a d.m. system must have his own decoder and encoder due to the feedback loop in the encoder. A d.m. encoder cannot be time shared with a number of subscribers as can a p.c.m. encoder. In the 1960's the main argument for d.m. was its cheapness per encoder, making it suitable when the number of subscribers was small, and the advantage of p.c.m. was the ability to time share the encoder when the number of subscribers was large. Large scale integration techniques have altered the situation, making encoder/decoder costs less important and introducing the idea of one p.c.m. encoder per subscriber to avoid analogue cross-talk involved in switching the p.c.m. encoder between channels. This change in events means that time sharing is not such a serious consideration in selecting an encoder as it used to be.

9.5.7. Filters

The final filter, F_0, in a d.m. decoder receives a digit stream at a rate equal to the transmission rate. This is in contrast to the situation in p.c.m. where the filters operate on p.a.m. samples occurring at the Nyquist rate, i.e. a much lower rate. As a consequence the filter design in d.m. is simpler than in p.c.m.

9.6. DITHER

Quantizers have errors in their output signals which correlate with their input signals. This correlation is greater for coarse quantization and becomes subjectively undesirable in both speech and television.

A reduction in this correlation, first described by Roberts[10] in connection with p.c.m. transmission of television pictures, adds pseudo-random noise to the input signal prior to quantization. At the receiver this same pseudo-random noise is subtracted from the recovered analogue samples to give the decoded analogue signal. This application of pseudo-random noise is referred to as 'dithering the input signal' and its effect is to change the quantization error from one which is conditional on the input signal to one which approaches white noise. The variance of the quantization error is not increased, and the dithering effect improves the subjective quality of speech[12] and television reception.[10,11] In the latter case for the same subjective quality of the received picture, a 6 bits per sample without dither can be replaced by 3 bits per sample with dither, i.e., the transmitted bit rate is halved.

A one bit p.c.m. system has a quantizer in the form of a zero crossing detector, and the correlation between the quantization error and the input signal is high (0·5). The remarks made so far in this section refer to linear p.c.m. Consider now the case of linear d.m. It has the same zero crossing detector as the one bit p.c.m. system but because of the feedback loop and the time sampling the linear d.m. becomes equivalent to a linear p.c.m. encoder using a multi-level quantizer. The important parameter is the correlation between the error signal and the input signal ($x(t)$ in Figure 1.1) and is $\ll 0.5$.

The effect of adding a pseudo-random noise signal $P(t)$ to $x(t)$ and subtracting $P(t)$ from the signal at the output of the integrator in the decoder tends to cause the quantization noise to 'whiten'. It has been reported[12] that at high clock rates an improvement of the order of 2 dB can be achieved in the signal-to-noise ratio. No similar improvement has been observed in p.c.m.

A possible explanation for this improvement in signal to noise ratio in the linear d.m. case can be envisaged with reference to Section 3.11 and Figure 3.17. The quantization noise in d.m. is the noise components which fall in the message band. The spectral density function of the error signal is composed of two components which are located at multiples of f_p and $f_p/2$.

When the sampling rate is low the spectrum located at zero and $\pm f_p/2$ overlap, but when the clock rate is high they separate, as shown in Figure 3.17 for the case when $f_p = 64f_{c2}$. At these higher clock rates when the spectra do not significantly overlap the effect of dither is to whiten them, with the result that the inband noise is reduced as the spectrum centred at zero is flattened and widened. Of course, it would be more desirable if dither had this effect at low clock rates, but this is not the case.

For companded delta modulation, which is discussed in Chapters 7 and 8, dither has little effect.

9.7. EFFECT OF SUCCESSIVE MODULATIONS ON SIGNAL-TO-NOISE RATIO

If a signal is repeatedly encoded and decoded the signal-to-noise ratio is degraded due to the accumulation of quantization and overload noise. Jayant and Shipley[13] investigated this situation and produced the following expressions for signal-to-noise ratio after m successive modulations.

Linear d.m.: $\qquad (\text{s.n.r.})_m = 30 - 10\log_{10}(3\cdot 15 + 0\cdot 35m^2) \qquad (9.15)$

First order c.f.d.m.: $(\text{s.n.r.})_m = 40 - 10\log_{10}(0.9 + 3.5m + 1.4m^2)$ (9.16)

p.c.m.: $(\text{s.n.r.})_m = (\text{s.n.r.})_1 - 10\log_{10}(m)$ (9.17)

The above expressions for the d.m. systems, obtained by computer simulation, apply for a 2 s speech input signal band-limited to 3300 Hz; the clock rate is 60 kHz and the final low-pass filter F_0 in the decoder has a 12 dB per octave roll off. Equation 9.17 for p.c.m. is derived by assuming that the s.n.r. is high and that each additional encoding introduces equal amounts of encoder generated noise.

As the number m of successive modulations are increased the $(\text{s.n.r.})_m$ decreases initially more rapidly with adaptive d.m. than with linear d.m. Equations 9.15 to 9.17 are valid for $m = 6$; $(\text{s.n.r.})_6$ is still better for adaptive d.m. than for linear d.m. but is of course considerably reduced in value compared to the initial condition, $m = 1$. The rate at which the s.n.r. for both adaptive d.m. and p.c.m. is degraded decreases monotonically with m, i.e.

$$|(\text{s.n.r.})_m - (\text{s.n.r.})_{m+1}| < |(\text{s.n.r.})_{m-1} - (\text{s.n.r.})_m|$$

The rate is approximately a constant for linear d.m.

REFERENCES

1. Cattermole, K. W., *Principles of Pulse Code Modulation*, Iliffe, London (1969).
2. Reeves, A. H., French Pat. 852, 183 (1938). Brit. Pat. 535, 860 (1939). U.S.A. Pat. 2,272,070 (1942).
3. Goodman, D. J., 'The application of delta modulation to analogue-to-p.c.m. encoding', *Bell Systems Tech. J.* **48**, No. 2, 321–343, February (1969).
4. Goodman, D. J. and Greenstein, L. J., 'Quantizing noise of d.m./p.c.m. encoders', *Bell Systems Tech. J.*, **52**, No. 2, 183–204, February (1973).
5. Pennington, H. and Steele, R., 'A-law pulse code modulation by a delta-modulation technique', *Electronic Letters*, **9**, No. 8/9, 171–173, May (1973).
6. Cattermole, K. W., Contribution to discussion of paper by Purton, R. F., 'A survey of telephone speech-signal statistics and their significance in the choice of a p.c.m. companding law', *Proc. I.E.E.*, **109B**, 486, November (1962).
7. Passot, M., 'Normal spectra application to the study of delta modulation', *Masters Thesis*, Loughborough University of Technology, England (1973).
8. Schindler, H. R., 'Delta modulation', *I.E.E.E. Spectrum*, 69–78, October (1970).
9. Panter, P. F., *Modulation, Noise and Spectral Analysis*, McGraw Hill, New York (1965).
10. Roberts, L. G., 'Picture coding using pseudo random noise', *I.R.E. Trans. Info. Theory*, **IT-8**, No. 2, 145–154, February (1962).
11. Limb, J. O., 'Design of dithered waveforms for quantized visual signals', *Bell Systems Tech. J.*, **48**, No. 7, 2555–2582, September (1969).
12. Jayant, N. S. and Rabiner, L. R., 'The application of dither to the quantization of speech signals', *Bell System Tech. J.*, **51**, No. 6, 1292–1304, July-August (1972).
13. Jayant, N. S. and Shipley, K., 'Multiple delta modulation of a speech signal', *Proc. I.E.E.E.*, 1382, September (1971).
14. Greefkes, J. A. and Riemens, K., 'Code modulation with digitally controlled companding for speech transmission', *Philips Technical Review*, **31**, No. 11/12, 335–353 (1970).

15. Donaldson, R. W. and Douville, R. J., 'Analysis subjective evaluation, optimization and comparison of the performance capabilities of PCM, DPCM, Δ-M, AM and PM voice communication systems', *I.E.E.E. Trans. Com. Tech.*, **COM-17**, 421–431, August (1969).
16. O'Neal, J. B. Jr., 'Delta modulation quantizing noise analytical and computer simulation results for Gaussian and television signals', *Bell Systems Tech. J.*, **45**, No. 1, 117–141, January (1966).
17. Hills, M. T. and Evans, B. G., *Telecommunications Systems Design*, Vol. 1, *Transmission Systems*, Allen & Unwin, London (1973).
18. Rao, V. G., 'Comparison of p.c.m., d.p.c.m. and d.m. for coding speech signals,' *J. Electron. Eng.*, Bangalore, **XVI**, No. 2, 89–93 (1972).
19. —. 'Calculated *R* values for the A-law,' C.C.I.T.T. Report COMXII, No. 77-E, Annex. (to draft Recommendation G172).

Chapter 10
ENCODERS WITH MULTI-LEVEL QUANTIZERS

10.1. INTRODUCTION

The basic d.m. systems uses a quantizer which has two quantization levels. Such a quantizer is easy to implement because it is a simple zero-crossing detector. It is ideally suited for use in binary transmission systems. While delta modulators are closed loop systems having quantizers with two quantization levels, pulse code modulation systems are open loop having quantizers with many levels, typically 32 to 8192. Linear and adaptive p.c.m. codecs are discussed in Chapter 9.

In this Chapter codecs are examined whose encoders have a feedback loop and a quantizer in the forward path. The quantizer will have a number of levels which always exceeds the two levels used in delta modulators but are generally less than those used in p.c.m. encoders. The number of levels are typically 4 or 8 although they may read 256 in some applications.

10.2. MULTI-LEVEL D.M. AND DIFFERENTIAL P.C.M.

If the zero crossing quantizer in the linear d.m. encoder shown in Figure 1.1 is replaced by one having more than two levels the resulting system is called multi-level d.m., Figure 10.1. In any practical arrangement the multi-level output signal $q^*(t)$ would be held and suitably filtered prior to transmission. The decoder is just a simple integrator and filter.

The number of quantization levels is usually restricted to four or eight, in order that the receiver should be able to identify these levels in the presence of channel noise. A system has been proposed[1] where the quantizer has seven levels, i.e. the output of the quantizer is zero when its input is zero; this is called a septernary p.a.m. system. This system has been considered for the transmission of TV signals. In order to successfully transmit multilevel signals the quality of the transmission channels must be better than those channels which handle binary signals. As the amount of computer traffic carried over the commercial telecommunication networks increases the movement away from networks designed primarily for telephony to those designed for high speed data transmission will probably be accelerated. With the passing of time it

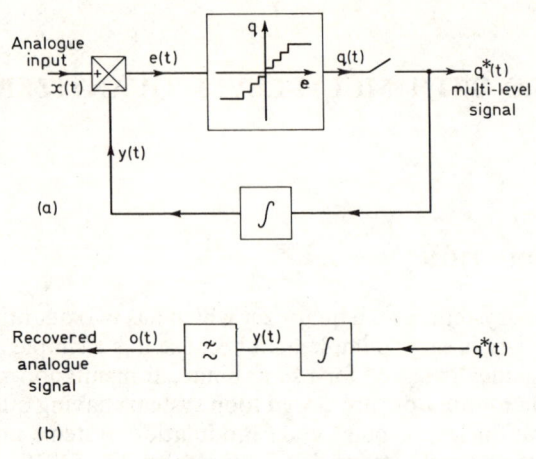

Fig. 10.1. Multi-level d.m. codec, (a) encoder; (b) decoder

seems likely that multi-level p.c.m. will replace binary p.c.m., and the introduction of multi-level d.m. and its derivatives will be serious rivals to multi-level p.c.m.

Although the discussion which follows is concerned with the binary transmission of multi-level d.m., the stability, quantization etc. of the encoder will be relevant for multilevel transmission.

If transmission is restricted to the binary type then the $q^*(t)$ samples obtained by sampling the quantized waveform $q(t)$ are binary encoded at a rate of nf_p, where n is the number of bits required to encode each of the N quantization levels and f_p is the clock rate of the multi-level d.m. encoder ($n = \log_2 N$).† The binary transmitted waveform $L(t)$ must be decoded at the receiver prior to integration and low-pass filtering. The arrangement is shown in Figure 10.2.

Suppose the n-bit encoder, i.e. p.c.m. word generator, is absorbed into the forward path of the multi-level d.m. encoder. In order to apply $q^*(t)$ to the integrator in the feedback path an n-bit decoder must be inserted in this path which decodes $L(t)$ into $q^*(t)$. The forward path now consists of a multi-level quantizer, sampler and n-bit encoder, and these three items constitute a p.c.m. encoder encoding error samples spaced $(1/f_p)$ seconds apart. The system is shown in Figure 10.3 and appears to be much more complicated than the arrangement of Figure 10.2. However, practical p.c.m. encoders[2] operate on a signal and produce the binary p.c.m. word in a single operation rather than sequentially quantizing sampling and encoding. For example, the A-law code generated by the A-law ΔM

---

† In the general case where the $q^*(t)$ samples are encoded into an m-level code, i.e. the transmitted $L(t)$ pulse can have any of m levels, $n = \log_m N$.

Encoders with multi-level quantizers

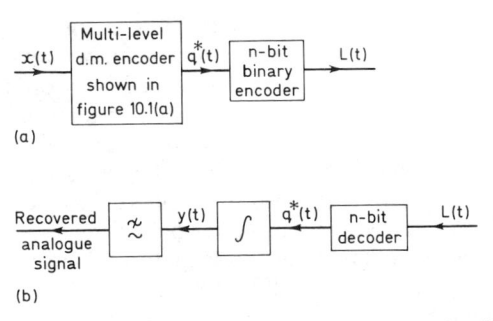

Fig. 10.2. Binary transmission of multi-level d.m. signal: (a) encoder, (b) decoder

described in Section 9.6 achieves the quantization with the aid of the d.m. section, and the A-code was effected by the adaptive counter. Consequently it may be easier to implement the arrangement shown in Figure 10.3 rather than that of Figure 10.2.

An alternative arrangement[3] is to use the system shown in Figure 10.2, but to change the quantizer in the multi-level d.m. encoder into two functions called the *classifier* and the *weighter*. The classifier contains the decision levels which divides the error signal into N ranges. The digital output from the classifier $C(t)$ is binary encoded by the external n-bit encoder for transmission. The signal $C(t)$ is also fed to the weighter which converts $C(t)$ in the quantization levels $q^*(t)$. The local decoder in the d.m. section is composed of the weighter and integrator, and consequently a weighter must be placed after the n-bit decoder in the receiver. In order to

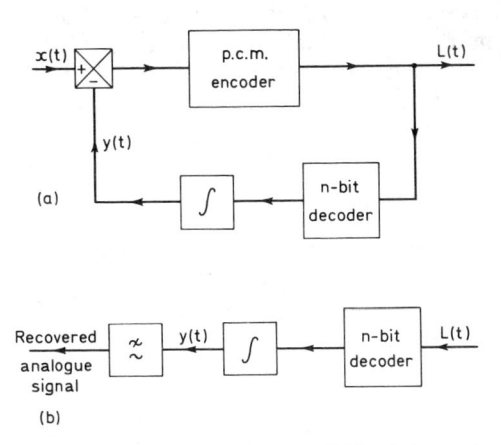

Fig. 10.3. Rearrangement of Fig. 10.2 to produce a multi-level d.m. codec having a binary output waveform L(t)

match the signals at the output of the integrators in the encoder and decoder the weighter and integrator circuits can be implemented digitally.

If the integrator in Figure 10.3 is replaced by a predictor[4,5] then the system is called differential p.c.m., or predictive quantizing.[4] A predictive system transmits the difference between the input signal and an estimate of this signal based on its history. If these differences are quantized prior to transmission the system is referred to as differential p.c.m. The predictors used in feedback encoders can be either linear[4] or non-linear.[6] However, differential p.c.m. encoders for example picturephone codec[7] often use the most simple of predictors–a leaky integrator.

In this chapter only the leaky integrator will be considered. This means that the multi-level d.m. using binary transmission is identical to differential p.c.m. using leaky integrators. In the literature d.m. is often described as one bit differential p.c.m. However, as differential p.c.m. may have complex predictors, the nomenclature used in this chapter to describe the codecs whose predictors are leaky integrators will be multi-level d.m. in preference to differential p.c.m. This nomenclature fits in with the remainder of the chapter when delta-sigma multi-level systems are considered.

10.2.1. Waveform tracking

The variation of the $y(t)$ signal at the output of the integrator in the multi-level d.m. encoder of Figure 10.1 is shown in Figure 10.4 for an arbitrary input signal $x(t)$. The quantizer has a uniform characteristic having eight levels, as illustrated by the solid line in Figure 10.5. In Figure 10.4 the corresponding $q^*(t)$ impulses are shown together with the $L(t)$ code assuming that each $q^*(t)$ impulse is encoded into a 3-bit binary code. The coding rules are given by Table 10.1. The $L(t)$ waveform is encoded prior to transmission using either alternate polarity coding or alternate digit inversion.

Table 10.1

Magnitude of $q^*(t)$	Binary code for $L(t)$
−3·5	1 1 1
−2·5	1 1 0
−1·5	1 0 1
−0·5	1 0 0
0·5	0 0 0
1·5	0 0 1
2·5	0 1 0
3·5	0 1 1

Encoders with multi-level quantizers 305

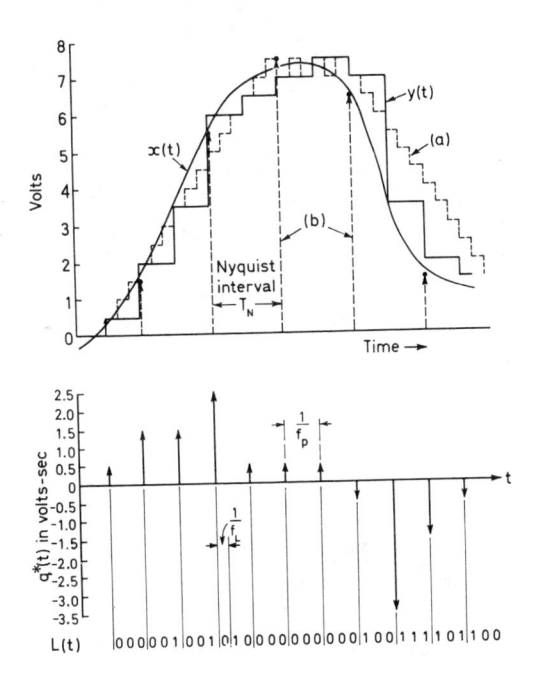

Fig. 10.4. For an arbitrary input signal x(t), the y(t) waveform and q*(t) samples apply for a multilevel d.m. encoder. Waveform (a) is the feedback waveform in a linear delta modulator, and the quantized samples in a linear p.c.m. encoder are labelled (b)

The clock rate f_p used in the multi-level d.m. encoder is a third of the bit rate of the line waveform $L(t)$, i.e. $f_L = 3f_p$. A linear d.m. encoder operating at a line rate of f_L and having a step size of 0·5 volts produces a response shown by Curve (a) in Figure 10.4 which is overloaded when the $x(t)$ input rapidly decreases.

Suppose $x(t)$ is pulse code modulated, i.e. it is uniformly quantized by the quantizer shown by the dotted line in Figure 10.5, then instantaneously sampled at the Nyquist rate and finally binary encoded into a six-bit word prior to transmission, where the line bit rate f_L is the same as that used for the three-bit multi-level d.m. As the p.c.m. system has twice as many bits in its code compared to multi-level d.m., f_p is twice the Nyquist rate. Hence the samples marked (b) in Figure 10.4, whose magnitudes are indicated by a circle, represent the quantized samples at the Nyquist rate. It can be seen that the $y(t)$ signal in the multi-level d.m. system differs from the quantized samples at the Nyquist instants spaced T seconds apart by less than 0·5 volts.

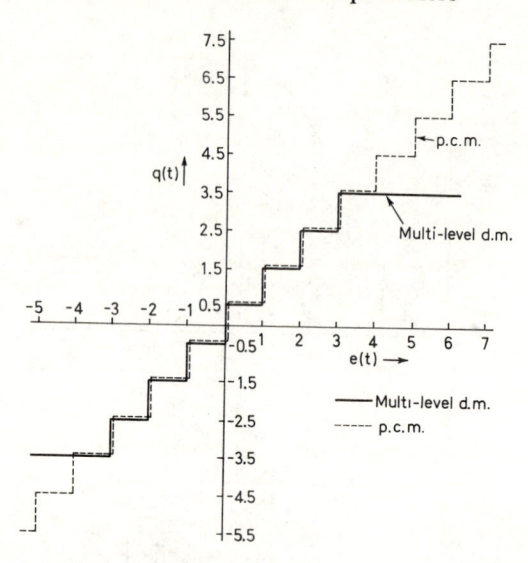

Fig. 10.5. *Quantizer characteristics; solid line applies to multi-level d.m. and the dotted line applies to p.c.m. (note, quantizers used in p.c.m. usually have the 'tread' rather than the 'riser' passing through the origin, as in the even staircase function shown in Fig. 1.13(a)*

In the case of the p.c.m. encoder overloading occurs when the amplitude of $x(t)$ exceeds the maximum quantization level, causing the peaks of $x(t)$ to be truncated, whereas the multi-level d.m. system is slope overloaded if the slope of $x(t)$ exceeds $3.5f_p$, as in the case of the 8-level quantizer shown in Figure 10.5. Consequently multi-level d.m. is well suited to input signals whose spectra match the overload characteristic, which decreases by 6 dB for every octave increase in frequency. Considered from the time rather than the frequency domain, because the multi-level d.m. system has an encoder which tracks the input signal, it is particularly useful when encoding signals which exhibit a high correlation between adjacent clock periods, i.e. speech and television signals.

10.2.2. Optimum quantizer characteristic

The act of quantizing an analogue signal $e(t)$ into N levels to produce a quantized signal $q(t)$ results in $e(t)$ being distorted. The distortion is some function of the difference between $e(t)$ and $q(t)$, and this section establishes the shape of a non-linear quantizer characteristic which minimizes this distortion when $e(t)$ is a signal having stationary statistical properties.

Bisection rule

Consider the non-uniform quantizer shown in Figure 10.6 which has been drawn from results derived by Max[8] when $e(t)$ is a Gaussian signal having zero mean and a mean square value σ_e^2 of unity. This characteristic has been optimised to give minimum mean square error between the input signal $e(t)$ and the quantized signal $q(t)$. It will be shown that this staircase characteristic has the property that if a point is placed at the centre of every vertical step a straight line can be drawn on the characteristic which passes through all of these points and through the origin,

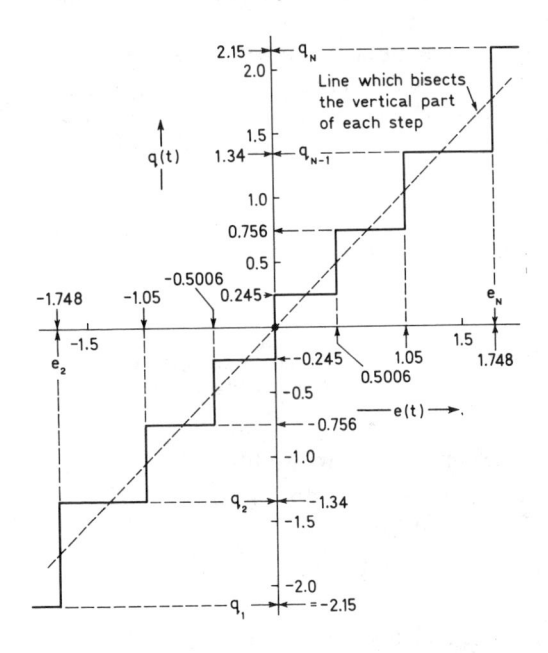

Fig. 10.6. *Optimum eight-level quantizer for Gaussian input signal having a mean square value of unity*

as shown in Figure 10.6. This feature of the optimum quantizer that a hypothetical straight line can be placed on its staircase function which bisects all the steps and has an inclination of 45° will be known here as the Bisection rule. This rule is independent of the number of quantization levels. If σ_e^2 is changed the values of $e(t)$ which cause changes in the quantized level have different values and the magnitude of these quantization levels are also altered, but nevertheless the Bisection rule is still applicable. In fact $e(t)$ need not be a Gaussian signal for the rule to apply, but it must have stationary statistical properties.

Theory of optimum quantizer

The Bisection rule will now be established, and the quantizers which conform to this rule will be used in the multi-level d.m. encoders whose performance is analysed in subsequent sections.

The procedure to derive the Bisection rule and to determine the minimum mean square error between the signals at the input and output of the quantizer is as follows.

Let the range of the input signal be divided up into N sections, where the end points of the sections are $e_1, e_2, \ldots, e_{N+1}$, and the constant output from each of these N sections is $q_1, q_2, \ldots, q_N$. The arrangement is shown in Figure 10.7.

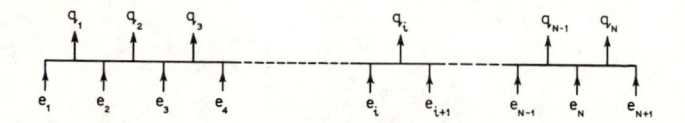

Fig. 10.7. The N sections of the amplitude range of the input signal $e(t)$ to the quantizer and the corresponding quantization levels

Let $f(e)$ be the amplitude probability density function of $e(t)$ and assume that it has stationary statistical properties. The distortion D will be defined as the mean square error

$$D = \langle \{e(t) - q(t)\}^2 \rangle$$

where $\langle \rangle$ means expected value. Let the expected value of D be D_i where $e_i < e(t) < e_{i+1}$, i.e. over the region where $e(t)$ is quantized to q_i. Then

$$D_i = \int_{e_i}^{e_{i+1}} (e(t) - q_i)^2 f(e) \, de(t)$$

and the total distortion is

$$D_t = \sum_{i=1}^{N} D_i$$

provided that the end points e_1 and e_{N+1} shown in Figure 10.7 are made to approach $\pm \infty$, respectively as $f(e)$ is a Gaussian probability density function.

$$D = \int_{e_1}^{e_2} \{e(t) - q_1\}^2 f(e) \, de(t) + \cdots + \int_{e_{j-1}}^{e_j} \{e(t) - q_{j-1}\}^2 f(e) \, de(t)$$
$$+ \int_{e_j}^{e_{j+1}} \{e(t) - q_j\}^2 f(e) \, de(t) + \cdots + \int_{e_N}^{e_{N+1}} \{e(t) - q_N\}^2 f(e) \, de(t)$$
(10.1)

To find the minimum value of distortion for a fixed number of quantization steps, as D is a function of two variables $e(t)$ and $q(t)$, D is partially differentiated with respect to the various values of e_i and q_i and the results equated to zero.

$$\frac{\partial D}{\partial e_j} = \frac{\partial}{\partial e_j} \left[\int_{e_{j-1}}^{e_j} \{e(t) - q_{j-1}\}^2 f(e) \, de(t) \right.$$
$$\left. + \int_{e_j}^{e_{j+1}} \{e(t) - q_j\}^2 f(e) \, de(t) \right]$$

$$\frac{\partial D}{\partial e_j} = (e_j - q_{j-1})^2 f(e_j) - (e_j - q_j)^2 f(e_j) \tag{10.2}$$

where $j = 2, 3, \ldots, N$. Also

$$\frac{\partial D}{\partial q_j} = \frac{\partial}{\partial q_j} \left[\int_{e_j}^{e_{j+1}} \{e(t) - q_j\}^2 f(e) \, de(t) \right]$$
$$= -2 \int_{e_j}^{e_{j+1}} \{e(t) - q_j\} f(e) \, de(t) \tag{10.3}$$

where $j = 1, 2, 3, \ldots, N$.

For a minimum, Equations 10.2 and 10.3 must be equated to zero. There are other conditions for the critical point depending on the higher derivatives but from physical considerations the point found in this way will give minimum distortion.

Because the amplitude of the error is of importance and not its sign, Equation 10.2 when equated to zero can be written

$$|e_j - q_{j-1}|^2 = |e_j - q_j|^2$$

and as $q_{j-1} \neq q_j$

$$e_j = \frac{q_j + q_{j-1}}{2}; \quad j = 2, 3, \ldots, N \tag{10.4}$$

This result establishes the Bisection rule.

Setting $\partial D / \partial q_j$ to zero gives

$$\int_{e_j}^{e_{j+1}} e(t) f(e) \, de(t) = q_j \int_{e_j}^{e_{j+1}} f(e) \, de(t) \quad j = 1, \ldots, N \tag{10.5}$$

Equation 10.5 can be stated as

mean value of $e(t)$ in range $e_j \leq e(t) \leq e_{j+1}$
$= q_j \times$ probability of $e(t)$ being in the range $e_j \leq e(t) \leq e_{j+1}$

Therefore, q_j is the centroid of area $f(e)$ between e_j and e_{j+1}. In order to determine the optimum e_i's and q_i's for a given $f(e)$ and N a particular value is ascribed to q_1, and the subsequent e_i's and q_i's are calculated using Equations 10.4 and 10.5. If q_N comes out as the centroid of the area of $f(e)$ between e_N and ∞, then the optimum e_i's and q_i's are determined. However, if q_N does not fulfill this condition q_1 is changed and the procedure repeated. Should the criterion for q_N stated above still not be fulfilled then the magnitude of q_1 is again altered. By repeatedly changing the magnitude of q_1 (with the aid of a computer) the criterion for q_N is eventually satisfied, whence the quantization characteristic is completely specified.

When $e(t)$ has a Gaussian probability density function

$$D_i = \frac{1}{\sigma(2\pi)^{1/2}} \int_{e_i}^{e_{i+1}} \{e(t) - q_i\}^2 \exp\left(-\frac{e^2(t)}{2\sigma^2}\right) de(t)$$

Suppose $e(t)$ is scaled by σ, then $q_i(t)$ must be similarly scaled, i.e.

$$e(t) = \sigma\varepsilon(t)$$

and

$$q_i = \sigma Q_i$$

Consequently

$$D_i = \frac{\sigma^2}{(2\pi)^{1/2}} \int_{\varepsilon_i}^{\varepsilon_{i+1}} \{\varepsilon(t) - Q_i\}^2 \exp\left(-\frac{\varepsilon^2(t)}{2}\right) d\varepsilon(t)$$

and therefore the mean square value of the error signal is proportional to the mean square value of the input signal, i.e. $D \propto \sigma^2$.†

Max[8] quotes results in tabular form for a Gaussian probability density function whose standard deviation σ is unity and where $e_{(N/2)+1} = 0$ for N even and $q_{(N+1)/2} = 0$ for N odd. The characteristic displayed in Figure 10.6 is drawn from these results with $N = 8$. The relationship between the mean square error D and the number of quantization levels N for $\sigma^2 = 1$ is

$$D = dN^{-k}$$

where d and k have values which depend on the number N of quantization levels. For example in multi-level d.m. encoders the quantizers often have only a small number of levels, say 4 to 8. For these quantizers the

† The property, $D \propto \sigma^2$, was pointed out to the author by his colleague, H. W. Pakes.

Encoders with multi-level quantizers 311

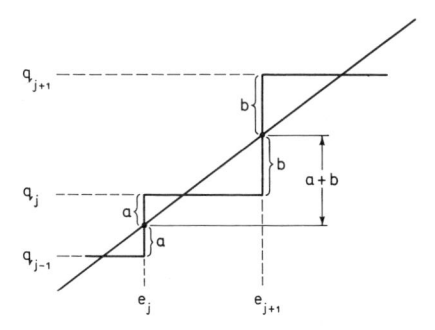

Fig. 10.8. *Super-position of a straight line on the optimum quantization characteristic*

approximate values of d and k are 1·3 and 1·76, respectively. However, for larger values of N, say 36, $d = 2 \cdot 21$ and $k = 1 \cdot 96$. As expected D decreases with increasing N. For the characteristic shown in Figure 10.6, $D \simeq 0 \cdot 033$.

Because D is proportional to σ^2, Max's result for Gaussian signals having $\sigma^2 = 1$ can, for the general case when $\sigma^2 \neq 1$, be amended to

$$D = dN^{-k} \cdot \sigma_e^2 \qquad (10.6)$$

Figure 10.8 shows a part of the quantization characteristic and complies with the previous theory, and in particular with Equation 10.4. A straight line $q(t) = me(t)$ has been placed on the characteristic which bisects the vertical part of each step. The slope of this straight line can be seen to be from the figure

$$m = \frac{a+b}{e_{j+1} - e_j} = \frac{\{(q_j - q_{j-1})/2\} + \{(q_{j+1} - q_j)/2\}}{e_{j+1} - e_j}$$

$$= \frac{\{(q_{j+1} + q_j)/2\} - \{(q_j + q_{j-1})/2\}}{e_{j+1} - e_j}$$

and from Equation 10.4

$$m = \frac{e_{j+1} - e_j}{e_{j+1} - e_j} = 1$$

Hence the optimum quantizer having more than two output levels has the characteristic that a straight line drawn through the mid-point of each step always has a slope of unity.

10.2.3. Quantizer and encoder distortion, s.q.n.r.

Having decided that the transmission is to be of the binary type and that output samples will be encoded into n bit words, then the number of steps N in the quantizer is fixed at 2^n. Suppose that the input signal to the encoder is of the Gaussian type and that its power level is adjusted to avoid overloading the multi-level d.m. encoder. The error signal $e(t)$ will be a close approximation to a Gaussian signal and will have its amplitude restricted by the feedback action of the encoder to avoid overloading the quantizer except on rare occasions. Consequently the mean square value of $e(t)$ is close to the value used in designing the quantizer, i.e. by adjusting the input level, the minimum quantization distortion is obtained. There is no need for the system to be over-sensitive to the input level as the feedback loop will endeavour to maintain $e(t)$ close to the optimum value, although only over a limited range.

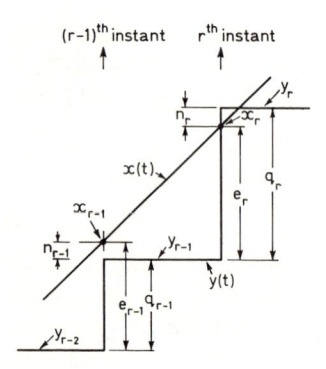

Fig. 10.9. Feedback signal $y(t)$ in a multi-level d.m. encoder

Figure 10.9 shows an arbitrary input signal $x(t)$ and the corresponding feedback, error and quantized signals for the multi-level d.m. encoder shown in Figure 10.1. It is assumed that this encoder has an output signal $q^*(t)$ which is a sequence of impulses and that an ideal integrator is used in its feedback loop. Between the $(r-1)$th and the rth sampling instants $y(t)$ is constant and equal to the value it had at the $(r-1)$th instant, namely y_{r-1}; consequently as $x(t)$ is increasing $e(t)$ and $q(t)$ increase. At the rth sampling instant the $y(t)$ signal is still y_{r-1} and the error signal

$$e_r = x_r - y_{r-1}$$

For this value of e_r an impulse of magnitude q_r is generated, i.e.

$$q(t) = q^*(t) = q_r$$

at the rth sampling instant; because of the ideal integrator in the feedback loop, $y(t)$ increases immediately to y_r after the rth sampling instant.

Encoders with multi-level quantizers

The output q_r of the quantizer at the rth sampling instant is $e_r + n_r$, where n_r is the error component in the quantizing process. Thus at the rth instant the input, error and quantized and feedback signals have values x_r, e_r, q_r, and y_{r-1}, respectively. Immediately after the rth sampling instant where x_r has increased by only an infinitesimal amount, $y(t)$ has increased by q_r

$$y_r = y_{r-1} + q_r$$

and $e(t)$ and $q(t)$ make rapid changes. However, these changes do not effect $y(t)$ because the sampling switch in Figure 10.1 remains open until the $(r+1)$th sampling instant.

The difference equations for the encoder shown in Figure 10.1 are

$$e_r = x_r - y_{r-1}$$

$$q_r = e_r + n_r$$

$$y_{r-1} = y_{r-2} + q_{r-1}$$

Rearranging these equations yields

$$x_r - y_{r-2} = q_{r-1} + q_r - n_r$$

but

$$q_{r-1} + q_r = y_r - y_{r-2}$$

and consequently

$$n_r = q_r - e_r = y_r - x_r$$

or

$$y_r = x_r + n_r$$

In the above equation n_r can have positive or negative values. Consequently the distortion samples $\{n_r\}$ at the output of the quantizer in the encoder are equal to the noise samples at the output of the integrator in the decoder, provided there are no transmission errors. When the encoder is correctly tracking a Gaussian input signal $x(t)$, which results in the signal $n(t)$ having a mean square value D defined by Equation 10.6, the noise power at the output of the decoder is D. If $x(t)$ has a mean square value of σ_x^2, the s.q.n.r. for the codec is

$$\text{s.q.n.r.} = \frac{\sigma_x^2}{D} = \frac{N^k}{d}\left(\frac{\sigma_x}{\sigma_e}\right)^2$$

and by putting $N = 2^n$ the s.q.n.r. can be represented by

$$\text{s.q.n.r.} = 3nk - 10\log_{10} d + 10\log_{10}\left(\frac{\sigma_x}{\sigma_e}\right)^2 \text{ dB} \qquad (10.7)$$

When N is small, say 4 to 8 levels, $d \doteqdot 1\cdot3$ and $k \doteqdot 1\cdot76$. The s.q.n.r. becomes

$$\text{s.q.n.r.} = 5\cdot3n - 1\cdot14 + 10 \log_{10} \left(\frac{\sigma_x}{\sigma_e}\right)^2 \text{ dB}$$

When N is 36, d and k have values of $2\cdot2$ and $1\cdot96$, respectively and

$$\text{s.q.n.r.} = 5\cdot9n - 3\cdot44 + 10 \log_{10} \left(\frac{\sigma_x}{\sigma_e}\right)^2 \text{ dB}$$

This latter equation is a close approximation to the one given by O'Neal.[10,11]

The spectral density function of the quantization error signal $n(t)$ is relatively flat over the message band provided $e(t)$ is continuously varying over its complete range when the encoder is tracking the input signal $x(t)$. This spectral density function has a peak of $f_p/2$ and a minimum in the region $f_p/4$ when the full range of the quantizer levels are not being continuously covered. At the other extreme when overloading occurs there is a considerable rise in the distortion at the low frequencies.

When the adjacent sample correlation of the input signal $x(t)$ is normalised (by dividing it by its mean square value) and its magnitude exceeds $0\cdot5$, then multi-level d.m. has a superior signal-to-noise ratio than p.c.m.[5] Multi-level d.m. having a 16 level quantizer requiring 4 bits per word transmission when used for the binary transmission of speech has a signal-to-noise ratio which is about 6 dB better than μ-law logarithmic p.c.m.,[5] when μ has a value of 100 and the sampling rate is 8 kHz. μ-law logarithmic p.c.m.[14] is used in the U.S.A. for the transmission of speech. Its companding law is similar to the A-law shown in Figure 9.5. Alternatively, for the same signal-to-noise ratio multi-level d.m. has one bit less than μ-law logarithmic p.c.m. This multi-level d.m., i.e. sub-optimal differential p.c.m., has an inferior signal-to-noise ratio of about 4 dB for the above speech transmission compared to optimal differential p.c.m.

10.2.4. Linear model of quantizer

The optimum quantization characteristic has been shown to have a slope of unity. Consequently, a convenient model for an optimum quantizer is a saturated amplifier having unity gain over its linear region,

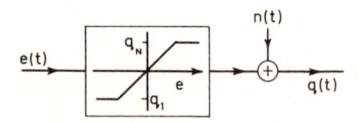

Fig. 10.10. Linear model of the quantizer

followed by a noise source $n(t)$ which allows for quantization noise. The model is shown in Figure 10.10. The saturation levels of the amplifier have been added to emphasise that the mean slope of the characteristic having a value of unity only applies over the range from q_1 to q_N. Assuming that the linear region of the amplifier is used, i.e. there is no peak clipping of the input signal $e(t)$, then the output signal $q(t)$ from the linear model is deemed to be a continuous signal.

$$q(t) = e(t) + n(t), \qquad q_1 \le q(t) \le q_N$$

where $n(t)$ is a function of $e(t)$.

The model assumes that all the quantization levels are used according to the amplitude probability density function of $e(t)$. For example, it cannot be used for a d.c. signal as $q(t)$ has then only one value and differs by a fixed amount from $e(t)$.

10.2.5. Linear analysis

The multi-level d.m. system shown in Figure 10.1 will now be analysed for input signals which do not overload the encoder.[9] It will be assumed that the model of the quantizer proposed in Figure 10.10 is valid and the amplifier operates in its linear region. By representing the multi-level quantizer having more than two levels by the simple linear equivalent

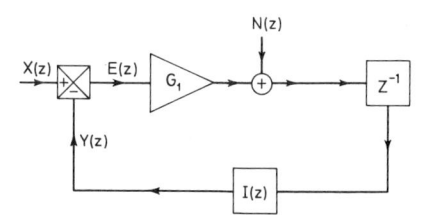

Fig. 10.11. Linear representation of the multi-level d.m. encoder

circuit of Figure 10.11 a linear representation of the encoder is obtained, where $X(z)$, $E(z)$, $Q(z)$, $Y(z)$ and $N(z)$ are the z-transforms of the signals $x(t)$, $e(t)$, $q(t)$. $y(t)$ and $n(t)$, respectively. It will further be assumed that the encoder is stable. The conditions governing stability are stated in Section 10.2.6.

When compared to Figure 10.1 it will be seen that Figure 10.11 contains an additional amplifier in the forward path having gain G_1. This has been included to determine if any benefit can be obtained by amplifying the error signal.

The z^{-1} block is necessary in Figure 10.11 to account for the one bit delay produced by the sampling process. As the analysis to be used involves the use of z-transforms, it does not matter if the encoder uses a zero order hold circuit after the sampler because the z-transform of such a circuit is unity. The integrator is represented by the pulse transfer function $I(z)$.

Writing the z-transform from Figure 10.11

$$Y(z) = \{E(z)G_1 + N(z)\}z^{-1}I(z) \qquad (10.8)$$

$$\frac{Y(z)}{E(z)} = G_1 I(z) z^{-1} + \frac{N(z)z^{-1}I(z)}{E(z)}$$

Let the open-loop gain in the absence of noise be

$$J(z) = \frac{Y(z)}{E(z)} = G_1 I(z) z^{-1} \qquad (10.9)$$

Since $E(z) = X(z) - Y(z)$ and using Equations 10.8 and 10.9

$$Y(z) = \frac{J(z)}{1+J(z)}\left\{X(z) + \frac{N(z)}{G_1}\right\} \qquad (10.10)$$

Suppose the integrator is the leaky type having

$$I(s) = \alpha/(s+\beta)$$

and its corresponding z-transform function

$$I(z) = \frac{\alpha}{1 - z^{-1}\exp(-\beta T)} \qquad (10.11)$$

β is zero for an ideal integrator and $1/(RC)$ for an RC-type integrator: α is a gain factor. Let

$$B = \exp(-\beta T) \qquad (10.12)$$

Generally, B has a value which is close to unity. Substituting $I(z)$ in Equation 10.9

$$J(z) = \frac{\alpha G_1 z^{-1}}{1 - z^{-1} B} \qquad (10.13)$$

Frequency domain

Writing $z = \exp(j2\pi f T)$, Equation 10.13 becomes

$$J \exp(j2\pi f T) = \frac{\alpha G_1}{\exp(j2\pi f T) - B}$$

and confining the analysis to frequencies $\leq f_{c2}$ where f_{c2} is the frequency

Encoders with multi-level quantizers 317

in the message input signal $x(t)$

$$|J(j2\pi f)| = \frac{\alpha G_1}{[(1-B)^2 + 4B \sin^2(\pi f/f_p)]^{1/2}}$$

$$\simeq \frac{\alpha G_1}{2\sqrt{B} \sin(\pi f/f_p)} \qquad (10.14)$$

since $(1-B) \simeq 0$. In the frequency domain the $J(z)/(1+J(z))$ term in Equation 10.10 becomes

$$\left|\frac{J(j2\pi f)}{1+J(j2\pi f)}\right| = \frac{1}{1+\left(\frac{2\sqrt{B}}{\alpha G_1}\right)\sin(\pi f/f_p)} \qquad (10.15)$$

The signal $y(t)$ at the output of the integrator in the decoder is filtered to produce the final decoded signal $o(t)$. The r.m.s. power spectral density function of the $y(t)$ signal which resides in the message band is therefore equal to the r.m.s. power spectral density function $|o(j2\pi f)|$ of the recovered signal $o(t)$. Consequently, from Equations 10.10 and 10.15

$$|o(j2\pi f)| = \frac{1}{1+\left(\frac{2\sqrt{B}}{\alpha G_1}\right)\sin(\pi f/f_p)}\left\{|X(j2\pi f)| + \frac{N(j2\pi f)}{G_1}\right\} \qquad (10.16)$$

where $|X(j2\pi f)|$ and $|N(j2\pi f)|$ are the r.m.s. power spectral densities at the baseband of $x(t)$ and $n(t)$, respectively.

For the recovered signal $o(t)$ to have no frequency distortion, Equation 10.15 is required to have a value of unity. This occurs only when $f = 0$, and is approximately valid when

$$f \ll \frac{f_p}{2} \quad \text{and} \quad f \ll \frac{\alpha G_1 f_p}{2\pi\sqrt{B}} \qquad (10.17)$$

It will be shown that $\alpha G_1 = 2$ is the highest value which will ensure the encoder is stable; for optimum idling $\alpha G_1 < \frac{4}{3}$. Hence for no frequency distortion in $o(t)$, $f_p \gg 4.7f$ since $B^{1/2} \simeq 1$; this cannot always be satisfied due to restrictions on channel bandwidth.

The r.m.s. quantization noise spectral density function

$$|N_Q(j2\pi f)|_\Delta = \frac{|N(j2\pi f)|}{G_1\left\{1+\frac{2\sqrt{B}}{\alpha G_1}\sin(\pi f/f_p)\right\}} \qquad (10.18)$$

If Inequality 10.17 is satisfied

$$|N_Q(j2\pi f)|_\Delta \simeq \frac{|N(j2\pi f)|}{G_1} \tag{10.19}$$

Equation 10.19 means that if G_1 is unity the r.m.s. noise component in the signal $o(t)$ at the output of the decoder approximates to the r.m.s. value of the noise signal generated in the multi-level quantization process. The effect is mentioned in Section 10.2.3. However, by introducing an amplifier in the forward path of the encoder having a gain G_1 the decoded noise signal is attenuated. However, G_1 cannot be unduly large and is in fact restricted as discussed in Section 10.2.8.

Signal-to-quantization noise ratio

The input Gaussian $x(t)$ applied to the multi-level d.m. encoder having a gain G_1 in its forward path has a mean square value of

$$\sigma_x^2 = \int_{-f_{c2}}^{f_{c2}} |X(j2\pi f)|^2 \, df$$

where $X(j2\pi f)$ is the Fourier transform of $x(t)$. The mean square value of the noise component in the decoded signal is obtainable from Equation 10.18

$$N_q^2 = \int_{-f_{c2}}^{f_{c2}} |N_Q(j2\pi f)|_\Delta^2 \, df$$

Assuming that $|N_Q(j2\pi f)|_\Delta^2$ is flat over the message band the s.q.n.r. becomes

$$\text{s.q.n.r.} = \frac{S^2}{N_q^2} = \frac{\int_{-f_{c2}}^{f_{c2}} |X(j2\pi f)|^2 \, df}{|N_Q(j2\pi f)|_\Delta^2 2f_{c2}} \tag{10.20}$$

If Inequality 10.17 applies and using Equation 10.19 the above expression for the s.q.n.r. becomes

$$\text{s.q.n.r.} = \frac{S^2}{N_q^2} = \frac{G_1^2 \int_{-f_{c2}}^{f_{c2}} |X(j2\pi f)|^2 \, df}{|N(j2\pi f)|^2 2f_{c2}}$$

or

$$\text{s.q.n.r.} = \frac{G_1^2 \sigma_x^2}{|N(j2\pi f)|^2 \cdot 2f_{c2}} \tag{10.21}$$

Encoders with multi-level quantizers

When G_1 and α are unity the denominator in the above equation is equal to D, i.e. Equations 10.7 and 10.21 are equivalent representations of the s.q.n.r.

It will be recalled that in the opening remarks of this section, on linear analysis, it was decided to place an amplifier having a gain G_1 between the output of the subtractor and the input to the quantizer to see if this strategy would produce any desirable features. Equation 10.21 suggests that a large value of G_1 will give the beneficial effect of a large s.q.n.r. However, the linear model (shown in Figure 10.11) of the multi-level d.m. encoder is based on the assumption that slope overloading does not occur, i.e. the variations of the signal $e(t)$ are confined to the linear part of the quantization characteristic shown in Figure 10.10. For a quantizer having a small number of levels the encoder may become slope overloaded if G_1 is in excess of unity for smaller slopes of $x(t)$. Consequently if G_1 is made large with the object of substantially increasing the s.q.n.r. the linear model is invalidated and Equation 10.21 will not apply. It will be shown in subsequent sections that the gains G_1 and α in the forward and feedback paths respectively of the multi-level d.m. encoder determine the encoder's stability and the minimum idling amplitudes of the feedback waveform $y(t)$. The restrictions on G_1 and α are discussed in Section 10.2.8.

10.2.6. Stability of the multi-level d.m. codec

To determine the stability of a multi-level d.m. codec the roots of the characteristic equation obtained from Equation 10.10 must be examined. The characteristic equation is

$$1 + J(z) = 0$$

and if the roots of this equation lie inside the unit circle the encoder is stable. Substituting for $J(z)$ from Equation 10.13 in the above equation yields

$$1 + J(z) = \frac{z + \alpha G_1 - B}{z - B} = 0$$

This equation has only one zero, $z = B - \alpha G_1$. The encoder is therefore stable if αG_1 satisfies the inequality

$$B - 1 < \alpha G_1 < 1 + B \tag{10.22}$$

As B is close to unity, even for a leaky integrator, the range of αG_1 which will ensure that the encoder is stable is

$$0 < \alpha G_1 < 2 \tag{10.23}$$

Encoders with multi-level quantizers

If αG_1 complies with the above inequality when encoding $x(t)$ the oscillations in the feedback waveform will not increase in amplitude.

10.2.7. Idle channel condition

When the encoder is idling it is desirable that the output from the quantizer oscillates between the lowest quantization levels at a frequency equal to half the clock rate. For a discussion of idle channel noise see Section 5.1.

Consider that the integrator has no leakage, i.e. $B = 1$ in Equation 10.12. At a particular sampling instant the value of the signal at the output of the integrator is

$$y_{r-1} = y_{r-2} + \alpha q_{r-1} \qquad (10.24)$$

where α is the gain of the integrator, and q_{r-1} is the output from the quantizer after it has been delayed by one clock period due to the sampling process. As the input signal $x(t)$ is zero in the idling condition $x_r = 0$ and $e_r = -y_{r-1}$. An amplifier having a gain of G_1 is located between the output of the subtractor in the encoder where the error signal $e(t)$ is produced and the input to the multi-level quantizer. The input sample to the quantizer is therefore

$$\varepsilon_r = -G_1 y_{r-1}$$

At the previous sampling instant the magnitude of the sample presented to the quantizer was

$$\varepsilon_{r-1} = -G_1 y_{r-2}$$

The change in the magnitude of these samples is

$$\varepsilon_r - \varepsilon_{r-1} = -G_1(y_{r-1} - y_{r-2})$$

and from Equation 10.24

$$\varepsilon_r - \varepsilon_{r-1} = -\alpha G_1 q_{r-1} \qquad (10.25)$$

The input sample to the quantizer therefore changes at each sampling instant by αG_1 multiplied by the quantized level produced at the previous sampling time.

If the quantizer has a uniform characteristic, as shown in Figure 10.12, and initially $q_r = q_B$, then at the next clock instant the input to the quantizer becomes $-\alpha G_1 q_B$. For the encoder to idle with $q(t)$ oscillating between $\pm q_A$

$$\alpha G_1 q_B \leq 2\varepsilon_A \qquad (10.26)$$

The equality sign in Equation 10.26 will just ensure minimum oscillations in the idling condition.

Encoders with multi-level quantizers

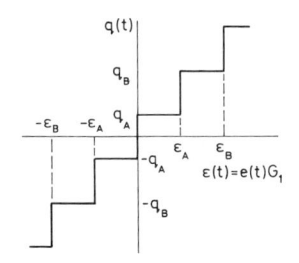

Fig. 10.12. Uniform quantization characteristic in the vicinity of the origin

In Figure 10.12, $q_B = 3q_A$ and $\varepsilon_A = 2q_A$ so that Inequality 10.26 becomes

$$\alpha G_1 < \tfrac{4}{3} \qquad (10.27)$$

This restriction overrides that imposed by the stability consideration of Inequality 10.23 which prevents the $y(t)$ waveform from exhibiting violent oscillations. If αG_1 is prevented from exceeding $\tfrac{4}{3}$ the oscillations will be a minimum when the encoder is in its idling state and the oscillations will decay when encoding a signal $x(t)$.

For the non-uniform quantizer of the type shown in Figure 10.6 the restrictions on αG_1 are found from Equations 10.26 and 10.4, i.e.

$$\alpha G_1 \leq \frac{2}{q_B}\left(\frac{q_A + q_B}{2}\right) = \frac{q_A}{q_B} + 1 \qquad (10.28)$$

For the 8-level quantization characteristic shown in Figure 10.6

$$\frac{q_A}{q_B} = 0\cdot 324$$

10.2.8. Limitations on gain

G_1 is the gain of the amplifier which is placed between the output of the subtractor and the input to the quantizer, while α is the gain of the integrator in the feedback loop. The previous two sections showed that the stability of the encoder and the correct idling conditions are dependent on the product of α and G_1. The maximum value of αG_1 is determined by either Equations 10.27 or 10.28 depending on whether the quantizer is uniform or non-uniform, respectively.

It is shown in the next section that for $G_1 \gg 1$ and the number N of quantization levels limited to 8, for example the encoder will be overloaded when the slope of $x(t)$ is less than the value which causes slope overload to occur when $G = \alpha = 1$. When G_1 is made greater than unity and α is selected to give minimum idling oscillations, as described in Section 10.2.7, then the improved s.q.n.r. should be obtained according to Equation 10.21 provided the input signal does not cause slope overload to occur.

Consider a simple example to illustrate why the s.q.n.r. is enhanced by making $G_1 > 1$ and $\alpha < 1$. Suppose that the multi-level d.m. encoder has a quantizer whose characteristic is as shown in Figure 10.5. Let G_1 be 2 and from Equation 10.27 the corresponding value of α is $\frac{2}{3}$. Consider two values of the error signals. First when $e(t) = 0.4$ the input signal to the quantizer is 0.8, and from Figure 10.5 this signal is quantized to 0.5. The resulting change in $y(t)$ is $0.5\alpha = \frac{1}{3}$. Secondly when $e(t) = 0.8$, after amplification by G_1 the quantized signal has a magnitude of 1.5 which causes $y(t)$ to change by unity. When compared to the encoder having $G_1 = \alpha = 1$, errors of 0.4 and 0.8 result in the same quantized level of 0.5, and therefore the same change in $y(t)$. Thus by making $G_1 = 2$, $\alpha = \frac{2}{3}$; for a range of $e(t)$ amplitudes which do not produce slope overload a larger number of quantization levels occur compared to $G_1 = \alpha = 1$. The changes in $y(t)$ per clock period are αq_{r-1}, where q_{r-1} is the value of $q(t)$ at the previous sampling time, and are not excessive as $\alpha < 1$. Because the subtractor is a linear network the principle of superposition can be applied to note that at a sampling instant the change in the signal level at the input to the quantizer, which is due solely to the change in $y(t)$, is $\alpha G_1 q_{r-1}$, i.e. $(\frac{4}{3})q_{r-1}$. Using the symbols displayed in Figure 10.12, if the error signal $\varepsilon(t)$ at the input to the quantizer is $<\varepsilon_A/G_1$ the change in $y(t)$ is αq_A; as $\alpha < 1$, $y(t)$ tracks $x(t)$ with smaller error compared to $G_1 = \alpha = 1$, when the slope of $x(t)$ is small. However, when the encoder is overloaded and the largest quantization level q_N occurs, the signal responds to the overload condition faster if $\alpha > 1$. This leads to the notion of producing an adaptive multi-level d.m. system where the signal $q^*(t)$ at the output of the encoder shown in Figure 10.1 is connected to a logical control system in the feedback loop. The size of the steps in the feedback waveform $y(t)$ are determined in accordance with the algorithm employed in the logical control system. As an illustration, consider that G_1 is fixed and α varies. When the encoder is overloaded and the maximum level q_N is repeatedly generated the largest value of α is produced. For sequences of $q^*(t)$ other than $\pm q_N$, α is reduced compared to its overload value in a way which depends on the encoding algorithm; when the smallest quantization levels $\pm q_A$ occur α is set to its minimum value. Other adaptation strategies may involve the simultaneous adjustment of G_1 and α and/or the values of the quantization thresholds and levels.

10.2.9. Overload characteristic of multi-level d.m. codec

Multi-level d.m. systems experience slope overload when the slope of the input signal $x(t)$ exceeds the maximum slope which the feedback

Encoders with multi-level quantizers

waveform $y(t)$ can have. For the uniform quantizer having N quantization levels uniformly spaced by $2q_A$ volts the maximum quantization level is $(N-1)q_A$. If this level is continually produced due to the existence of an overload condition $y(t)$ increases at its maximum rate of $(N-1)q_A f_p$ volts/second. The overload characteristic applies when

$$x(t) = E_s \cos 2\pi f_s t$$

and is governed by the equation formed by equating the maximum slope of $x(t)$, namely $2\pi f_s E_s$ to the maximum slope of $y(t)$. For $B = 1$

$$2\pi f_s E_s = (N-1)q_A f_p$$

When the quantization levels in the quantizer are not uniformly spaced, and the maximum level is q_N, the overload characteristic is determined by

$$2\pi f_s E_s = q_N f_p$$

The overload characteristic therefore has the same shape as the characteristic of the linear d.m. system discussed in Section 1.2.5, see Equation 1.4, except that V is replaced by q_N. Over the message band the overload characteristic falls by 6 dB per octave increase in frequency, irrespective of whether the quantizer levels are uniformly or non-uniformly spaced.

When an amplifier having a gain G_1 is used after the subtractor in the multi-level d.m. encoder then all error signals $e(t)$ which are greater than or equal to e_m and less than e_N (where e_N is the decision value of $e(t)$ which gives the maximum quantization level q_N) will be quantized to q_N if

$$G_1 = e_N/e_m$$

Because of the requirements for minimum idling amplitude in the $y(t)$ waveform the 'gain' α in the feedback loop is restricted according to Inequality 10.27 and Equation 10.28. This means that if G_1 is sufficiently large, say above 1·4, then $\alpha < 1$. Consequently when the encoder is overloaded the change in $y(t)$ per clock period is αq_N, and the overload characteristic is therefore

$$2\pi f_s E_s = \alpha q_N f_p$$

For the uniform quantization characteristic the maximum value of α is just less than $4/(3G_1)$ and the above equation becomes

$$2\pi f_s E_s = \frac{4}{3} \cdot \frac{q_N f_p}{G_1}$$

10.3. MULTI-LEVEL D.S.M.

The multi-level d.s.m. encoder is formed by placing an integrator before the input to the multi-level d.m. encoder shown in Figure 10.1(a).

The decoder is simply a low-pass filter. The multi-level d.s.m. system is the same as the d.s.m. system described in Section 1.4 and shown in schematic form in Figures 1.9 and 1.10, with the important exception that the quantizer has more than two levels. As with the multi-level d.m. encoder described in Section 10.2, an amplifier having a gain of G_1 is included in series with the integrator in the forward path of the encoder; the output signal from the encoder is considered to be fedback directly to the error point. When used with pre-emphasising and de-emphasising filters at the encoder and decoder respectively it becomes Brainard and Candy's direct-feedback encoder.[9]

Using the same method of analysis as that employed in Section 10.2.5

$$|o(j2\pi f)| = \left|\frac{J(j2\pi f)}{1+J(j2\pi f)}\right| \cdot |X(j2\pi f)| + \frac{|N(j2\pi f)|}{|1+J(j2\pi f)|}$$

The noise r.m.s. spectral density function at the output of the decoder is

$$|N_Q(j2\pi f)|_{\Delta-\Sigma} = \frac{G_1}{|J(j2\pi f)|} \cdot |N_Q(j2\pi f)|_\Delta \qquad (10.29)$$

where the subscripts $\Delta - \Sigma$ and Δ applied to a symbol means that the symbol relates to a multi-level delta-sigma modulation codec and a multi-level d.m. codec respectively. For no frequency distortion Equation 10.15 must be equal to unity, i.e. Inequality 10.17 applies. If these conditions are satisfied

$$|N_Q(j2\pi f)|_{\Delta-\Sigma} = |N(j2\pi f)|/|J(2\pi f)|$$

and from Equation 10.14

$$|N_Q(j2\pi f)|_{\Delta-\Sigma} = \frac{2\sqrt{B}\sin(\pi f/f_p)}{\alpha G_1} \cdot |N(j2\pi f)|$$

The variation of the r.m.s. spectral density function as a function of frequency is approximately sinusoidal. At low frequencies

$$|N(j2\pi f)|_{\Delta-\Sigma} = \frac{2\pi\sqrt{B}}{\alpha G_1} \cdot \frac{f}{f_p} \cdot |N(j2\pi f)| \qquad (10.30)$$

where

$$f \ll \frac{\alpha G_1 f_p}{2\pi\sqrt{B}}$$

The spectral density function of the decoded noise is obtained by squaring Equation 10.30. A simplified expression for $|N(j2\pi f)|^2_{\Delta-\Sigma}$ may be obtained by noting that for a leaky integrator in the feedback loop of the encoder a

Encoders with multi-level quantizers 325

typical value of $\sqrt{B}$ is 0·98, while in order to ensure that the encoder idles with minimum quantization levels of $\pm q_A$, as described in Section 10.2.7, we require $\alpha G_1 = 1\cdot3$, approximately. Using these values of $\sqrt{B}$ and αG_1 the spectral density function of the noise at the output of the decoder is

$$|N(j2\pi f)|^2_{\Delta-\Sigma} = 22 \cdot \frac{f^2}{f_p^2} \cdot |N(j2\pi f)|^2 \qquad (10.31)$$

where $f \ll 0\cdot21 f_p$. This equation shows that the spectral density function of the quantization noise depends on the square of the frequency, a situation previously encountered in Section 2.4 in d.s.m. codecs.

The noise power is the integral over the baseband frequencies of the square of Equation 10.29 or if applicable, Equation 10.30. The s.q.n.r. is obtained as in Section 10.2.5.

The same conditions of stability and idling apply as for the multi-level d.m. system, namely that αG_1 is governed by Equation 10.28. In multi-level d.s.m. codecs the quantization noise can be reduced by increasing αG_1 but this appears to be of little advantage as this product cannot be increased above 1·3, approximately.

10.3.1. Pre- and de-emphasis

Suppose a baseband signal $b(t)$ is pre-emphasised by a filter $H_1(j2\pi f)$ to produce the signal $x(t)$ at the input to the encoder. Let the signal $y(t)$ in the decoder be de-emphasised by a filter $H_2(j2\pi f)$ prior to the final filtering process. $H_1(j2\pi f)$ and $H_2(j2\pi f)$ have relatively short time constants compared with that of the integrator in the encoder, and have been discussed in Section 8.5.

The linear analysis is modified by replacing

$$|X(j2\pi f)| \quad \text{by} \quad |H_1(j2\pi f) \cdot B(j2\pi f)|$$

where $B(j2\pi f)$ is the Fourier transform of $b(t)$ and

$$|o(j2\pi f)| \quad \text{by} \quad |o(j2\pi f) \cdot H_2(j2\pi f)|$$

The wanted components of the r.m.s. spectral density function $|o(j2\pi f)|$ at the output of the decoder are unchanged provided the product

$$H_1(j2\pi f) \cdot H_2(j2\pi f) = 1$$

However, the unwanted noise component of $|o(j2\pi f)|$ is modified by the de-emphasising filter and becomes

$$|N_Q(j2\pi f)|_{\Delta-\Sigma} \cdot |H_2(j2\pi f)|$$

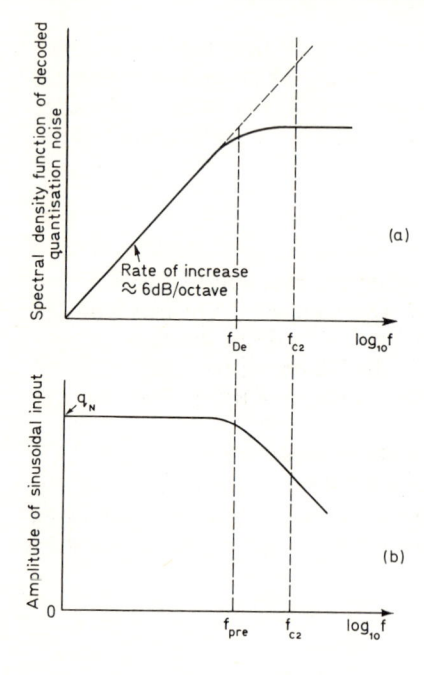

Fig. 10.13. Multi-level d.s.m. Effect of pre- and de-emphasis on (a) spectral density function of decoded quantization noise, (b) overload characteristic

Let $H_2(j2\pi f)$ be an *RC* integrator

$$H_2(j2\pi f) = \frac{f_{De}}{(f_{De}^2 + f^2)^{1/2}} \qquad (10.32)$$

where

$$f_{De} = \frac{1}{2\pi R_2 C_2} \gg f_1$$

and f_1 is the break frequency of the integrator in the encoder, $f_1 = (\beta/2\pi)$. The effect of $|H_2(j2\pi f)|$ only becomes significant for frequencies in the message band which are in the vicinity of or greater than f_{De}. This is illustrated by the noise characteristic shown in Figure 10.13(a).

A pre-emphasising *RC* circuit having a break frequency $f_{pre} = f_{De}$ reduces the overload characteristics at high frequencies as shown in Figure 10.13(b). Thus for frequencies greater than f_{pre} multi-level d.s.m. with pre- and de-emphasising filters behave as multi-level d.m. systems.

More complex filters[9] for $H_1(j2\pi f)$ and $H_2(j2\pi f)$ will give greater control of the overload and noise characteristics. The simple *RC* filters have practical value; they illustrate that if the noise characteristics are improved the overload characteristic deteriorates and vice versa. For signals whose spectrum falls at high frequencies the use of pre- and de-emphasis is desirable as by this means the overload condition is avoided and the noise is reduced where it matters most.

10.4. SLIDING SCALE SYSTEM

The encoders described in the last section can be companded by using two quantizers. The characteristics of these quantizers are shown in Figure 10.14, where the fine quantizer has been drawn with seven levels, and the coarse quantizer, with three.

The behaviour of the sliding scale encoder is readily appreciated by referring to Figure 10.15(a) which shows a delta-sigma type representation; to simplify the description switches S_1 to S_5 have been included. The integrator is assumed to be ideal and to receive impulse samples. At the commencement of a clock period S_1 closes and allows a sample of $x(t)$, say X_j, to be passed to the integrator to produce a signal at its output

$$E_{j-1} + X_j = I_j$$

where E_{j-1} is the integrated error from the previous clock period. Switch S_2 samples I_j and if $|I_j| < i_Q$ the output $Q(t)$ from the coarse quantizer

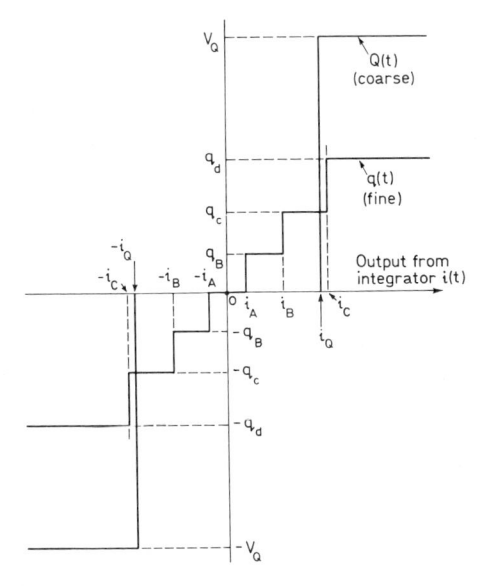

Fig. 10.14. Coarse and fine quantizers in sliding scale system. (After Brown[12])

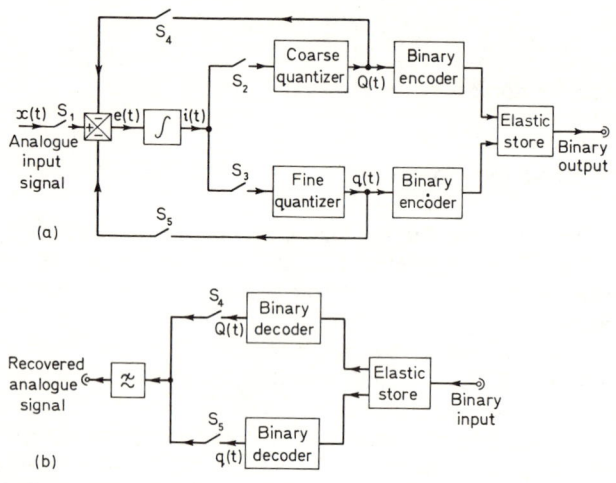

Fig. 10.15. *Sliding scale system:* (a) *encoder,* (b) *decoder.* (*After Brown*[12])

is zero, but if $|I_j| \geq i_Q$ a sample $\pm V_Q$ is fedback to the error point via switch S_4.

Suppose that $I_j \geq i_Q$: the sample $+V_Q$ is inverted at the error point and the signal at the output of the integrator becomes $E_{j-1} + X_j - V_Q$. This signal is now sampled by switch S_3 and the fine quantizer is brought into operation. Suppose that $E_{j-1} + X_j - V_Q$ lies between $-i_B$ and $-i_A$ (see fine quantizer characteristic). This means that when the coarse quantizer was brought into action it changed the sign of the integrated error. The output of the fine quantizer becomes $-q_B$; after sampling by switch S_5, polarity inversion at the error detector and subsequent integration the error

$$E_j = E_{j-1} + X_j - V_Q + q_B$$

The next clock period commences with switch S_1 closing and so on.

Without the coarse quantizer the largest amplitude $x(t)$ can have without causing peak clipping is $\pm q_d$. The introduction of the coarse encoder extends this value to $\pm(V_Q + q_d)$ and the effective quantization levels from 7 to 19. The values of the quantization levels in this encoder are

$$V_Q \pm q_d, \quad V_Q \pm q_c, \quad V_Q \pm q_B, \quad \pm V_Q,$$
$$\pm q_c, \quad \pm q_B, \quad 0,$$
$$-(V_Q \pm q_d), \quad -(V_Q \pm q_c), \quad -(V_Q \pm q_B)$$

The name 'sliding scale' is given to this encoder because the effective

Encoders with multi-level quantizers

quantization levels are those of either the fine quantizer which are around zero ($i(t) < i_Q$) or the levels of the fine quantizer slid to either $\pm V_Q$.

If the transmitted signal is in a binary form then the levels of the coarse quantizer can be encoded with two bits and those of the fine quantizer by three bits; only one binary encoder need be used. These binary signals must be multiplexed for transmission and de-multiplexed prior to decoding at the receiver. A five-bit multi-level d.m. system has a quantizer in its encoder with 32 levels which is more than the 23 available with the sliding-scale encoder. Consequently, from both performance and implementation criteria the five-bit multi-level d.s.m. is superior to the sliding scale encoder. However, the sliding-scale encoder only needs the services of the coarse quantizer when it is experiencing an overload condition. This encoder is therefore particularly suited to signals which use the fine quantizer for most of the time, but occasionally need both the fine and coarse quantizers. Because the receiver must distinguish between the serially transmitted two- and five-bit words additional word identification bits must also be transmitted. However, a superior lower bit-rate self-synchronizing encoding strategy[12] is to generate a 111 or 000 pattern if V_Q or $-V_Q$ occurs respectively followed by a 3-bit word representing the levels of the fine quantizer. Thus when both quantizers are used a 6-bit word is generated, but if only the fine quantizer is required a 3-bit word is formed. The decoder upon detecting a 111 or 000 code knows that the succeeding three bits apply to the fine quantizer and consequently there is no necessity to transmit synchronising information. As the $\pm q_d$ levels are only used in conjunction with the coarse quantizer since $i_Q < i_c$, they are allowed to have the same code words as $\pm V_Q$, e.g. $V_Q - q_d$ is coded as 111000.

The data rate of this encoder per sample period is not constant, and an elastic store is provided to regulate this variable encoding rate to give a constant transmission bit rate.

The receiver shown in Figure 10.15(b) has an elastic store to receive the constant data rate. If the coarse binary decoder identifies the presence of $\pm V_Q$, switch S_4 is closed and the data is passed to the filter for decoding. A second word is then removed from the store and decoded by the fine binary decoder to one of 7 possible levels before being fed to the filter via switch S_5. By proceeding in this manner for each encoded sample of $x(t)$ the latter signal is recovered.

A sliding scale encoder has been designed for encoding television signals.[12] For ease of implementation the value of i_Q is made just less than i_c which leads to 19 distinguishable quantization levels. When encoding television signals the system can function on a three bit per sample transmission channel if the sample rate of the sliding scale encoder is reduced by 5%. Its subjective performance approaches that of a seven-bit p.c.m. system.

10.5. SYLLABICALLY COMPANDED TWO-BIT P.C.M.

An uncompanded two-bit p.c.m. system has a quantizer with four levels, as shown in Figure 10.16. The quantizer characteristic together with p.c.m. code is shown in Table 10.2.

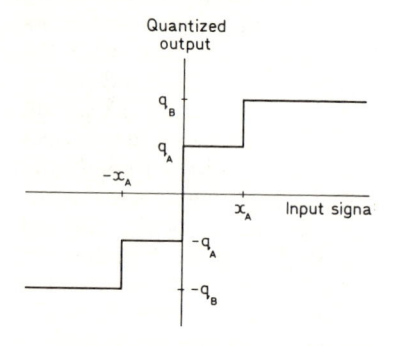

Fig. 10.16. Uniform four-level quantizer

The range of $x(t)$ which can be encoded is small, and generally the distortion is unacceptably high. However, the transmitted bit-rate is very low, i.e. twice the Nyquist rate, and for narrow-band channels this feature is highly desirable. In order to maintain the low bit rate, but considerably reduce the distortion, the syllabically companded two-bit p.c.m. system has been developed by Wilkinson.[13] It is designed for speech signals band-limited between 250 and 2,400 Hz and operates at a bit rate of 9.6 kbits/s.

The encoder is shown in Figure 10.17. Three simple quantizers are required which can be implemented with two i.c. operational amplifiers. Comparators C_1 and C_2 decide if the magnitude of the input $x(t) > X_t$ or $< -X_t$, respectively, and the final comparator C_3 decides the polarity of $x(t)$.

The outputs from C_1 and C_2, if any, are applied to an OR gate whose output is sampled at the Nyquist rate f_N and passed to a zero order hold circuit to produce the amplitude digits $L_a(t)$. The output from comparator C_3 is also sampled at the Nyquist rate to produce the polarity digits

Table 10.2

Range of input signal	Quantized output	p.c.m. Code Polarity	Magnitude
$-\infty$ to $-x_A$	$-q_B$	0	1
$-x_A$ to 0	$-q_A$	0	0
0 to $+x_A$	$+q_A$	1	0
$+x_A$ to ∞	$+q_B$	1	1

Encoders with multi-level quantizers 331

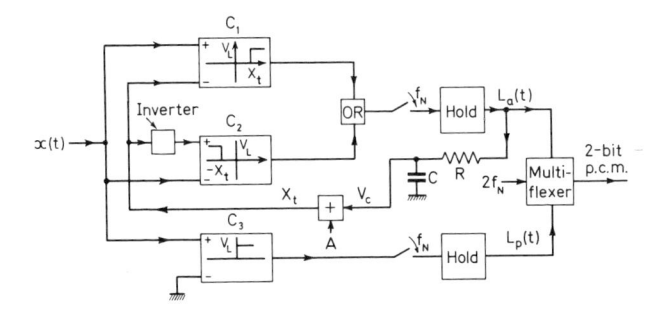

Fig. 10.17. *Syllabically companded 2-bit p.c.m. encoder.* (*After Wilkinson*[13])

$L_p(t)$. As $L_a(t)$ and $L_p(t)$ are binary signals the multiplexer receives these signals once every Nyquist interval and produces a two-bit word in accordance with the rules of Table 10.2.

This table represents the requirements for two-bit p.c.m. In the syllabically companded encoder the threshold levels are not fixed at $\pm x_A$ but vary in accordance with the ratio of logical ones to zeros in $L_a(t)$ and the time constant of the syllabic RC filter. The actual threshold X_t in Figure 10.17 is seen to be the sum of the voltage across the capacitor and a d.c. voltage A. If the peak amplitude of $|x(t)| < A$ the binary signal $L_a(t)$ consists of a sequence of zeros, $V_c = 0$ and $X_t = A$. The maximum value of V_c occurs when $x(t)$ overloads the coder causing long sequences of logical ones. This causes V_c to approach, exponentially, the binary level of $L_a(t)$, i.e. $\pm V_L$. Consequently the amplitude range of the coder is A to $V_L + A$.

When encoding $x(t)$ the value of the threshold X_t follows the variations in the amplitude of the envelope extracted from $L_a(t)$ by the syllabic RC filter. For example, if $x(t)$ is a sinusoid $E_s \sin \omega_s t$, the X_t threshold in the steady-state makes small variations due to the charging and discharging of the capacitor C, as illustrated in Figure 10.18. $L_a(t)$ is periodic as shown; when

$$X_t = X_{t_1}, \quad L_a(t) \text{ changes from } V_L \text{ to } 0$$

and when

$$X_t = X_{t_2}, \quad L_a(t) \text{ changes 0 to } V_L$$

Hence, $L_a(t) = 0$ in one cycle of the input for a duration

$$\lambda = \frac{1}{2\pi f_s} \left\{ \sin^{-1}\left(\frac{X_{t1}}{E_s}\right) + \sin^{-1}\left(\frac{X_{t2}}{E_s}\right) \right\} \quad (10.33)$$

Encoders with multi-level quantizers

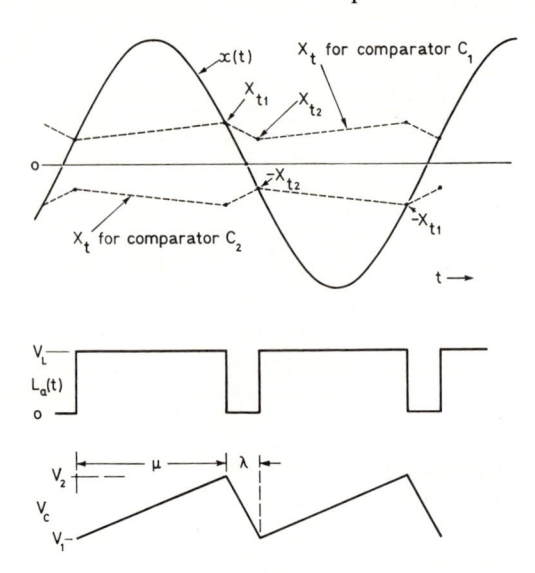

Fig. 10.18. Syllabically companded 2-bit p.c.m. encoder–waveforms when encoding a sinusoidal signal $x(t)$

Applying Equations 2.55 and 2.56 enables the maximum and the minimum value of V_c to be calculated, V_2 and V_1, respectively. Hence

$$X_{t1} = A + V_2 = A + V_L \left\{ \frac{1 - \exp(-\mu/RC)}{1 - \exp[-(\lambda+\mu)/RC]} \right\} \quad (10.34)$$

and

$$X_{t2} = A + V_1 = A + V_L \left\{ \frac{(1 - \exp(-\mu/RC))\exp(-\lambda/RC)}{1 - \exp[-(\lambda+\mu)/RC]} \right\}$$

where $\mu = (1/2f_s) - \lambda$, X_{t1}, X_{t2} and λ can be found uniquely by graphical means. This is apparent by considering the simple case when the time constant RC is very long, i.e.

$$RC \gg (\lambda + \mu)$$

and the voltages X_{t1} and X_{t2} shown in Figure 10.18 approximate to the same value to give a threshold voltage

$$X_t = A + V_L \frac{\mu}{\lambda + \mu} = V_L(1 - 2f_s\lambda) + A$$

where

$$\lambda = \left(\frac{1}{\pi f_s}\right) \sin^{-1}\left(\frac{X_t}{E_s}\right)$$

Substituting for λ in the equation for X_t gives

$$X_t \simeq E_s \cos\left\{\frac{\pi}{2}\left(\frac{X_t - A}{V_L}\right)\right\} \tag{10.35}$$

This equation when plotted shows that when X_t is small it has an approximately linear relationship with E_s, and the rate of increase of X_t with E_s diminishes with increasing E_s.

The comparators C_1 and C_2 shown in Figure 10.17 adjust the threshold values X_t for input sinusoids having different values of E_s and f_s such that the binary waveform $L_a(t)$ representing the magnitude of the 2-bit p.c.m. word has a binary shape typified by Figure 10.18; λ varies in the way described in the previous analysis.

By arranging for the encoder to vary X_t according to the value of E_s the companding action is equivalent to replacing a static level x_A (Figure 10.16) by a variable level X_t. It is left to the decoder to perform the cardinal task of estimating the appropriate values of q_A and q_B.

10.5.1. Decoder

The decoder must first identify which bit in each two-bit word contains amplitude information and which bit contains polarity information.

Assuming that the input signal is band limited between frequencies f_{c1} and f_{c2}, i.e. contains no d.c. component, the mean value of $L_p(t)$ over a long time is half the logic rail voltage. However, the mean value of the amplitude digits $L_a(t)$ varies considerably from approximately zero to the logic rail voltage V_L. This characteristic enables the de-multiplexing, i.e. the separation of $L_a(t)$ and $L_p(t)$ in the received 2-bit p.c.m. signal $m(t)$ to be easily accomplished. Alternate bits in $m(t)$ are directed into two RC circuits; the voltage at the output of these circuits which is closest to $V_L/2$ is the one which receives the $L_p(t)$ signal. Hence the relative time locations of the $L_a(t)$ and $L_p(t)$ signals are established. When speech is transmitted, band-limited between 250 Hz and 2,400 Hz, a suitable value for the RC time constant in the circuits used to identify the $L_a(t)$ and $L_p(t)$ signals is 50 ms.

After de-multiplexing the p.c.m. signal it is digital-to-analogue converted to produce a signal $q(t)$ which in any Nyquist interval can have one of four different values. The digital-to-analogue conversion is made on the basis that an uncompanded 2-bit p.c.m. signal has been received. The difficulty is in deciding which values to assign to the q_A and q_B values shown in Figure 10.16 as they depend on the statistical properties of $x(t)$ which may be stationary, for example Gaussian, or non-stationary as in the case of speech signals. It is found empirically that for speech signals

$$\pm q_A \simeq \pm 0.75 \quad \text{and} \quad \pm q_B \simeq \pm 1.5$$

Having discussed how $L_p(t)$ and $L_a(t)$ are separated, and that $q(t)$ is produced with levels $\pm q_A$ and $\pm q_B$ which are determined empirically for a given type of $x(t)$ signal, the remainder of the decoding process will now be described. The $L_a(t)$ signal is reconverted to X_t. This is achieved by passing $L_a(t)$ through a syllabic filter to give a voltage V_c to which is added a d.c. voltage A. This defines the minimum value of X_t when $V_c = 0$. Thus for errorless transmission the $L_a(t)$ and X_t signals in the decoder are

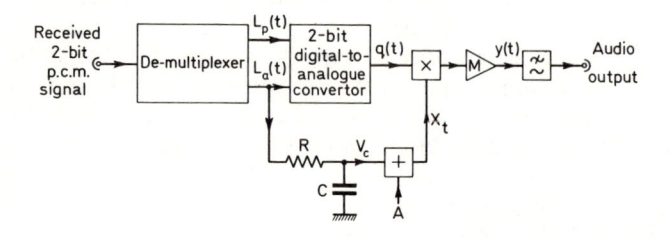

Fig. 10.19. *Syllabically companded 2-bit p.c.m. decoder. (After Wilkinson[13])*

identical to those at the encoder, see Figures 10.17 and 10.19. The quantized signal $q(t)$ now has its levels multiplied by X_t and the product amplified by an amplifier having a gain M. Thus X_t varying at the syllabic rate corresponds to x_A and the quantization levels $\pm q_A$ and $\pm q_B$ (in Figure 10.16) varying at the syllabic rate. The syllabically companded 2-bit p.c.m. codec therefore behaves as a p.c.m. system whose quantizer has many levels. Finally the signal $y(t)$ at the output of the amplifier need only be filtered to remove frequency components outside of the message band to yield the recovered speech signal.

10.5.2. Decoder output

The waveform $y(t)$ at the output of the amplifier in the decoder is shown in Figure 10.20. The peak level is $|q_B M X_t|$ and the lower level is $|q_A M X_t|$, where M is the gain of the amplifier.

Encoders with multi-level quantizers

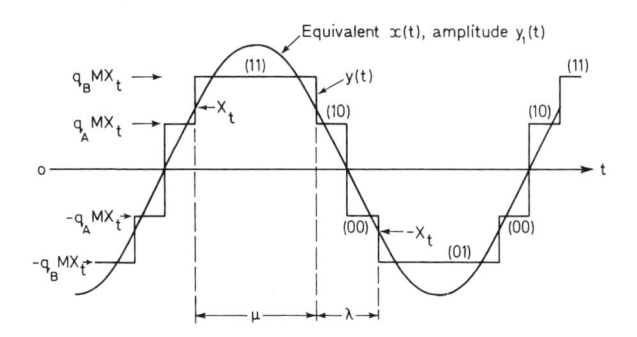

Fig. 10.20. Waveform $y(t)$ at the output of the amplifier in the decoder when a sinusoid is encoded

By applying Fourier analysis the amplitude of the fundamental component of $y(t)$ can be shown to be

$$y_1(t) = \frac{4}{\pi}q_B M X_t \left\{ \frac{q_A}{q_B} + \left(1 - \frac{q_A}{q_B}\right) \sin\left(\frac{\pi(X_t - A)}{2V_L}\right) \right\} \quad (10.36)$$

If $q_B = 2q_A$

$$y_1(t) = \frac{2}{\pi}q_B M X_t \left\{ 1 + \sin\left(\frac{\pi(X_t - A)}{2V_L}\right) \right\} \quad (10.37)$$

For $X_t \ll V_L$, $y_1(t) \propto X_t$. From Equation 10.35, $X_t \doteq E_s$, thus

$$y_1(t) \doteq \frac{2}{\pi}q_B M E_s \quad (10.38)$$

The amplitude of the output sinusoid $y_1(t)$ is proportional to the amplitude of the input sinusoid E_s, when $X_t \ll V_L$. Under these conditions, if $y_1(t) = E_s$, $q_B M = \pi/2$. By selecting $q_B M$ to be slightly less than $\pi/2$, say $\frac{3}{2}$, $y_1(t) \doteq E_s$ over a 35 dB variation in E_s.

10.5.3. Comparison with the SCALE system

Because of the ease of de-multiplexing the received p.c.m. signal the simplicity of the circuitry and the flat overload characteristic, it is reasonable to compare this system with the SCALE system described in Chapter 7. Wilkinson finds that when $x(t)$ is a white noise signal band-limited between 450 and 550 Hz the two-bit adaptive p.c.m. codec operating with a bit rate of 9.6 kbits/s has a peak signal-to-noise ratio of 13 dB and a value exceeding 10 dB over an input signal range of 30 dB. For the SCALE system encoding the same input signal and functioning at the same bit rate the peak s.n.r. is 10 dB and the s.n.r. is greater than 7 dB over an input signal range of 15 dB.

When speech has its higher frequencies pre-emphasised and band-limited between 250 Hz and 2,400 Hz, and both the two-bit adaptive p.c.m. codec and the SCALE system encode this speech signal at 9·6 kbits/s, then in the presence of zero errors, or when not more than 5% of the received bits are received in error, the decoded signals have similar subjective appeal. The transmission errors are assumed to be randomly distributed in time.

The resulting quality of these two systems which are performing at a bit rate which is only twice the Nyquist rate is quite striking.

REFERENCES

1. Hullett, J. L., 'Impairment of differentially quantized T.V. signals by transmission errors', *Australian Telecomm. Research*, **4**, No. 2, 28–35, November (1970).
2. Cattermole, K. W., *Principles of Pulse Code Modulation*, Iliffe, London (1969).
3. Limb, J. O. and Mounts, F. W., 'Digital differential quantizer for television', *Bell Systems Tech. J.*, **48**, 2583–2596, September (1969).
4. O'Neal, J. B. Jr., 'Predictive quantizing systems, (differential pulse code modulation) for the transmission of television signals', *Bell Systems Tech. J.*, **45**, 689–721, May–June (1966).
5. McDonald, R. A., 'Signal-to-noise and idle channel performance of differential pulse code modulation systems—particular applications to voice signals', *Bell Systems Tech. J.*, **45**, No. 7, 1123–1151, September (1966).
6. Fine, T., 'Properties of an optimal digital system and applications', *I.E.E.E. Trans. Info. Theory*, **IT-10**, 287–296, October (1964).
7. Millard, J. B. and Maunsell, H. I., 'Digital encoding of the video signal', *Bell Systems Tech. J.*, **50**, 459–479, February (1971).
8. Max, J., 'Quantizing for minimum distortion', *I.R.E. Trans. Info. Theory*, **IT-6**, 7–12, March (1960).
9. Brainard, R. C. and Candy, J. C., 'Direct-feedback coders: design and performance with television signals', *Proc. I.E.E.E.*, **57**, No. 5, 776–786, May (1969).
10. O'Neal, J. B. Jr., 'A bound on signal-to-quantizing noise ratios for digital encoding systems', *Proc. I.E.E.E.*, **55**, No. 3, 287–292, March (1967).
11. O'Neal, J. B. Jr., 'Signal-to-quantizing-noise ratios for differential p.c.m.', Correspondence in *I.E.E.E. Trans. Com. Tech.*, **COM-19**, 568–569, August (1971).
12. Brown, E. F., 'A sliding-scale direct-feedback p.c.m. coder for television', *Bell Systems Tech. J.*, **48**, No. 5, 1537–1553, May–June (1969).
13. Wilkinson, R. M., 'An adaptive pulse code modulator for speech signals', Report No. 72001, *Signals Research and Development Establishment*, Christchurch, Hants, January (1972).
14. Smith, B., 'Instantaneous companding of quantized signals', *Bell Systems Tech. J.*, **36**, 653–709, May (1957).

Chapter 11

DELTA MODULATION DIGITAL FILTERS

11.1. INTRODUCTION

For the purpose of this chapter, the linear delta modulator is the encoding device which transforms the analogue signal into a suitable binary form for digital filtering. Usually digital filters are presented with linear p.c.m. 'words', i.e. analogue samples which have been encoded into binary numbers. The coefficients of the weighting response are normally represented digitally and digital multiplication is used. The nature of the binary signal at the output of the linear delta modulator results in a somewhat different approach to digital filtering. However, the same performance can be achieved as for filters operating on p.c.m. signals, but the implementation is different.

Linear d.m. has been described in Section 1.2. It consists of a delta modulator which encodes analogue signals $x(t)$ into binary signals $L(t)$ and a decoder which is simply an integrator followed by a low pass filter F_0. The received $L(t)$ waveform is integrated and then filtered to produce the decoded signal $o(t)$ which is a close replica of $x(t)$. The function of the filter F_0 is to exclude quantization noise above the bandwidth of interest. Because the decoder is linear and time invariant the decoded output signal $o(t)$ is unchanged if the integrator and F_0 are interchanged, as shown in Figure 11.1. The input to the filter is now a binary signal $L(t)$

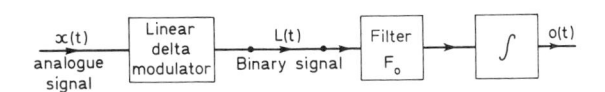

Fig. 11.1. Linear d.m. system. Integrator in the decoder positioned after the low-pass filter

which is appropriate if F_0 is realised as a digital rather than an analogue filter. If the d.m. system is correctly designed slope overload can be avoided and quantization noise reduced to negligible proportions: consequently $o(t)$ can be assumed equal to $x(t)$. If F_0 is a low pass filter $x(t)$ will be recovered after integration. In general, F_0 can be designed to perform some filtering operation on $x(t)$ before recovery since the d.m. encoder and integrator effectively cancel.

337

11.2. D.M. NON-RECURSIVE FILTER WITH ANALOGUE COEFFICIENTS

Consider the linear delta modulator shown in Figure 11.2. Suppose that the encoder is not slope overloaded by the input signal $x(t)$ and that the

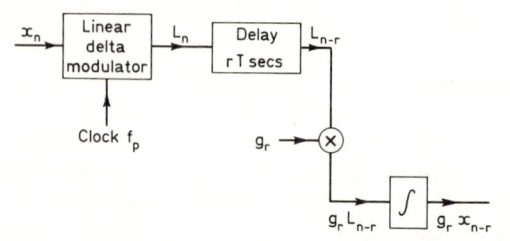

Fig. 11.2. Effect of delaying the binary signal L(t) by rT seconds and multiplying by a constant g_r

quantization noise is negligible. This requires a high ratio of clock rate f_p to step-size γ in the feedback signal $y(t)$, and $f_p \gg f_{c2}$ where f_{c2} is the highest frequency in $x(t)$. For these conditions $y(t)$ closely approximates $x(t)$.

If the signal $L(t)$ at the output of the delta modulator is delayed rT seconds, where $T = 1/f_p$, before being passed to the local decoder, i.e.

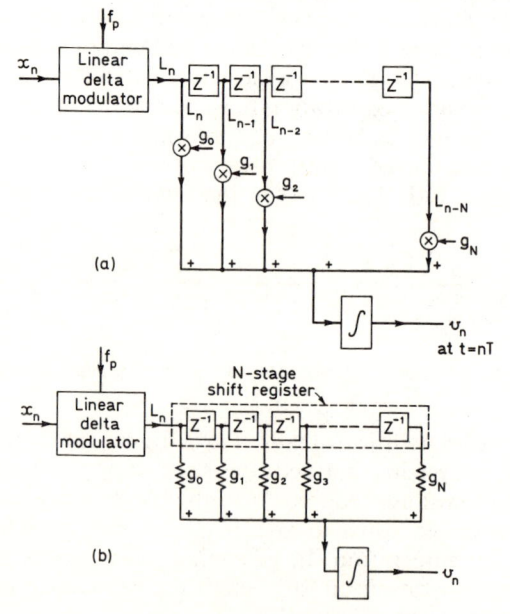

Fig. 11.3. d.m. binary transversal filter: (a) using multipliers, (b) using weighted resistors

Delta modulation digital filters

the integrator, then the signal at the output of the integrator is $x(t-rT)$. Suppose $t = nT$, the nth sampling instant, and that the output of the rT delay is multiplied by a constant g_r, then the signal at the output of the integrator is $g_r x_{n-r}$. The arrangement is shown in Figure 11.2. The input to the integrator $g_r L_{n-r}$, which may be positive or negative, consists of narrow pulses which approximate to impulses of strength $V\tau g_r$, where V is the magnitude and τ the duration of L_{n-r}.

If the output of the delta modulator is applied to a shift register enabling $L_n, L_{n-1}, L_{n-2}, \ldots, L_{n-N}$ signals to be produced, i.e. the shift register has $N+1$ outputs spaced by delays of T seconds and if each of these outputs are multiplied by coefficients $g_0, g_1, g_2, \ldots, g_N$, respectively, added together and then integrated, the resulting signal at time nT is

$$v_n = \sum_{r=0}^{N} g_r x_{n-r} \qquad (11.1)$$

If the values of the coefficients g_r are suitably selected a non-recursive filter[3] having a required frequency response can be obtained. The arrangement of the filter to achieve Equation 11.1 is shown in Figure 11.3(a). The multipliers and coefficients can be replaced by suitable resistors, as shown in Figure 11.3(b), having conductances proportional to the weighting coefficients $g_0, g_1, \ldots, g_N$. The arrangement of a binary shift register and set of weighting resistors is known as a binary transversal filter.[1]

11.2.1. Pulse transfer function

The pulse transfer function of this non-recursive filter is found[3-6] by taking the z-transform of v_n

$$V(z) = \sum_{n=0}^{\infty} v_n z^{-n} = \sum_{n=0}^{\infty} \left\{ \sum_{r=0}^{N} g_r x_{n-r} \right\} z^{-n} \qquad (11.2)$$

Let $k = n-r$

$$V(z) = \sum_{k=-r}^{\infty} \sum_{r=0}^{N} g_r x_k z^{-(k+r)}$$

$$= \sum_{k=-r}^{\infty} x_k z^{-k} \sum_{r=0}^{N} g_r z^{-r}$$

and with $x_k = 0$ for negative values of k

$$V(z) = X(z) \sum_{r=0}^{N} g_r z^{-r}$$

where $X(z)$ is the z-transform of the input sequence. Therefore

$$G(z) = \frac{V(z)}{X(z)} = \sum_{r=0}^{N} g_r z^{-r} \qquad (11.3)$$

If a sequence containing a solitary 'one', i.e. 0, 0, 1, 0, 0, 0, 0, 0 is applied to the input of the shift register the v sequence at the output of the filter is $g_0, g_1, g_2, \ldots, g_N$. This output sequence is called the weighting sequence although it is also frequently referred to as the impulse response. The pulse transfer function of the filter is therefore the z-transform of the weighting sequence. Should a negative weighting be required the associated resistor is connected to the inverted output of the flip-flop in the shift register.

11.2.2. Frequency response of the filter

Now $z = e^{sT}$, where s is the complex frequency variable used in Laplace transformations. The frequency response is found by replacing z in Equation 11.3 by $\exp(j2\pi f T)$, i.e. by putting $s = j2\pi f$.

$$G[\exp(j\pi f T)] = \sum_{r=0}^{N} g_r \exp(-j2\pi f Tr) \qquad (11.4)$$

Consideration will now be given to determine what latitude exists in obtaining a suitable impulse response from some desired frequency domain specification for the case of an N-stage non-recursive filter. Now a unique relationship exists between the weighting sequence $\{g_r\}$ and a sequence of frequency samples $\{G_m\}$. This relationship is found by evaluating the frequency response $G[\exp(j2\pi f T)]$ at $(N+1)$ discrete frequencies, i.e. it is sampled at frequencies spaced $1\{(N+1)T\}$Hz apart. Replacing f in the above equation by $m/\{(N+1)T\}$ and to conform with conventional representation of Equations 11.5 and 11.6, $(N+1)$ will be written as N in this section.

$$G_m = \sum_{r=0}^{N-1} g_r \exp\left\{-j2\pi\left(\frac{m}{NT}\right)Tr\right\} \qquad (11.5)$$

where $m = 0, 1, 2, \ldots, (N-1)$.

The frequency samples $G_0, G_1, G_2, \ldots, G_{N-1}$ are collectively called the discrete Fourier transform (DFT) of the weighting sequence $g_0, g_1, g_2, \ldots, g_{N-1}$; G_m is a periodic sequence with period N.

For the weighting sequence to be real, a condition which produces a physically realizable filter, it is necessary for G_0 and $G_{N/2}$ to have a phase shift of 0° or 180°. The remaining frequency samples can be chosen as required, except that the pairs G_1 and G_{N-1}, G_2 and G_{N-2}, etc. must be complex conjugates. Samples corresponding to frequencies above f_{c2}

may be usefully set to zero to alternate quantization noise above the bandwidth of $x(t)$.

When the frequency samples have been selected, the weighting sequence can be obtained by taking the inverse discrete Fourier transform (IDFT) of $G_0, G_1, \ldots, G_{N-1}$

$$g_r = \frac{1}{N} \sum_{m=0}^{N-1} G_m \exp\left\{ j2\pi \frac{rm}{N} \right\}, \quad r = 0, 1, \ldots, (N-1) \quad (11.6)$$

Equations 11.5 and 11.6 show that the resolution, i.e. the spacing between independently selected frequency samples, depends on N. Therefore any design technique[2] is constrained by the setting of only $N/2$ independent complex frequency samples which clearly sets a limit on the frequency resolution obtainable for a given number of stages.

11.2.3. Noise in d.m. non-recursive filter

The frequency response of the non-recursive d.m. filter shown in Figure 11.3(b) is symmetrical about d.c. from $-f_p/2$ to $+f_p/2$; it is periodic outside these limits. In order to prevent the input signal $x(t)$ from overloading the delta modulator $f_p/2$ must be considerably in excess of the highest frequency f_{c2} in $x(t)$. A typical response is shown in Figure 11.4.

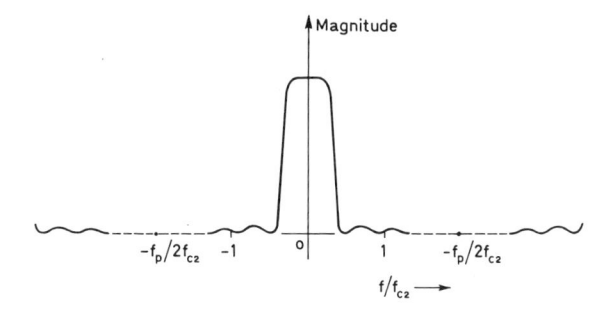

Fig. 11.4. Typical low pass frequency response showing f_{c2} considerably less than 'folding frequency' $f_p/2$

The only source of noise in the filter[2] is due to quantization noise in the encoder provided f_p/f_{c2} is sufficiently high for a given d.m. step-size. Consequently Figure 11.3(b) can be rearranged as shown in Figure 11.5. The integrator has been placed after the output of the delta modulator to produce $o(t)$. This signal only differs from $x(t)$ by the d.m. error signal $e(t)$. As $o(t)$ is an analogue signal the shift register has been replaced by an analogue delay with N taps spaced T seconds apart. The input and output

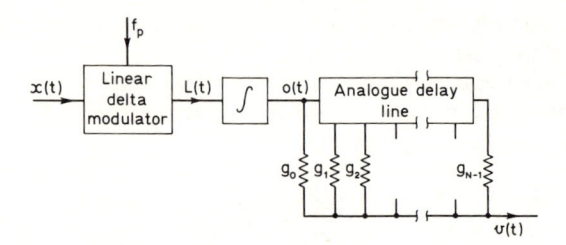

Fig. 11.5. Equivalent arrangement of Fig. 11.3(b)

signals in Figures 11.3(b) and 11.5 are then identical since the rearrangement involves only linear elements.

By replacing the filter of Figure 11.3(b) by the equivalent filter of Figure 11.5 the noise performance of the filter can be determined. This is because the spectral density function of $o(t)$ is the sum of the spectral density function of the input signal and that due to the quantization effects, as discussed in Section 3.10.

The noise in the filter is the noise component $n(t)$ in $o(t)$ filtered by the binary transversal filter. For band-limited white noise the mean square value of $n(t)$, with the aid of Equations 3.85, 3.75 and 11.4, becomes

$$\langle \{n(t)\}^2 \rangle = \int_{-\infty}^{\infty} S_{nn}^*(f) \cdot |H(j2\pi f)|^2 \cdot |G[\exp(j2\pi f T)]|^2 \, df \qquad (11.7)$$

11.3. D.M. RECURSIVE FILTER WITH ANALOGUE COEFFICIENTS

A recursive filter employs feedback to obtain the required filter characteristics. This section describes a recursive d.m. filter[2]–the advantages and disadvantages of non-recursive and recursive filters are discussed in the next section.

If the analogue output signal $v(t)$ in the non-recursive d.m. filter is subtracted from the input signal $x(t)$ such that the delta modulator now encodes the difference signal $d(t)$ then the arrangement is a recursive filter. The analogue and digital outputs of the filter are $d(t)$ and the binary signal $L(t)$ at the output of the encoder respectively. In this direct form instability will almost certainly occur with typical values of weighting coefficients. This is because of a high probability that $v(t)$ will frequently be increased per clock period by an amount considerably greater than the step size γ resulting in the delta modulator being slope overloaded.

In order not to restrict the values of the weighting coefficients it is necessary to band-limit $v(t)$ to the same pass-band as $x(t)$, thereby ensuring that $d(t)$ does not contain any frequencies higher than those of $x(t)$. This

Delta modulation digital filters 343

is achieved by placing a low pass filter F_u in the feedback path of the recursive d.m. filter. The restrictions on the clock rate f_{c2} and γ values which prohibit the encoder from being overloaded have already been discussed in the case of the d.m. non-recursive filter; if the slope of $d(t)$ satisfies Equation 1.4, the filter noise will be due only to quantization effects. It is desirable that the low pass filter F_u which follows $v(t)$ should have a gain of unity and zero phase shift for frequencies up to f_{c2}. Above this frequency the gain of the filter F_u should be zero. A filter such as F_u does not effect the frequency response in the frequency band which is of interest but improves the noise performance of the recursive d.m. filter. To meet these conditions the pulse transfer function of F_u should be of the form

$$u(z) = u_0 + u_1(z + z^{-1}) + u_2(z^2 + z^{-2}) + \ldots + u_N(z^N + z^{-N}) \quad (11.9)$$

From the above equation it can be seen that F_u has an impulse response which commences before the impulse has been applied, i.e. it has a 'non-causal' response.

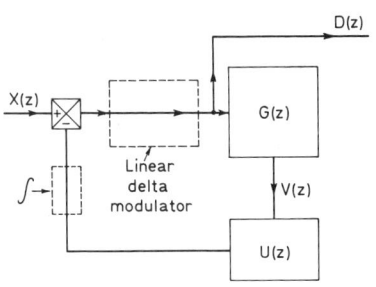

Fig. 11.6. D.M. recursive filter at sampling instants

Figure 11.6 shows the d.m. recursive filter as a discrete time network. The delta modulator is assumed to have nullified the integrator. The loop equation is

$$X(z) - D(z)G(z)u(z) = D(z)$$

The pulse transfer function of the non-recursive d.m. filter is therefore

$$\frac{D(z)}{X(z)} = \frac{1}{1 + u(z)G(z)} = \frac{1}{1 + u(z)\sum_{r=0}^{N-1} g_r z^{-r}} \quad (11.10)$$

Figure 11.7 illustrates one possible form of the d.m. recursive filter. The delay per stage in the shift register is T seconds and is represented by z^{-1} in the figure. In a practical arrangement there would be one resistor from the output of each stage in the shift register connected to the input of the integrator. The non-causal response of F_u is taken into consideration by the use of the a_0, a_1 and a_2 resistors. The a_0 and a_1 resistors are arranged

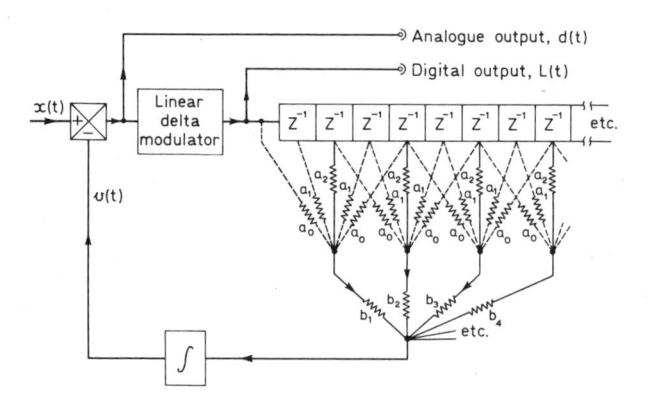

Fig. 11.7. D.M. recursive filter. (After Lockhart[2])

symmetrically about each a_2 resistor in order to conform with Equation 11.9. If F_u had not been incorporated in the feedback path of the filter $G(z)$ would have been realised by resistors a_2+b_1, a_2+b_2 and a_2+b_3, etc. For performance comparable with that of p.c.m. digital filters the delay between these resistors must be less than $1/f_{c2}$; in this example, $g_r = 0$ for all r except 2, 4, 6, ... The number of the a resistors associated with each b resistor and the magnitude of these resistors can be varied according to the specification of the filter.

11.3.1. Noise in d.m. recursive filter

The equivalent circuit of the d.m. recursive filter has the shift register and resistor networks shown in Figure 11.7 replaced by an equivalent analogue transversal filter of the type shown in Figure 11.5. The output of the transversal filter is subtracted from the input signal to give the signal $d(t)$ at the output of the filter, which is also the signal connected to the input of the transversal filter. The delta modulator and the integrator are removed, but the quantization effects of the delta modulator are represented by adding a noise signal $e(t)$ to the input signal $x(t)$. Assuming that $x(t)$ and $e(t)$ are statistically independent of each other the noise component $n(t)$ of the signal $d(t)$ can then be determined by convolving $e(t)$ with the impulse response of the filter. The mean square value of $n(t)$ namely $\langle\{n(t)\}^2\rangle$ can then be found. An alternative method of calculating $\langle\{n(t)\}^2\rangle$ is to multiply $S_{nn}(f)$, the spectral density function of $e(t)$, by the power transfer function of the filter and then integrate over all frequencies.

In order to apply this latter method the transfer function of the d.m. recursive filter described in the last section will now be considered. The pulse transfer function of this filter is given by Equation 11.10 and its frequency response is determined by replacing z with $\exp(j2\pi fT)$. For the purpose of explanation suppose that the frequency response is as shown in Figure 11.4, and repeats at integer multiples of the clock rate f_p. The modulus of this transfer function is squared and then multiplied by $S_{nn}(f)$, the spectral density function of the signal $e(t)$, to yield the spectral density function of $n(t)$. Now from Section 3.11.1

$$S_{nn}(f) = S_{nn}^{*}(f)|H(j2\pi f)|^2$$

The first term, $S_{nn}^{*}(f)$, in this product is composed of spectra whose centres are located at d.c., $\pm f_p/2$, $\pm 3f_p/2$, $\pm 5f_p(2, \ldots,$ etc. as indicated by Equation 3.93 and whose shape is shown in Figure 3.16 over the frequency range 0 to $f_p/2$ for various conditions. The second term $|H(j2\pi f)|^2$ is the power transfer function of a hold circuit and is obtainable from Equation 3.75. $|H(j2\pi f)|^2$ is zero at integer multiples of f_p and has peaks at odd integer multiples of $f_p/2$ which decreases rapidly with increasing odd integer values, and are very small at $\pm 3f_p/2$. Consequently the spectra due to the digital filter and to $S_{nn}^{*}(f)$, whose centres are located at integer multiples of $\pm f_p$, are reduced to insignificant proportions by $|H(j2\pi f)|^2$. Consequently the value of $\langle \{n(t)\}^2 \rangle$ is mainly due to the quantization noise in the delta modulator whose frequency components reside in the spectrum of the frequency response of the digital filter which is located about d.c.

From the above discussion the expression for the mean square value of the noise at the output of the filter is

$$\langle \{n(t)\}^2 \rangle = \int_{-\infty}^{\infty} \frac{S_{nn}(f)}{|1 + u[\exp(j2\pi fT)] \cdot G[\exp(j2\pi fT)]|^2} \, df$$

11.4. DISCUSSION

The advantages and disadvantages of non-recursive and recursive filters are well documented in the literature.[3-6] For those readers unfamiliar with this subject it should be noted that:

1. The non-recursive filter, unlike the recursive filter, does not have any closed loops. As a consequence the non-recursive filter does not suffer from stability problems.

(2) The recursive filter is said to be unstable if its weighting sequence does not asymptotically decay to zero. The recursive filter may be unstable if one or more of its poles reside outside the unit circle.

(3) Because of the feedback the weighting sequence of a recursive filter may be infinite in length. The non-recursive filter, however, has a finite length weighting sequence which is an advantage when spurious inputs are applied to the filter in that the effects of these inputs vanish

after a period of time which is dependent on the length of the weighting sequence and the clock rate.

(4) The recursive filter usually causes significant phase distortion, but the non-recursive filter can easily be designed to have a linear phase/frequency characteristic. As a result the non-recursive filter causes a minimum of distortion to the signal waveform which it passes, subject to the amplitude characteristic not introducing amplitude distortion.

(5) As a general rule it is easier to achieve a given arbitrary response with a non-recursive filter than with a recursive filter.

(6) Normally, for a given specification, a recursive filter can be constructed with less components than a non-recursive filter.

More comparisons can be made, but those above are some of the more salient ones. These remarks apply for both the conventional filters using p.c.m. and d.m. filters. The differences between the conventional digital filters and the d.m. filters described above are:

(7) A conventional digital filter can usefully filter frequencies up to the 'folding frequency' $f_p/2$. The d.m. filters are restricted to frequencies less than f_{c2} which do not overload the delta modulator. For the case of linear d.m. operating with low quantization noise $f_p/2 \doteqdot 10f_{c2}$.

(8) With conventional digital filters a p.c.m. encoder is used to encode the analogue imput signal into a binary one. This means that care must be taken to handle a p.c.m. word (see Chapter 9) because of its hierarchical structure. The coefficients are in a digital form and digital multipliers must be used. If an analogue output is required a p.c.m. decoder must be used. All these factors mean that it is more expensive to construct a conventional digital filter than one using a delta modulator. The multiplication summing in d.m. filters can be achieved by resistors, and the decoding by an integrator and low-pass filter.

(9) The noise performance of a d.m. filter depends only on the quantization noise produced in the delta modulator. A conventional digital filter has noise due to the quantization effects in the encoding, and round-off errors occur in the processes of digital multiplication and summation.

(10) The accuracy of the designed impulse response in a d.m. filter depends on the tolerance of the resistor values, the voltage tolerance of the shift register outputs and the accuracy of the integration. These disadvantages can be virtually overcome as described in Section 11.5.2. The corresponding accuracy in the p.c.m. filter is dependent on the number of bits assigned to the weighting coefficients.

Eventually large-scale integration techniques will enable the weighting sequence to be generated with a negligible error by using digital multiplication. Digital filters programmed on a digital computer can already achieve this. The binary transversal filter using resistors to form the weighting sequence can introduce errors due to the accuracy of these resistors, as mentioned in (10) above. In order to remove the resistors one approach is to use the fact that a delta modulator encodes on a bit-by-bit basis unlike the word-by-word encoding used by p.c.m. If the weighting coefficients are specified by only two values then the multiplication of these coefficients by the ones or zeros stored in the shift register becomes a simple task. The advantages and disadvantages arising from the use of these binary coefficients will be discussed in the next section.

11.5. PROGRAMMABLE BINARY TRANSVERSAL FILTER

If the weighting sequence $g_0, g_1, \ldots, g_N$ of the binary transversal filter shown in Figure 11.3(b) is replaced by a sequence having binary coefficients then it can be stored in an $(N+1)$-bit shift register as shown in Figure 11.8.

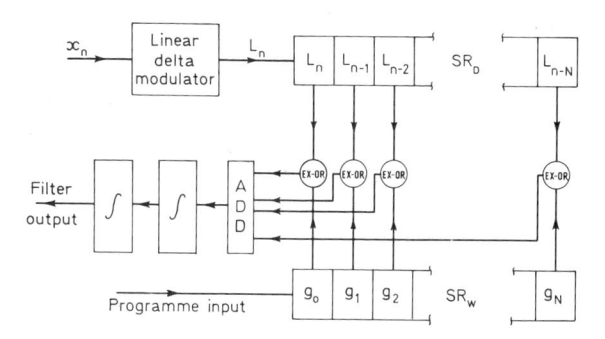

Fig. 11.8. Programmable binary transversal filter having binary weighting coefficients. (After Lockhart and Babary[8])

The change to binary coefficients advocated by Lockhart[7,8] means that the weighting sequence can be simply changed by resetting the shift register and reading-in a new set of data. In this way the weighting sequence is programmable.

This digital filter, in addition to being easily programmable, is simple to implement because the weighting coefficients are binary valued and these values are the same as the binary levels $\pm V$ say, of the signal at the output of the delta modulator. Consequently the multiplication of L_{n-r} and g_r can then be accomplished by using an Exclusive-OR circuit which will produce a level $-V$ if L_{n-r} and g_r have the same polarity, and $+V$ if they are different. The outputs of all the exclusive-OR's are added together and then integrated to give the filtered signal at the output of a binary transversal filter. However, as the binary signal $L(t)$ presented to the binary transversal filter is generated by a delta modulator, a further integrator is required to recover the filtered output. This is because $L(t)$ is a binary representation of the differential of $x(t)$.

11.5.1. Selecting the binary weighting sequence

Given a required continuous impulse response $g(t)$ the corresponding weighting sequence $\{g_r\}$ for analogue coefficients are the values of $g(t)$ at the sampling instants spaced T seconds apart.

The procedure to determine the weighting sequence $\{g_r\}$ where each coefficient has a binary number is very simple. The required $g(t)$ is applied

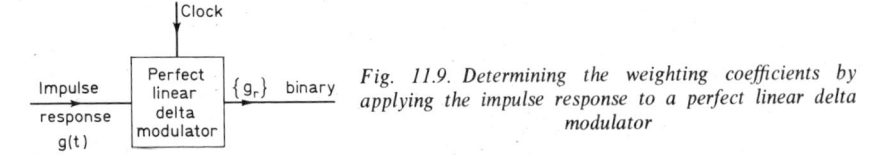

Fig. 11.9. Determining the weighting coefficients by applying the impulse response to a perfect linear delta modulator

to a hypothetical linear delta modulator as shown in Figure 11.9. The binary output sequence is the required weighting sequence $\{g_r\}$. This statement can be easily verified by noting that if the $\{g_r\}$ sequence is used in the binary transversal filter shown in Figure 11.3(b), then the application of a logical 'one' to this filter will yield an output signal v_n which is the impulse response of the filter and is identical to the required impulse function $g(t)$.

The determination of $\{g_r\}$ can be accomplished with the aid of a computer. However, it can easily be established by drawing the required $g(t)$ response and assuming that this response is the input to the delta modulator. The encoder's feedback waveform $y(t)$ and binary signal $L(t)$ are then constructed in a manner similar to that used in deriving the waveforms in Figure 1.2. The signal so produced is identical to $\{g_r\}$.

11.5.2. Recirculating shift registers

The parallel processing used in Figure 11.8 can be replaced by a serial method using recirculating shift registers as shown in Figure 11.10. The only disadvantage of this method is that the maximum processing rate is reduced. The advantage of using recirculating resistors are that the summing resistors are no longer required, thereby reducing cost if large-scale integration is used. In this case the *ends* of the shift registers rather than each stage are accessed, and only one exclusive-OR gate is required.

The shift register SR_B which stores the weighting sequence is one bit longer than the shift register SR_A which stores the last N bits of data from the encoder. Suppose that these registers are of the same length, say $N+1$ bits. The contents of these registers can only be obtained at the output of the last stage and because in one d.m. clock period it is necessary to execute $N+1$ multiplications and additions, see Figure 11.8, it follows that the

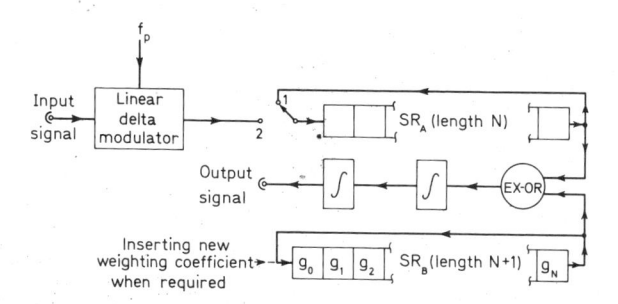

Fig. 11.10. *Programmable binary transversal filter having binary weighting coefficients where the processing is done serially. (After Lockhart and Babary[8])*

shift registers shown in Figure 11.10 must be clocked at

$$f_R = (N+1)f_p$$

At the end of a d.m. period the clock associated with SR_B must be frozen by moving the switch to position 2. All the digits in SR_A are then moved one stage to the right and the longest stored data sample is rejected. The switch is now in position 2 and the current delta modulator output sample is inserted into SR_A. The switch is now moved back to position 1 and both shift registers have their contents recirculated at f_R bits/s and multiplication and addition is achieved by the exclusive-OR and the first integrator, respectively. The first integrator in Figure 11.10 acts as both the adder and first integrator in Figure 11.8, providing the required sum at the end of each d.m. period. The second integrator performs the d.m. decoding, i.e. the same function as the integrator in Figure 11.8. Interruption of the recirculating digits whenever a new binary sample is inserted into SR_A can be prevented by arranging SR_B and SR_A to have $N+1$ and N stages, respectively. The clock rate is still $f_R = (N+1)f_p$ bits/s and consequently after every $1/f_p = T$ seconds the weighting sequence returns to the location shown in Figure 11.10.

The behaviour of the filter can be seen with reference to the following example. Suppose that initially the switch in Figure 11.10 is in position 2, the delta modulator data L_6 is inserted into SR_A and all the bits move one stage to the right, with the result that bit L_{10} is rejected, i.e. $N=4$ in this

Table 11.1

d.m. period T	Recirculating clock period	Contents of shift registers	Output of exclusive-OR	Changes
1	1	$SR_A =\ \ \ \ L_6\ \ L_7\ \ L_8\ \ L_9$ $SR_B = g_4\ \ g_0\ \ g_1\ \ g_2\ \ g_3$	$g_3 L_9$	$\begin{cases} L_6 \text{ inserted} \\ L_{10} \text{ rejected} \end{cases}$
1	2	$SR_A =\ \ \ \ L_9\ \ L_6\ \ L_7\ \ L_8$ $SR_B = g_3\ \ g_4\ \ g_0\ \ g_1\ \ g_2$	$g_2 L_8$	
1	3	$SR_A =\ \ \ \ L_8\ \ L_9\ \ L_6\ \ L_7$ $SR_B = g_2\ \ g_3\ \ g_4\ \ g_0\ \ g_1$	$g_1 L_7$	
1	4	$SR_A =\ \ \ \ L_7\ \ L_8\ \ L_9\ \ L_6$ $SR_B = g_1\ \ g_2\ \ g_3\ \ g_4\ \ g_0$	$g_0 L_6$	
1	$N+1 = 5$	$SR_A =\ \ \ \ L_6\ \ L_7\ \ L_8\ \ L_9$ $SR_B = g_0\ \ g_1\ \ g_2\ \ g_3\ \ g_4$	$g_4 L_9$	
2	$1+(N+1) = 6$	$SR_A =\ \ \ \ L_5\ \ L_6\ \ L_7\ \ L_8$ $SR_B = g_4\ \ g_0\ \ g_1\ \ g_2\ \ g_3$	$g_3 L_8$	$\begin{cases} L_5 \text{ inserted} \\ L_9 \text{ rejected} \end{cases}$
2	7	$SR_A =\ \ \ \ L_8\ \ L_5\ \ L_6\ \ L_7$ $SR_B = g_3\ \ g_4\ \ g_0\ \ g_1\ \ g_2$	$g_2 L_7$	
2	8	$SR_A =\ \ \ \ L_7\ \ L_8\ \ L_5\ \ L_6$ $SR_B = g_2\ \ g_3\ \ g_4\ \ g_0\ \ g_1$	$g_1 L_6$	
2	9	$SR_A =\ \ \ \ L_6\ \ L_7\ \ L_8\ \ L_5$ $SR_B = g_1\ \ g_2\ \ g_3\ \ g_4\ \ g_0$	$g_0 L_5$	
2	10	$SR_A =\ \ \ \ L_5\ \ L_6\ \ L_7\ \ L_8$ $SR_B = g_0\ \ g_1\ \ g_2\ \ g_3\ \ g_4$	$g_4 L_8$	
3	$1+2(N+1) = 11$	$SR_A =\ \ \ \ L_4\ \ L_5\ \ L_6\ \ L_7$ $SR_B = g_4\ \ g_0\ \ g_1\ \ g_2\ \ g_3$	$g_3 L_7$	$\begin{cases} L_4 \text{ inserted} \\ L_8 \text{ rejected} \end{cases}$

example. The subsequent behaviour of the filter is shown in Table 11.1 for each period $T_R = 1/f_R$. In the first recirculating period the output of the exclusive-OR is $g_3 L_9$, and $g_2 L_8$, $g_1 L_7$, $g_0 L_6$ and $g_4 L_9$ in subsequent periods. When the weighting sequence has returned to its original position, L_5 is inserted into SR_A and L_9 is rejected; the process continues without any interruption in the clock rate. The output of the first integrators in Figures 11.8 and 11.10 are identical except for a phase difference of T_R seconds.

11.5.3. Hybrid arrangement

The parallel and serial configuration shown in Figures 11.8 and 11.10, respectively can be incorporated in a hybrid digital filter which is able to exchange the advantages and disadvantages of these configurations in a flexible manner. This hybrid filter incorporates recirculating shift

Delta modulation digital filters

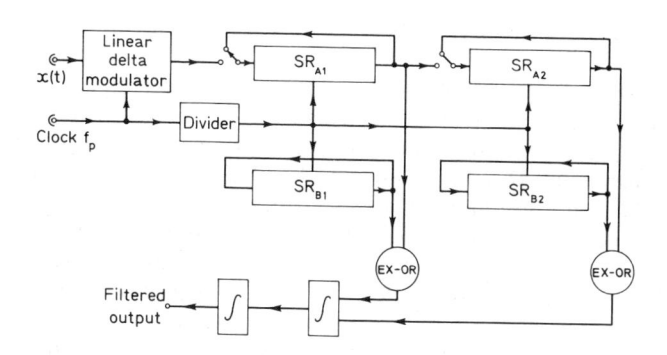

Fig. 11.11. Hybrid d.m. binary transversal filter with binary weighting coefficients. (After Lockhart and Babary[8])

register and parallel multiplication, as illustrated in Figure 11.11. Each recirculating shift register in Figure 11.10 has been replaced by two smaller ones with the object of reducing the clock rate at the expense of additional logic. Shift register SR_A has been supplanted by SR_{A1} and SR_{A2}, both of length $N/2$, and SR_B has been changed to SR_{B1} and SR_{B2} where these registers have $\{(N/2)+1\}$ stages. The weighting sequence is stored in these latter shift registers, i.e. there are $N+2$ coefficients in this sequence. By dividing SR_A and SR_B in this way the recirculating shift registers are clocked at $\{(N/2)+1\}f_p$ bits/s, a reduction of approximately $\frac{1}{2}$ when compared with the arrangement of Figure 11.10, i.e. $(N+2)/2(N+1)$, for $N \gg 2$. The method of operation is identical to that of the previous filter except that the d.m. data is inserted into SR_{A1} and SR_{A2} every $(N/2)+1$ clock periods. As the number of shift registers is increased their length and clock rate is reduced. Eventually the filter of Figure 11.7 is reached.

11.6. RECURSIVE FILTER USING RECIRCULATING SHIFT REGISTERS

The recursive d.m. filter shown in Figure 11.7 and discussed in Section 11.3 is implemented by using one register from each stage of the shift register. The weighting sequence due to these resistors can be replaced by an equivalent sequence $\{g_r\}$ composed of binary coefficients using the selection technique described in Section 11.5.1. This binary sequence is housed in a shift register enabling a similar arrangement to be used as in Figure 11.8 where the signal at the output of the 2nd integrator is a close approximation to the signal $v(t)$ in Figure 11.7. Hence a recursive filter using recirculating shift registers is formed.

11.7. ELEMENTARY FILTER SECTION

High order digital filters are usually implemented by a network of elementary sections.[4] Such a filter section[9] can be constructed for d.m. filtering using the recirculating shift register technique just described. The arrangement is shown in Figure 11.12 where a delta-sigma modulator instead of a delta modulator is used to obtain a delta modulated output thereby avoiding the necessity of using a double integrator (Figure 11.10). The non-recursive sub-section accepts a delta modulated signal, which may be the output from a previous section, and produces a filtered analogue signal which is applied to the recursive section. The data in the shift registers is recirculated in one d.m. clock period T, and once every T seconds the delta-sigma modulator accepts an analogue sample and outputs a binary sample into SR_c. The sequence of events in this elementary section is readily apparent from Sections 11.5 and 11.6.

11.8. DIGITAL IMPLEMENTATION OF D.M. FILTERS

There are two subtractors, one in the recursive filter part of the elementary filter section and the other in the delta-sigma modulator. These can be combined and implemented by a parallel adder. The integrators shown in Figure 11.12 are supplanted by up-down counters which receive bits at

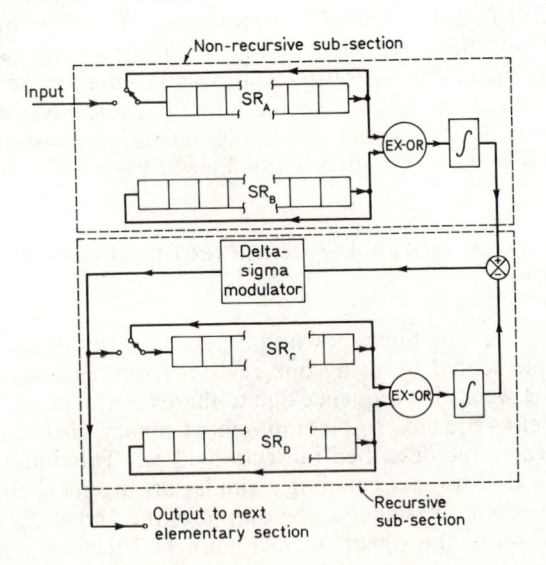

Fig. 11.12. Elementary filter section

the recirculating clock rate, and convey their number in a parallel form to the adder at the lower clock rate of the delta-sigma modulator. The parallel adder therefore adds the contents of the up-down counter to the number fed back from the output of the delta-sigma modulator using two's complement arithmetic. The output of the adder is fed in parallel form to a set of JK flip-flops which accumulates the numbers. The JK accumulator, replaces the integrator in the delta-sigma modulator. The comparator action is achieved by noting the sign digit in the accumulator which is also the bit at the output of the elementary filter section.

The filter section requires two exclusive-OR's and two up-down counters. By increasing the recirculating clock rate, generally by a factor of two, only one exclusive-OR and up-down counter need be used. In the first half of the d.m. period the exclusive-OR and up-down counter service the recirculating shift registers in the nonrecursive section: in the second half of the d.m. period the remaining shift registers in the recursive section are attended to.

The elementary filter section is completely digital, and the contents of the SR_B and SR_D shift registers are readily changed since the weighting sequence is programmable. This arrangement exploits the basic simplicity of d.m.

REFERENCES

1. Voelcker, H. B., 'Generation of digital signalling waveforms', *I.E.E.E. Trans. Com. Tech.*, **COM-16**, 81–93, (1968).
2. Lockhart, G. B., 'Digital encoding and filtering using delta modulation', *Rad. and Elec. Engnr.*, **42**, No. 12, 547–551, December (1972).
3. Rabiner, L. R., Gold, B. and McGonegal, C. A., 'An approach to the approximation problem for non-recursive digital filters', *I.E.E.E. Trans. Audio and Electroacoustics*, **AU-18**, No. 2, 83–106, June (1970).
4. Gold, B. and Rader, C. M., *Digital Processing of Signals*, McGraw Hill (1969).
5. Lindorff, D. P., *Theory of Sampled-data Control Systems*, Wiley (1965).
6. Ackroyd, M. H., *Digital Filters*, Butterworths (1973).
7. Lockhart, G. B., 'Binary transversal filters with quantized coefficients', *Electronic Letters*, **7**, No. 11, 305–307, June 3 (1971).
8. Lockhart, G. B. and Barbary, S. P., 'Binary transversal filters using recirculating shift registers', *Rad. and Elec. Engn.*, **43**, No. 3, 224–226, March (1973).
9. Lockhart, G. B., 'Digital filtering and delta modulation encoder', *I.E.E. Colloquium on Digital and Distributed Filters*, Digest No. 1973/6, 7/1–7/2, March 15 (1973).

Chapter 12
INSTRUMENTATION USING D.M.

12.1. INTRODUCTION

Although the vast majority of delta modulation systems have been designed for telecommunication applications they have considerable potential in other fields; in particular instrumentation relies increasingly on digital techniques. The delta modulator is an attractive alternative to p.c.m. for encoding analogue signals into binary form, not only on a cost basis which is becoming less important due to large scale integration techniques, but because the encoding is done on a bit-by-bit basis and the decoder is inherently simple.

At the beginning of the chapter methods for measuring noise in d.m. systems are described, and then some examples of applications of d.m. in the field of instrumentation are considered.

12.2. MEASUREMENT OF NOISE IN A D.M. CODEC

The distortion introduced by a delta modulation system when accommodating signals having non-stationary probability density functions, and in particular speech signals can be determined by a method due to Kikkert.[1] Depending on the power level of the signal being encoded this method can be used to measure either quantization noise or quantization plus slope overload noise. The block diagram is shown in Figure 12.1.

The delta modulation system under investigation, denoted by A, introduces delay[2] and distortion between signals $x(t)$ and $o(t)$. The object of the equipment shown in the dotted lines is to measure the distortion, i.e. noise, and to do this it must eliminate the effect of the time delay in the system under test. This is accomplished by encoding say a speech signal $S(t)$ in another delta modulator E whose clock rate is very high and whose step-size is small. This is decoded by the decoder E to give $x(t)$ which has negligible quantization noise, i.e. approximates to $S(t)$. The binary stream at the output of the encoder E is delayed by the shift register prior to being decoded to give $x(t)_d$. This signal is identical to $x(t)$ but delayed by $n/f_{p(E)}$, where n is the number of stages in the shift register.

The d.m. system A whose noise performance is being evaluated operates at its practical clock rate $f_{P(A)}$. The clock rate $f_{P(E)}$ of the d.m. codec E

Instrumentation using d.m.

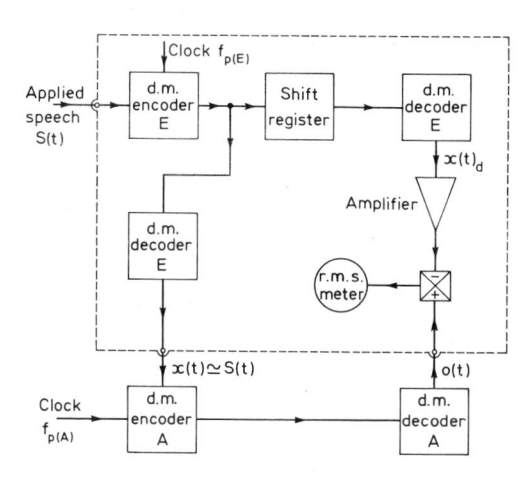

Fig. 12.1. Arrangement for measuring distortion in a d.m. codec. (After Kikkert[1])

is arranged so that $f_{P(E)} \gg f_{P(A)}$. The decoded analogue signal $o(t)$ differs from the signal $x(t)$ applied to the input of the d.m. codec A due to the noise generated in the codec. The noise generated is large compared to that in the d.m. system E and consequently $o(t)$ differs from $x(t)$ by a much greater extent than does $x(t)$ from $S(t)$. The signal $o(t)$ is compared with $x(t)_d$, the delayed version of the signal $x(t)$, to form the difference signal $o(t) - x(t)_d$ whose value is indicated by an r.m.s. meter. The measurement procedure for a given signal $S(t)$ consists of varying the length of the shift register and the gain of the amplifier until the meter reading is a minimum. This reading is a representation of the noise produced in the d.m. codec A.

12.2.1. Uncorrelated noise power in d.m. codec

This method known as the Noise Loading Test[3,4] is applicable for measuring the quantization noise component $n(t)$ in the error signal $e(t)$ when the delta modulator is frequently overloaded.

The spectral density function $S_{yy}(f)$ of the waveform $y(t)$ at the output of the local decoder in a linear d.m. system can be expressed with the aid of the model of a delta modulator shown in Figure 4.5. In this figure the equivalent linear filter allows for slope overloading effects as described in Section 4.3.1. As the input signal $x(t)$ has a spectral density function $S_{xx}(f)$ and the filter has a power transfer function $|G(j2\pi f)|^2$ the signal at the output of the filter has a spectral density function which is the product of these two functions. Consequently,

$$S_{yy}(f) = S_{xx}(f)|G(j2\pi f)|^2 + S_{nn}(f) \qquad (12.1)$$

where $S_{nn}(f)$ is the spectral density function of the noise signal $n(t)$.

In order to simplify the description of the Noise Loading Test the following discussion will assume that the above spectral density functions are defined for only positive frequencies rather than the usual definition which applies to all frequencies. Consider the frequency band of $x(t)$ divided into N frequency slots of width Δf where

$$N\Delta f = f_{c2} - f_{c1} \qquad (12.2)$$

f_{c2} and f_{c1} are the highest and lowest positive frequencies in $x(t)$ respectively and N is integer valued and large compared to unity.

Suppose that $x(t)$ is passed through a filter which suppresses the signal in one of these slots

$$S_{xx}(f) = 0, \qquad f_0 - \frac{\Delta f}{2} \le f \le f_0 + \frac{\Delta f}{2} \qquad (12.3)$$

where

$$f_0 = f_{c1} + n\Delta f + (\Delta f/2) \qquad (12.4)$$

and n can only have one value during a particular measurement, but over a series of measurements will have values from 0 to $N-1$. For frequencies other than those defined by Equation 12.3 the value of $S_{xx}(f)$ is generally finite over the band f_{c1} to f_{c2}.

Provided Equation 12.3 applies, $S_{nn}(f) = S_{yy}(f)$ over the frequency band Δf, determined by putting $S_{xx}(f) = 0$ in Equation 12.1. The reason for the value $S_{yy}(f)$ over the frequency band Δf, centred at f_0, is because the effects of quantization cause a continuous spectrum to exist over this frequency band which does not correlate with the input spectrum.

The noise power in nth frequency slot is therefore

$$N_{qn}^2 = \int_{f_{c1} + n\Delta f}^{f_{c1} + (n+1)\Delta f} S_{yy}(f) \, df \qquad (12.5)$$

If f_0 is varied over the frequency band of the input signal $x(t)$ in steps of Δf, where the initial and final values of f_0 correspond to $n = 0$ and $n = N-1$, respectively then the noise power due to quantization is

$$\langle \{n(t)\}^2 \rangle = \sum_{n=0}^{N-1} N_{qn}^2 = \sum_{n=0}^{N-1} \int_{f_{c1} + n\Delta f}^{f_{c1} + (n+1)\Delta f} S_{yy}(f) \, df \qquad (12.6)$$

When $S_{yy}(f)$ is defined for both positive and negative frequencies the right-hand side of Equation 12.6 is multiplied by two.

Instrumentation using d.m. 357

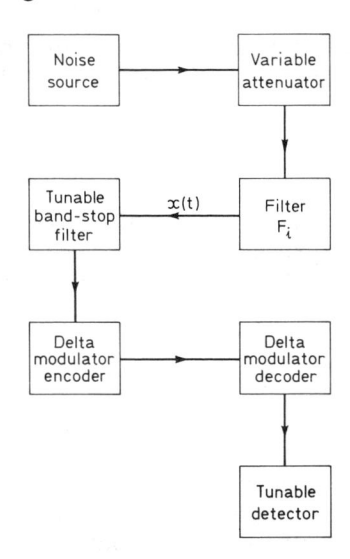

```
        ┌──────────┐          ┌──────────┐
        │  Noise   │─────────▶│ Variable │
        │  source  │          │attenuator│
        └──────────┘          └──────────┘
                                    │
                                    ▼
        ┌──────────┐               ┌──────┐
        │ Tunable  │    x(t)       │Filter│
        │band-stop │◀──────────────│  F_i │
        │  filter  │               │      │
        └──────────┘               └──────┘
             │
             ▼
        ┌──────────┐          ┌──────────┐
        │  Delta   │          │  Delta   │
        │modulator │─────────▶│modulator │
        │ encoder  │          │ decoder  │
        └──────────┘          └──────────┘
                                    │
                                    ▼
                              ┌──────────┐
                              │ Tunable  │
                              │ detector │
                              └──────────┘
```

Fig. 12.2. *Arrangement for measuring uncorrelated noise power in a d.m. system*

The experimental arrangement for measurement of uncorrelated noise power $\langle\{n(t)\}^2\rangle$ is shown in Figure 12.2. A noise source which produces nearly white noise over a band of frequencies in excess of the pass-band of the filter F_i is attenuated to produce a required long term mean square power level. The output of the attenuator is then band-limited by F_i to produce a Gaussian signal $x(t)$ which is passed through a tunable bandstop filter having a stopband of Δf Hz to produce the input signal to the delta modulator. This signal is then processed by the delta modulation system and a tuned detector, connected to the decoded signal, is tuned to f_0 with an effective pass-band of Δf. The detector records the mean square value of the noise signal in the band Δf, represented by Equation 12.5.

The centre frequency f_0 of the tunable bandstop filter and the tunable detector are then moved in the same direction by Δf and the noise power is again recorded. By moving f_0 in steps of Δf over the entire frequency band of the filter F_i the shape of the spectral density function $S_{nn}(f)$ is determined over this frequency band and the mean square value of the uncorrelated noise component $n(t)$ found.

The procedure is then repeated for different values of the attenuator setting to give the value of $\langle\{n(t)\}^2\rangle$ over a range of mean square power levels of the signal $x(t)$.

The measurement of signal power plus total noise power is achieved by repeating the procedure, except that the tunable bandstop filter is removed allowing $x(t)$ to proceed directly to the input of the delta modulator. The tuned detector now records both signal and noise

358 Instrumentation using d.m.

components in the frequency band Δf where the noise components are due to both quantization and slope overload effects.

12.3. D.M. TIME SCALER

The system to be described will store a transient, or part of an analogue signal, and display it repeatedly with time scaling.† The transient is encoded by a suitable delta modulator into a binary signal which is sequentially stored in a long shift register. The data is removed periodically from this register at a bit rate k times slower than that used in the encoding process, and thence decoded into a time scaled version of the original transient. The ratio of the encoding to decoding bit rates is the scale factor k.

The system diagram of the d.m. time scaler is shown in Figure 12.3. The band-limited analogue signal $x(t)$ is applied to both the delta modulator for encoding into a binary waveform $L(t)$ and to a quantizer Q whose output is a logic zero if $|x(t)| < V_t$ or a logic 1 if $|x(t)| \geq V_t$. Therefore Q performs as a threshold detector observing signals whose instantaneous magnitudes are $\geq V_t$. The value of V_t can be selected by the operator.

The first time that $|x(t)| \geq V_t$ a memory is set; the output $m(t)$ signal of the memory M is then changed from a logic zero to a logic one. The $m(t)$ signal remains at this logic level, even if $|x(t)| < V_t$, until the memory receives a reset signal. Even after the reset signal has occurred the output

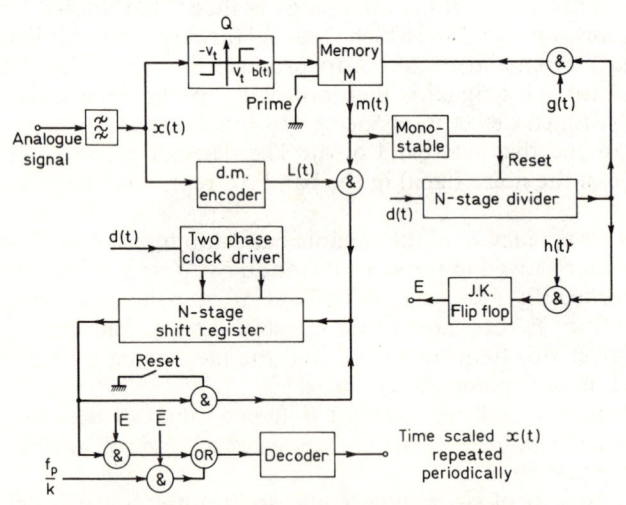

Fig. 12.3. System arrangement of d.m. time scaler

† The assistance of G. M. Kennard and L. Leong in the realisation of this system is gratefully acknowledged by the Author.

of the quantizer is prohibited from again setting the memory until the latter has been manually primed.

When $m(t)$ is a logic one the binary signal $L(t)$ is allowed to pass into the N-stage shift register. This shift register is manually reset prior to the reception of the analogue signal $x(t)$. The arrival of the $m(t)$ signal is used to trigger a monostable whose output pulse has a duration which is small compared to $1/f_p$, where f_p is the clock rate associated with the encoder and the rate at which data is inserted into the shift register. The monostable is used to reset an N-stage divider. As shown in Figure 12.3 the input to this divider is a signal $d(t)$, where

$$d(t) = g(t) + h(t) \tag{12.7}$$

$$g(t) = m(t) \cdot f_p \tag{12.8}$$

$$h(t) = \overline{m(t)} \cdot (f_p/k) \tag{12.9}$$

The plus sign in Equation 12.7 refers to the logic OR operation, the dot sign in Equations 12.8 and 12.9 refers to the logic AND operation and $\overline{m(t)}$ is the logical inversion of $m(t)$.

Thus prior to the arrival of a transient $\geq V_t$, i.e. when $\overline{m(t)}$ is a logic 1, the input to the divider consisted of a pulse train having a repetition frequency of f_p/k bits/s. Directly the $m(t)$ signal changes to a logic 1 the divider is reset to zero via the monostable, and the higher clock rate f_p bits/s is applied to the input divider. In addition the encoded data from the delta modulator is allowed to pass into the N-stage shift register at f_p bits/s.

Directly N bits of data from the delta modulator have been inserted into the shift register, the N-level divider produces an output pulse. This pulse together with the signal $g(t)$ produces a signal which causes the memory to be reset resulting in the signal $m(t)$ returning to the logic 0 level. As a consequence no more data is inserted into the shift register.

The $d(t)$ signal now reverts to its form prior to the memory being set, i.e. to a clock rate of f_p/k. The contents of the filled shift register are now removed from its output and re-inserted into its input at the bit rate of the signal $d(t)$. This recirculation process is continually repeated. Directly the memory is reset in addition to the initiation of the recirculation of the data in the shift register at f_p/k the N-stage divider changes from dividing at the bit rate f_p to a slower one of f_p/k. Consequently after every recirculation of the data in the shift register a pulse is generated at the output of the divider. The time duration between the pulses at the output of the divider is Nk/f_p and each of these pulses are allowed to trigger a flip-flop whose output E is applied to an AND gate. The recirculating data in the shift register is also applied to this AND gate and provided E is a logical one the data is allowed to pass to the decoder which produces an analogue version of the encoded transient stored in the shift register.

During the next kN/f_p seconds E is a logical zero preventing the stored data from being decoded. During this interval $\bar{E}$, the logical inversion of E, allows the $d(t)$ signal to pass into the decoder which enables the signal at the output of the decoder to recover to a zero level. The process of passing data to the decoder for Nk/f_p seconds followed by an equal period when the decoder is allowed to recover to a zero state in order to receive the data once more, is continually repeated. Consequently the transient $x(t)$ is transformed into a time scaled periodic signal. Due to the output of the decoder being allowed to recover to zero in between the appearances of the time scaled transient it is subjected to a d.c. shift $\mp V_t$. This level can be removed by adding $\pm V_t$ to the output of the decoder.

12.3.1. Frequency limitations

It is necessary when encoding transients having high frequency components to operate the delta modulator at a high clock rate. Delta modulators have been operated at clock rates in excess of 100 MHz, but MOST shift registers have difficulty at the present time in operating with clock rates well below this figure. As the clock rate f_p must be the same for both the encoder and shift register when data is being encoded and stored it follows that the limiting factor governing the storage of the fastest transient is the performance of the shift register. This limitation can be mitigated by employing a number of shift registers, say μ, and inserting the data from the delta modulator into these registers in a sequential fashion. This means that the clock rate for the encoder is μ times greater than the clock rate for the shift register.

The lowest frequency that the system can accommodate is unlimited if a static shift register is employed, but is limited when dynamic shift registers are used. This is because there is a minimum clock frequency associated with dynamic shift registers. Systems using dynamic shift registers can only achieve a time scaling factor of about 100 to 1,000, depending on the quality of the time-scaled transient.

12.3.2. Applications of the d.m. time scaler

Ultra violet and pen recorders

Briefly the principle of the ultra violet (u.v.) recorder or oscillograph is as follows. The electrical signal to be recorded is applied to a moving coil galvanometer. To this galvanometer is attached a mirror whose angular position is proportional to the instantaneous value of the applied signal. A u.v. light source is directed onto this mirror and the reflected beam traverses a moving paper which is sensitised to u.v. light enabling the position of the reflected beam to be instantly visible. The combined movements of the mirror and the paper means that the visible path of the u.v. light beam on the paper represents the variation of the amplitude of the applied signal as a function of time.

These oscillographs have a number of independent recording channels, typically between 12 and 50. They are generally employed to record signals whose highest frequency is in the range of 40 to 1,000 Hz, although the maximum frequency that can be recorded is of the order of 15 kHz.

The pen recorder differs from the oscillograph in that the response of an electrical signal is not produced by the movement of a light beam on u.v. sensitised paper, but by a pen which makes an ink drawing on a paper surface. The ink writing system usually has less independent recording channels and is cheaper than the u.v. oscillograph. However, an important distinction between these two types of recording instruments is that the u.v. oscillograph has a much greater bandwidth than the ink writing pen recorder which can only respond to frequencies up to a few Hz.

When the d.m. time scaler is used to time scale a transient having frequency components well above 15 kHz, the time scaled transient can now be recorded by the u.v. oscillograph. By using a static rather than a dynamic shift register in the d.m. time scaler, data can be removed from the shift register at a speed which is so slow that when decoded the time stretched transient can be recorded by the ink writing pen recorder. Thus a transient that was so fast that it was beyond the capability of a u.v. oscillograph, can with the aid of the d.m. time scaler be recorded by the relatively inexpensive ink writing recorder.

Peripheral for hybrid computer

A hybrid computer is the combination of an analogue and a digital computer. The digital computer is basically used to control the analogue computer whose function is to process or generate the analogue data. It is sometimes necessary[5] for the hybrid computer to be remotely located from the operator, and the analogue signals entering and emanating from this computer may have frequency components which are in excess of the bandwidth of the available telephone circuits thereby prohibiting the use of these circuits for transmission of these signals. However, by using a d.m. time scaler at the computer and operator sites the analogue signals can be 'time stretched' which reduces the bandwidth of these signals and enables them to be satisfactorily transmitted over telephone circuits. Because the digital part of the computer requires binary data it may be convenient to transmit the 'time stretched' binary signals from the d.m. time scaler rather than the corresponding time stretched analogue versions.

The transmission of binary signals has the advantage that after the decoding process at the end of a noisy transmission channel the recovered analogue signals will be less corrupted by noise than if they were transmitted in an analogue form.

d.m. time scaler used as an oscilloscope aid

The use of a dynamic shift register requires that data, if it is not to be lost, must be recirculated through the shift register above some minimum clock rate.

A consequence of using dynamic shift registers is therefore that the time scaled transient is continually repeated (as described in Section 12.3), i.e. it is transformed into a periodic signal.

If the original transient occurs only once and is to be observed on an oscilloscope, careful adjustments of this instrument must be made. However, by using the d.m. time scaler the transient is time stretched and becomes periodic. It can be displayed by an oscilloscope whose triggering capability is irrelevant and whose bandwidth is small compared to the oscilloscope requirements for viewing the original transient.

12.4. DISTANCE DOMAIN D.M.

The delta time scaler has the ability to encode transients into binary signals and convey them at a suitable bit rate and with a relatively small time delay to a computer for analysis. However, if the transient is very fast it cannot be encoded and time scaled by this means and other methods such as the use of a sampling-oscilloscope in conjunction with a computer can be employed.

Distance-domain d.m.[6] offers a method of encoding very fast transients provided they can be displayed on an oscilloscope and photographed by a polaroid camera. It is the photographic trace which is delta modulated into a binary waveform, and as this is a permanent record the speed of encoding is adjusted to suit particular requirements. The distance-domain d.m. does have other applications which are discussed in Section 12.4.1, such as profile encoding of certain three dimensional objects, but it has been primarily conceived for encoding photographic traces.

In the distance-domain delta modulator the photograph is laid on a transparent sheet and illuminated from below. A photo-detector located in a mobile head attempts to track the light passing through the white trace in an analogous way to the tracking of the input signal in a linear delta modulator. The mobile head makes two movements during each sampling period Δx. The first movement is $\pm \Delta y$ along the y-axis, where the polarity is selected such that the mobile head continues to track the light trace. The mobile head is then forced to move by $+\Delta x$ along the positive axis. Observe that Δx and Δy are analogous to the clock period T and the step height γ in the linear delta modulator described in Section 1.2.

There are however two important differences between time- and distance-domain d.m. The photographic trace may have a width greater than Δx in which case, unless care is taken, the mobile detector may step inside the trace for a number of sampling periods resulting in a considerable encoding error. Secondly, the time-domain delta modulator determines its binary output by the sign of the error at a sampling instant, whereas in distance-domain encoding the mobile head can only provide information that the error is zero. In order to deduce the sign of the error the presence of an optical detection of the trace in both the Δx and Δy movements must be interpreted by an encoding algorithm. The sign of the error is decided at the end of a Δx step; the appropriate electrical output binary signal $c(x)$ is generated and used to control the movement $f(x)$ of the mobile head in the y direction by either $\pm \Delta y$.

In order to avoid the mobile head wandering about inside the trace when the step sizes are very much smaller than the width of the trace an edge-tracking algorithm is used which confines the mobile head to track either edge of the trace as shown in Figure 12.4. This is achieved by the optical system observing if the mobile head crosses the trace during the Δy and Δx steps. The observations are stored until the end of the sampling period when two logic signals $P_{\Delta x}$ and $P_{\Delta y}$ are formed. If the trace is crossed the logic level of the signal is 1, and 0 when it is not detected. The binary signal $C(x)$ is then generated by taking into consideration the $P_{\Delta x}$ and $P_{\Delta y}$ signals and the polarity of the previous $C(x)$ signal. If $P_{\Delta x}$

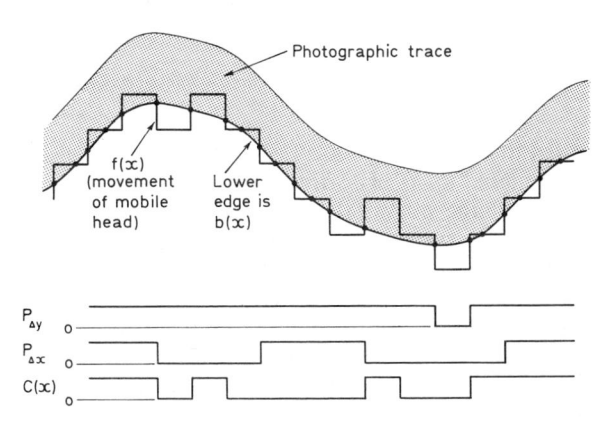

Fig. 12.4. Edge tracking a photographic trace

and $P_{\Delta y}$ have the same logic levels the new $C(x)$ has the same polarity as the previous one, but if the levels are different its polarity, i.e. its logical state, is reversed. The $C(x)$ so formed is applied to a circuit which drives the stepper motor which moves the mobile head $\pm \Delta y$ via a belt and pulley. If $C(x)$ is a logical 1 Δy is positive, and vice versa. Having completed the movement in the y direction the y gantry is moved Δx by applying a signal $g(x)$ to the x stepper motor. At the end of the Δx movement a new $C(x)$ signal is generated and so on. Figure 12.5 illustrates the mechanical arrangement.

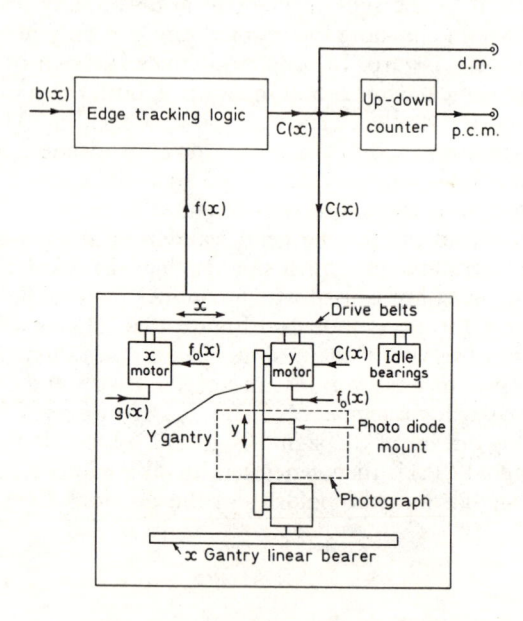

Fig. 12.5. Distance domain d.m. system

When the photograph has been positioned the search mode is initiated via the signal $f_0(x)$. The mobile head makes repeated Δy movements without change in Δx. If it traverses the photograph without crossing the trace it is stepped-on by Δx and the procedure of sweeping in the y direction for each Δx step continues until one edge of the trace is found when tracking and therefore encoding commences. Figure 12.4 shows the $f(x)$ and logic signals when encoding an arbitrary trace. For simplicity the movement of the mobile head has been drawn as a perfect step-type waveform, whereas in practice there would be oscillations due to imperfections in the mechanical system when the encoder is tracking at 60 to 100 bits/s.

12.4.1. Discussion

Distance-domain d.m. suffers from slope overload in the same way as linear d.m. This effect can be mitigated by reducing the slope of the trace on the oscilloscope prior to taking the photograph or by allowing the mobile head to make successive Δy steps during each sampling interval Δx until the edge of the trace is crossed. This latter method is a form of multi-level d.m. described in Section 10.2.

The signal $C(x)$ is the delta modulated binary output signal as shown in Figure 12.4. If it is connected to an up-down binary counter then the number in the counter at any time represents the magnitude of the edge of the trace $b(x)$ in binary numbers. Data can be removed from this counter in a form and at a speed suitable for computer peripherals to handle.

There are numerous applications for this type of encoder. The edge tracking algorithm enables profiles of objects to be encoded and analysed by computer. It is suitable for most of the applications mentioned in Section 12.3.2. In addition is can be used as an instantaneous wattmeter to measure the power in very fast transients, such as switching in transistor circuits. The voltage and current transients are photographed, encoded and the results fed to the computer for analysis.

12.5. DELTA-SIGMA WATTMETER

Instantaneous power in a load is measured by multiplying the voltage $v(t)$ across it by the current $i(t)$ flowing through it. The multiplication is achieved in the delta-sigma wattmeter by using two samplers rather than a multiplier. Sampling and multiplication are closely related. For example, double sideband suppressed carrier modulation can be produced by either multiplying an analogue signal with a sinusoidal carrier, or by sampling the analogue signal with a binary pulse train $p(t)$ and filtering these samples. Alternatively these samples can be produced without a sampler if a multiplier is used. This is achieved by multiplying $p(t)$, whose binary values are arranged to be zero and unity, with the analogue signal.

The delta-sigma wattmeter[7] is given this name because the voltage $v(t)$ developed across the load in which the power is to be measured is encoded by a delta-sigma modulator, d.s.m. into a binary waveform $L(t)$. The d.s.m. decoder is just a low-pass filter F_0 as described in Section 2.4. By connecting the $L(t)$ signal to the input of the filter F_0 the voltage plus a quantization error is obtained at the output of this filter. Although a delta modulator can be used in place of the delta-sigma modulator the latter is more suitable if the spectrum of the voltage signal $v(t)$ is flat.

Having argued that multiplication of two signals can be achieved by one signal sampling another, it might be anticipated that if the load current $i(t)$ is sampled by $L(t)$, and $L(t)$ is a binary waveform which is modulated by $v(t)$, the product $v(t).i(t)$ will be formed by passing the sampled $i(t)$ waveform through the filter F_0. However, one sampler is insufficient for if $v(t)$ is so small that the d.s.m. encoder idles $L(t)$ is a squarewave and if $i(t)$ is finite the current samples after passing through the filter will produce a voltage proportional to $i(t)$. This is undesirable because viewed as a multiplier the delta-sigma modulator idles if $v(t)$ is zero; as $i(t)$ is greater than zero the product $w(t) = v(t).i(t)$ should be zero. In order to ensure that the wattmeter satisfies the requirements of a multiplier, that $w(t)$ is zero when either $i(t)$ or $v(t)$ are zero, a condition which cannot occur in a load, two samplers are used as shown in Figure 12.6. The principle of the wattmeter shown in this figure will now be described.

The binary waveform $L(t)$ and its logical inverted signal $\overline{L(t)}$ have a phase difference of 180° and are arranged to sample the scaled analogue current signal $i(t)$ to produce the sampled signals $s_1(t)$ and $s_2(t)$, respectively.

$$s(t) = s_1(t) - s_2(t)$$

is the signal which when passed through the filter F_0 yields the instantaneous power signal $w(t)$. For the condition when $v(t)$ is below the threshold of encoding and therefore $L(t)$ is a squarewave whose period is $2/f_p$ (where f_p is the clock rate used in the d.s.m. encoder and $i(t)$ is finite and band limited to frequencies below f_{c2}, $f_p \gg f_{c2}$) and $i(t)$ is a d.c. signal, then $s_1(t)$ and $s_2(t)$ are squarewaves which are 180° out of phase with each other and have the same pulse repetition frequency as $L(t)$. $s(t)$ is therefore a squarewaveform of period $2/f_p$ and is rejected by the filter F_0 since its pass-band only extends up to the frequency f_{c2} which is small compared to $f_p/2$; consequently $w(t)$ is zero. If $i(t)$ is a random signal $s_1(t)$ and $s_2(t)$ in adjacent clock periods are nearly equal in amplitude and

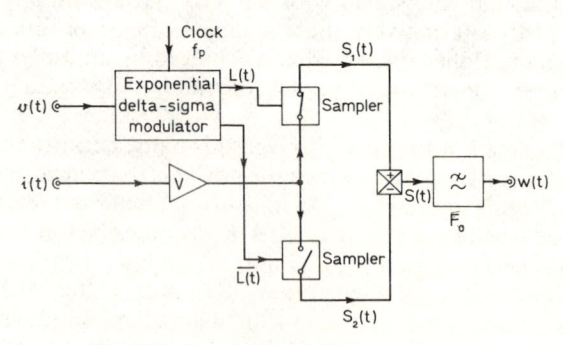

Fig. 12.6. Essential parts of the wattmeter

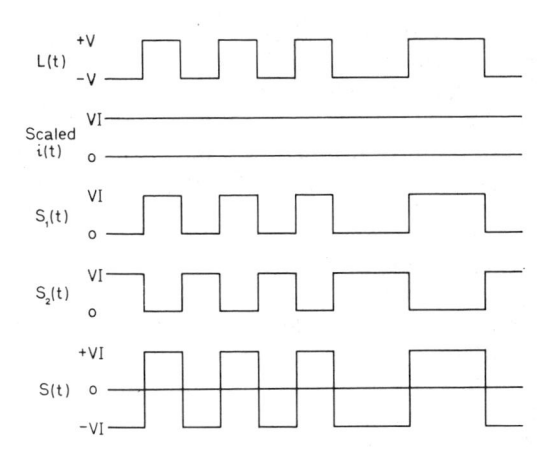

Fig. 12.7. Sampled waveforms for arbitrary $L(t)$ and $i(t)$ signals

opposite in polarity. Hence $s(t)$ is again a square waveform of period $2/f_p$, but its amplitudes in adjacent periods are nearly equal and opposite, and consequently the desired result that $w(t)$ is negligible is obtained.

The error in $w(t)$ is reduced if the d.s.m. encoder is operated at a high clock rate f_p compared to the highest frequency f_{c2} in the load waveforms, and changes at the output of the integrator due to the $L(t)$ waveform are arranged to be just large enough to prevent the encoder being overloaded by $v(t)$. Because $f_p \gg f_{c2}$ the current waveform $i(t)$ changes its amplitude by only a small amount over tens of clock periods. Figure 12.7 shows an arbitrary $L(t)$ pattern and a positive current signal VI. The current signal, of magnitude I, has been multiplied by a scaling constant V prior to its presentation at the samplers in order to ensure that $s(t)$ has the magnitude VI, rather than I. The resulting $s_1(t)$, $s_2(t)$ and $s(t)$ waveforms are also displayed–the latter waveform is the same as $L(t)$ but scaled by VI. When $s(t)$ is filtered $w(t)$ is recovered. If $i(t) = -I$, then $s(t)$ is the inverse of $L(t)$ scaled by VI. In Figure 12.7 $L(t)$ represents a positive voltage signal but the polarity of the $s(t)$ signal is determined by the normal algebraic laws of multiplication and depends on the sign of $i(t)$ and $v(t)$.

There may be errors in $v(t)$ and $i(t)$ because they have not been transduced correctly by the sensors prior to their presentation at the delta-sigma wattmeter. The wattmeter has a system error as distinct from a circuit error. Considering the encoder as delta modulator preceded by an integrator, the encoding error is rarely in excess of $\pm \gamma$, see Figure 1.2. For exponential d.m. the average step-size is D, rather than γ, and is given by Equation 2.33. The decoded voltage error is less than $|D|$ in d.m. because the filter F_0 removes the high frequency steps. However, in the d.s.m. encoder the absence of the integrator in the decoder means that some of these high frequency effects are re-introduced. For good performance a high V/D_{min} ratio is required which necessitates having $f_p \gg f_{c2} \gg f_1$; V is the peak signal that can be encoded by a delta-sigma modulator and D_{min} the minimum value of D due to the quantizer not having infinite gain and the existence of propagation delays around the feedback loop. The error in the encoder is modified by the sampling process, i.e. scaled by I, and is dependent on the phase relationship of $v(t)$ and $i(t)$.

The delta-sigma modulator is formed with two operational amplifiers, and a D-type flip-flop which acts as a sample and hold circuit. One operational amplifier is used in the integrator and the other as a comparator. The samplers s_1 and s_2 in the wattmeter are implemented with field effect transistors. These are preferred to bipolars because they have a low ratio of resistance when conducting to resistance when switched off. To realise the signal $s(t)$ the outputs of the samplers are applied to an operational amplifier having a differential input capability. The signal $s(t)$ is passed through a low-pass filter and the resulting instantaneous power signal $w(t)$ can be conveniently displayed on an oscilloscope. The mean power is obtained by averaging $w(t)$. If accuracy is not too important and mean power is required then $s(t)$ can be connected to a moving coil micro-ammeter which performs the dual functions of decoding and display. Range setting amplifiers ensure that the delta-sigma modulator is properly encoding, particularly when measuring power in low impedance loads.

12.6. SPEED CONTROL OF AN INDUCTION MOTOR

The decoder in a delta modulation system is a local decoder followed by a low-pass filter. The local decoder is the network in the feedback loop of the encoder. In a linear d.m. system the local decoder is an integrator but in companded d.m. codecs it may consist of multipliers, shift registers etc., as described in Chapters 7 and 8. The function of the local decoder is to convert the binary signal $L(t)$ at the output of the delta modulator to a form that when filtered is a close replica of the analogue signal $x(t)$ applied to the encoder.

Suppose the binary waveform $L(t)$ is transmitted through a channel whose characteristic is identical to the characteristic of the d.m. decoder. The channel will perform two functions, it will convey information and decode the signal. This concept of delta modulating an analogue signal and using a device with an appropriate characteristic for decoding the binary signal is used in the speed control of an induction motor, where the motor is assumed to act as the decoder. It is found[8] that a linear delta-sigma modulator is a suitable encoder for an 'induction motor decoder'. The motor is a type of low-pass filter and accepts the baseband

Instrumentation using d.m.

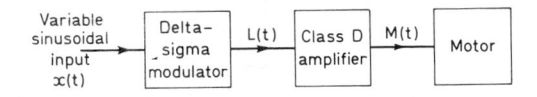

```
                ┌──────────┐        ┌──────────┐
 Variable       │ Delta-   │  L(t)  │ Class D  │ M(t)  ┌───────┐
 sinusoidal ───▶│ sigma    │───────▶│ amplifier│──────▶│ Motor │
 input          │ modulator│        │          │       └───────┘
 x(t)           └──────────┘        └──────────┘
```

Fig. 12.8. Speed control system

components, i.e. a sinusoid, in the binary signal connected to its stator windings and rejects the higher frequencies. It therefore behaves as if a sinusoid has been applied rather than a binary waveform. By varying the frequency of the sinusoid being encoded the speed of the motor can be controlled.

The speed control system is shown in Figure 12.8. The sinusoid $x(t)$ is encoded into a binary waveform $L(t)$ by the delta-sigma modulator. This encoding occurs at a low power level. $L(t)$ is taken from the output of a D flip-flop which acts as a sample and hold circuit and will typically have binary levels of zero and $+5$ V. $L(t)$ is amplified in a Class D amplifier to produce the higher power binary signal $M(t)$. When this signal is applied to the stator windings of the motor the torque developed by the motor depends on the amplitude and frequency of $x(t)$. The motor behaves as if $x(t)$ had been linearly amplified and applied to the motor, rather than the binary signal $M(t)$.

When $x(t)$ is zero the encoder idles and $L(t)$ and $M(t)$ are square waves of period $2/f_p$, where f_p is the clock rate of the encoder. As $f_p \gg f_{c2}$, where f_{c2} is the highest frequency to be encoded, the lowest line frequency $f_p/2$ in $M(t)$ is so high that the motor does not respond in spite of the large amplitude square wave presented to its stator windings. Suppose a sinusoid $x(t)$ having an amplitude E_s and frequency f_s is applied to the encoder. As E_s increases from zero the idle pattern is broken and the $L(t)$ and $M(t)$ waveforms have their binary levels sinusoidally distributed in time according to the periodicity of $x(t)$. The rotor will be accelerated and its speed increased as E_s is increased. The harmonics of $x(t)$ present in $M(t)$ are very small compared to the fundamental when the encoder is tracking and only become significant when E_s is increased above the overload level of the encoder. For the severest overload condition, and one to be avoided, $M(t)$ is a squarewave of period $1/f_s$. The overload condition can be delayed by increasing f_p, but this results in greater device dissipation in the driver stages of the Class D amplifier.

The relationship between E_s and f_s depends on the required motor performance. If the criterion is that the motor should operate over a wide speed range at constant torque and with constant temperature rise, there is a unique value of E_s for each f_s. The adjustment of the amplitude E_s for every frequency f_s of the sinusoid $x(t)$ in order to satisfy the above criterion is laborious. A preferred method is to use an oscillator whose sinusoidal output $a(t)$ can be varied over a wide range of frequencies but whose amplitude is always constant. The sinusoid $a(t)$ is passed through a network whose attenuation versus frequency characteristic is such that the emergent signal $x(t)$ has the correct value of E_s for each value of f_s. This means that by varying the frequency of the oscillator the desired value of E_s is automatically presented to the encoder; the motor when presented with the binary signal $M(t)$ will respond only to the fundamental frequency in $M(t)$ and will always have the correct amplitude to satisfy the performance criterion for the motor.

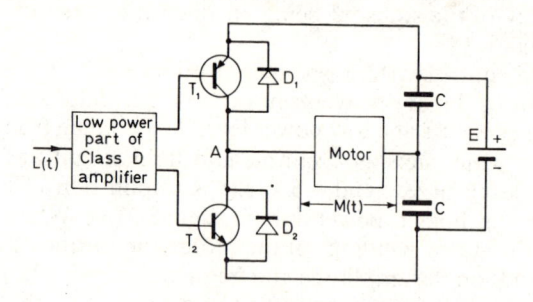

Fig. 12.9. *Class-D amplifier and motor*

For a single phase induction motor a suitable Class-D amplifier is shown in Figure 12.9. The function of the two capacitors is to make the d.c. potential $E/2$, approximately, at their junction in order that the binary levels of the voltage across the motor will be $\pm E/2$, and to provide smoothing to protect the power transistors. When $L(t)$ is a logic 1 transistor T_1 is switched-on and transistor T_2 is switched-off; when $L(t)$ is a logic 0 the order is reversed. When T_1 conducts point A is connected to $+E$ volts and the motor current can flow through either T_1 or D_1 depending on its direction of flow prior to the instant when $L(t)$ changed to a logic 1. When T_2 conducts point A is earthed and the motor current selects the appropriate path through T_2 or D_2. When $L(t)$ is a logic 1 the voltage across the motor

$$M(t) = E - E/2$$

When $L(t)$ is a logic 0 the voltage is reversed to $0 - (E/2)$, i.e. $M(t)$ has binary levels of $\pm E/2$.

12.6.1. Summary

In this speed control system the sinusoid $x(t)$ is encoded by a delta-sigma modulator into a binary waveform $L(t)$ which is then amplified by a Class-D amplifier to produce a signal $M(t)$ having binary levels of $\pm E/2$, where E is the d.c. supply voltage. If $M(t)$ is passed through a low-pass filter F_0 an amplified version of $x(t)$ plus some quantization noise results. If the amplitude E_s of $x(t)$ is increased the sinusoid $o(t)$ at the output of the filter is linearly increased, and if the frequency of $x(t)$ is changed so is the frequency of $o(t)$. The d.s.m. encoder, Class-D amplifier and filter F_0 act as and are easier to implement than a linear amplifier. In the speed control system the filter F_0 is replaced by an induction motor which acts as a filter removing $o(t)$ from the binary signal $M(t)$. The motor can be operated over a wide speed range at constant torque by varying the amplitude and frequency of the sinusoid $x(t)$. By using a variable frequency oscillator with constant amplitude followed by a shaping network to produce the $x(t)$ signal which has the correct values of amplitude for all frequency settings, according to the operation criterion, the speed control of the motor is obtained by merely varying the oscillator frequency.

Comparative measurements[8] of the torque developed by the motor as a function of its speed using this system and using a variable frequency power source connected to the stator windings exhibit only small differences which are of the order of the experimental accuracy.

This method of source encoding has particular advantages in control of motors in remote locations where the controller and motor are separated by a noisy channel. In this situation the $L(t)$ signal is generated at the output of the controller and on its recovery at the distant terminal it is used to produce $M(t)$ which in turn drives the motor.

REFERENCES

1. Kikkert, C. J., 'Digital techniques in delta modulation', *I.E.E.E. Trans. Com. Tech.*, **19**, 570–573, August (1971).
2. Handler, H. and Mangels, R. H., 'A delta-sigma modulation system for time delay and analogue function storage', *Proceedings Spring Joint Computer Conference*, 303–313, (1964).
3. Aaron, M. R., Fleischman, J. S., McDonald, R. A. and Protonotarios, E. N., 'Response of delta modulation to Gaussian signals', *Bell Systems Tech. J.*, **48**, Pt. 1, 1167–1195, May-June (1969).
4. Schwartz, M., Bennett, W. R. and Stein, S., *Communications Systems and Techniques*, McGraw-Hill, New York (1966).
5. Howe, R. M. and Hollstein, R. B., 'Time shared hybrid computers: a new concept in computer-aided design', *Proc. I.E.E.E.*, **60**, No. 1, 71–77, January (1972).
6. Steele, R. and Stevens, P., 'An automatic digital trace encoder for polaroid photographs', *Electronic Engineering*, 80–85, September (1973).
7. Dempster, I. A. and Steele, R., 'Delta-sigma wattmeter', *Electronics Letters*, **7**, No. 18, 519–520, September 9 (1971).
8. Creighton, G. K. and Steele, R., 'Source frequency adjustment controls motor speed', *Electronic Engineering*, 70–71, May (1973).

Appendix

DEFINITIONS OF QUANTIZATION NOISE

If the delta modulator is correctly tracking the input signal $x(t)$, i.e. without experiencing any slope overload, and there are no channel errors, then the decoded signal $o(t)$ can be represented as the sum of $x(t)$ and an unwanted noise signal. This noise signal is referred to as *quantization noise* or *granular noise*, and originates in the encoding process due to the existence of the error signal $e(t)$ which is the difference between $x(t)$ and the feedback signal $y(t)$. The value of $e(t)$ must be small if $o(t)$ is to be a close facsimile of $x(t)$, but cannot be zero for the delta modulator to function.

Different definitions of quantization noise are given in the literature relating to delta modulation systems. These definitions have often been chosen on a convenience basis to align with combinations of the theoretical approach, the method of computer simulation, and the measurement technique employed. Four of these definitions (perhaps descriptions is more apt) which are in vogue are given.

(1) Quantization noise is the difference between the final decoded signal $0(t)$ which occurs at the output of the filter F_0 and the input signal $x(t)$, where the relative delay and amplitude change (if any) of the decoded signal is compensated for prior to the computation.

(2) Quantization noise is the signal which occurs when $e(t)$ is passed through F_0, i.e. it is the component of $e(t)$ which is in the message band.

(3) Quantization noise is $e(t)$.

(4) Quantization noise is the difference between $x(t)$ delayed by one clock period and the signal at the output of the local decoder in the decoder. Here it is the residual error, i.e. the error which remains following the up-dating of the local decoder.

It may be noted that definitions (1) and (2) give the lowest values of quantization noise and have similar values. Definitions (3) and (4) yield higher values of quantization noise as the noise has not been band-limited by the filter F_0.

INDEX

Adaptation constants, first order constant factor delta modulation 246, 250
  second order constant factor delta modulation 253
Adaptive delta modulation 183, 217, 220, 223, 260
A-law, segmented 272, 290
A-law code 270, 276, 278, 302
A-law delta modulation 275–9
  multiplexing 280
A-law non-linearity 272
A-law pulse code modulation, using analogue compression 282
Amplitude quantization 105
Amplitude range, asynchronous delta modulation 169
  delta-sigma modulation 19
  exponential delta modulation 62
  exponential delta-sigma modulation 65
  linear delta modulation 11
Analogue compression 282
Analogue transmission 2
Analysis-synthesis methods 214
Asymmetry 152, 153, 154, 156
  effect on tracking 157
Asynchronous delta modulation 165–82
  amplitude range 169
  channel bandwidth 181
  minimum channel bandwidth 168
  overload characteristic 168
  overload condition 169
  quantization noise 169–71
  signal-to-quantization noise ratio 172, 181
  zero crossings 166–8, 171–2
Asynchronous pulse delta modulation 172–8
  idle channel state 177
  minimum channel bandwidth 177
  signal-to-quantization noise ratio 176
  zero crossings 174–6
Autoconvolution, nth order 147
  of band-pass white noise 88
  of low-pass white noise 85
Autocorrelation function 81, 83, 85, 91, 94, 96, 98–100, 159

Band-limited white noise 85, 88, 169, 186
  input signal 136
Band-pass filter 7, 202
Basis function 82
Binary coefficients 347
Binary impulse 36
Binary pulses 3, 8
Binary signal 1
Binary transmission 2
Binary transversal filter 339
Binary waveform 28, 191, 366, 368
Binomial theorem 86
Bisection rule 307–9
Bode plot 50, 54, 56
Bosworth Candy delta-sigma modulation 234–8
  adaptation algorithm 235
  step input 235–6
  television performance 262
  threshold control 237
Brainard and Candy's direct-feedback encoder 324
Break angular frequency 139
Break frequency 139, 267
Buffer 30

C.C.I.T.T. 213, 272, 290
Channel errors 295
Classifier 303
Closed loop system 132, 135, 139
Companded double integration delta modulation 241–2
  step response 241
Companding 183, 223, 233, 330
  high order 195
  instantaneous 223
  threshold of 208–9
Companding characteristic 183
Compression circuits 183, 219
Computer, hybrid 361
Continuous delta modulation 200–4
  dynamic range 204
  signal-to-quantization noise ratio 202
Cross-correlation function 122, 128–9
Cross-power spectra function 129

Index

Cross-spectral density function 131
  in terms of Normal spectra 101–3
Cross-talk 282

Decoder, adaptive local 228
  analogue s.c.d.s.m. 185, 187
  asymmetrical 152, 157
  asynchronous delta modulation 165, 166
  d.c. signals 20
  delta-sigma modulation 17
  double integration delta modulation 44
  exponential delta modulation 57
  externally companded delta modulation 205
  gain function of 52
  idling waveform 70
  linear delta modulation 3, 4, 6, 8, 11
  local 7, 50, 52, 132, 184, 185, 191, 202, 223, 239, 242, 303, 368
    RC arrangement 54
    signal from 98
    noise at output of 141
    optimum 225
  pulse delta modulation 66, 67
  rectangular wave modulation 178, 180
  SCALE 207
  syllabically companded two-bit p.c.m. 333
Decoder output, noise power 105
  syllabically companded two-bit p.c.m. 334
De-emphasis 325
Delta modulation, and pulse code modulation 283
  digital 243
  digital filters 337–53
  double integration, overload characteristic 50
  linear. *See* Linear delta modulation
  performance 110–16
Delta modulation filters, digital implementation 352
Delta modulation non-recursive filter 338–42
  advantages and disadvantages 345–7
  frequency response 339, 340, 341
  noise in 341
Delta modulation recursive filter 342–5
  advantages and disadvantages 345–7
  noise in 344
  using recirculating shift registers 351
Delta modulation–time–division multiplexed signal 29

Delta-sigma modulation 17–20
  amplitude range 19
  signal-to-quantization-noise ratio 19
  transmission errors in 163
Delta-sigma wattmeter 365–8
Differential pulse code modulation 304
Digital delta modulation 243
Digital errors 2, 233
Digital filters 337, 346–53
Digital implementation of delta modulation filters 352
Digital-to-analogue conversion 333
Digital transmission 2, 214
Digitally controlled delta modulation 213
Discrete adaptive delta modulation 227
Discrete Fourier transform 340
Discrete pulse-phase modulation model 21–5
Distance-domain delta modulator 362–5
Distortion 2, 28, 29, 158, 312
Distortion coefficients 111
Disortion noise power 114
Distortion power in message band 148
Distortion spectral density function 109, 110, 114, 148
DITHER 297–8
Double integration delta modulation 35–57
  idling modes 36
  overload characteristic 50
  signal-to-quantization noise ratio 54
'Down-pulses' 228, 230

Edge tracking algorithm 363, 365
Elementary filter section 352
Encoder, adaptive 15, 242
  always in overload 125
  analogue s.c.d.s.m. 185
  asymmetry 152–8
  asynchronous 132–3
  asynchronous delta modulation 152, 165, 169
  B.C. d.s.m. 236
  Brainard and Candy's direct-feedback 324
  companded 183
  continuous delta modulation 200
  d.c. signals 20
  difference quations 313
  direct feedback 234
  discrete delta modulation 227
  double integration delta modulation 35, 36, 42, 56
  exponential delta modulation 57, 60

Index

Encoder—*continued*
  externally companded delta modulation 205
  1st order c.f.d.m. 246
  gain factor 108
  h.i.d.m. 228
  linear 4, 295
  linear delta modulation 1–14, 17, 20–31, 56
  mapping delta modulation 226
  minimum step size 249
  modified robust delta modulation 260
  multi-level d.s.m. 323
  optimum 224
  p.c.m. 266, 270
  performance calculation 109
  pulse delta modulation 66, 67, 73
  rectangular wave modulation 178
  SCALE 207, 209
  sliding scale 327
  stabilising response by means of prediction 42
  tracking 29
  with multi-level quantizers 301–36
Error function 151
Error signal, energy density of 136
  spectral density 136
  spectral density function 140
Error waveform 127, 159, 292
  mean square value 142, 160
Exclusive–OR circuit 347
Expansion circuits 219
Exponential delta modulation 57–64
  amplitude range 62
  idling characteristic 58
  over-load characteristic 60
  quantization noise 63
  signal-to-quantization noise ratio 64
  slope overload with sinusoidal input signals 123–6
Exponential delta-sigma modulation 64–5
  amplitude range 65
  overload characteristic 64, 65
  signal-to-quantization noise ratio 65
Externally companded delta modulation 204–6

Feedback pulse, height of 190
Filter section, elementary 352
Filters 189, 297
  band-pass 202
  low-pass 337, 368

Finality of coefficients 82
First order constant factor delta modulation 243
  adaptation constants 246, 250
  overload condition 245
  step input response 246
  television performance 264
Flip-flop 7, 57, 60, 66, 219, 368
Frequency deviation 22
Frequency domain 316
Frequency response, delta modulation non-recursive filter 339, 340, 341

Gaussian probability density function 83, 134, 167, 310
Gaussian signal 135, 157
  slope overload 126–42
Granular noise 11, 79, 372

Hermite polynomial basis functions 82, 102
Hermite polynomial model 150–1
High information delta modulation 228
  television performance 262
Hold circuit 7, 94
Hunting amplitude 249, 250
Hunting pattern 247
Hybrid computer 361
Hybrid delta modulation binary transversal filter 350–1

Ideal filter 162
Idle channel condition, multi-level delta modulation 320
Idle channel noise 152–8
Idle channel state, asynchronous pulse delta modulation 177
Idling behaviour, linear delta modulation 8
Idling characteristic, exponential delta modulation 58
Idling error waveform 70
Idling modes, double integration delta modulation 36
Idling patterns, damping high order 44
  syllabically companded delta-sigma modulation 187
Idling state, pulse delta modulation 67, 74
Idling waveform at decoder 70
  pulse delta modulation 72

Index

Impulse function 36, 166
Impulse response 51, 94, 340
Impulse train 98–9
Induction motor, speed control 368–71
Infinite sawtooth functions 105–7
Infinite staircase characteristic 169
Infinite staircase functions 105–7
Infinite staircase model 25–7, 79, 80–1
Instantaneously adaptive delta modulation 223, 295
  performance criteria 223
  similarity between systems 261
Instantaneously companded delta modulation 223–65, 291, 296
Instrumentation 354–71
Integrator 17, 35, 50, 52, 204, 296, 303, 304, 320, 328
  leaky 304
  RC type 64
Inverse discrete Fourier transform 341

Jump phenomena 137–8

Linear delta modulation 1–34, 283
  amplitude range 11
  decoder 6
  definition 3
  delayed encoding 14
  description of system 4–14
  encoder 6
  idling behaviour 8
  models 20–7
  principle of 3
  quantization noise 11, 373
  sample and hold circuit 28
  signal-to-quantization noise ratio 13
  slope overload 9–11, 131
    with sinusoidal input signals 118–26
  with Gaussian signals 126–51
  time-division multiplexing 29–31
  zero order hold 7
Linear filter model 127
  quasi-linearisation 131–9
  transfer function of 139
Linear pulse code modulation 266, 283, 296
Long distance communication 2
Low-pass filter 259, 324, 337, 368

Mapping delta modulation 226
Mean square value of error waveform 160
Mehler's formula 84, 170

Message band 159
  distortion power in 148
Monostable 66, 74
MOST shift registers 360
Multi-level delta modulation 301
  idle channel condition 320
  limitations on gain 320
  linear analysis 315
  stability of codec 319
Multi-level delta-sigma modulation 323–7
Multi-level pulse code modulation 302
Multi-level quantizers 301, 306

Noise, at output of decoder 141
  in delta modulation non-recursive filter 341
  in delta modulation recursive filter 344
  measurement of 354
Noise characteristics 326
Noise-loading tests 355, 356
Noise power 13, 159
  decoder output 105
  delta modulation codec 355
  distortion 114
Normal signals 82
Normal spectra 100
Normal spectrum coefficients 84, 106
Normal spectrum technique 80–5
Nyquist interval 275, 276, 281, 285, 333
Nyquist rate 267, 275, 279, 305, 330

Open loop gain 316
Optimum quantizer 306–11, 314
Oscillograph 360, 362
Output power in message band 109
Overload characteristic 18, 162
  asynchronous delta modulation 168
  double integration delta modulation 50
  exponential delta modulation 60
  exponential delta-sigma modulation 64, 65
  multi-level delta modulation 322
  pulse delta modulation 71
  syllabically companded delta-sigma modulation 187
Overload condition, asynchronous delta modulation 109
  double integration delta modulation 36
  first order constant factor delta modulation 245
SCALE 209

Index

Pen recorder 361
Photo-detector 362
Pitch companded delta modulation 217
Pitch pulse extractor 216
Power in decoded signal 104
Power transfer function 18, 80, 355
Prediction, damping oscillatory idling response by 44
Prediction constant 43–5, 53
Prediction system 42
Pre-emphasis 325
Probability density function 83, 133–4
Programmable binary transversal filter 347–51
Propagation effects 60
Pseudo-random noise 297, 298
Pulse code modulation, A-law 270
   and delta modulation 283
   differential 304
   linear 266, 283, 296
      delta modulation technique 268
   logarithmic 288
   multi-level 302
   principle of block diagram 266
Pulse compression multiplexing 30
Pulse delta modulation 66–77
   idling state 67, 74
   idling waveform 72
   overload characteristic 71
   signal-to-quantization noise ratio 73
Pulse delta modulation codec 73
   performance 75
   setting-up procedure 75
Pulse group delta modulation 238–41
   step input response 240
Pulse-phase modulation 21
Pulse stretcher 21
Pulse transfer function 339, 343

Quantization characteristic 306, 311, 319, 321
Quantization distortion 312
Quantization effects 18
Quantization error 14, 297
Quantization error signal 314
Quantization levels 328, 334
Quantization noise 18, 28, 64, 78–117, 162–3, 188, 267, 315, 317
   as function of zero crossings 171
   asynchronous delta modulation 169–72
   exponential delta modulation 63
   linear delta modulation 11

Quantization noise—*continued*
   mean square value 140
Quantizer 3
   linear model 314
   multi-level 301
   optimum 306–11, 314
Quasi-linearisation, linear filter model 131–9

RC integrator 163
RC time constant 162
Rectangular wave modulation 178–81
   decoder 180
   signal-to-noise ratio 180
Repeaters 2
Robust delta modulation 260

SCALE 206–13, 287, 289, 335
   decoder 207
   dynamic range 213
   encoder 207, 209
   multiplier characteristic 207
   overload condition 209
   parameters 211
   performance 211
   signal-to-quantization noise ratio 213
   threshold of companding 208–9
Second order constant factor delta modulation 252, 287, 296
   adaptation constants 253
   pulse response 256
   signal-to-signal ratio 259
   step input 254
   step input response 253
   tracking an arbitrary signal 258
Segmented A-law 272, 290
Septernary p.a.m. system 301
Series resonant circuit 54
Shift-register 230, 343, 347, 359
   binary 339
   dynamic 362
   MOST 360
   recirculating 348, 350–1
Shift theorem 93
Signal-to-distortion ratio 290
Signal-to-noise ratio 8, 14, 115–16, 142, 147, 158, 183, 290, 295, 298, 314
   effect of successive modulations 298–9
   rectangular wave modulation 180
   second order constant factor delta modulation 259
   syllabically companded delta-sigma modulation 187

Index

Signal-to-quantization noise ratio 109–10, 285, 286–7, 291, 292, 318, 322
  asynchronous delta modulation 172, 181
  asynchronous pulse delta modulation 176
  continuous delta modulation 202
  delta-sigma modulation 19
  double integration delta modulation 54
  exponential delta modulation 64
  exponential delta-sigma modulation 65
  linear delta modulation 13
  pulse delta modulation 73
  SCALE 213
  syllabically companded delta-sigma modulation 187, 188
Sinusoidal input signal 157
  slope overload 118–26
Sliding scale system 327–9
Slope-limiter model 143–50
Slope overload 14, 15, 79, 80
  delta modulator 131
  Gaussian signals 126–42
  linear delta modulation 9–11
  sinusoidal input signals 118–26
    exponential delta modulation 123–6
    linear delta modulation 118–26
Spectral density function 12, 18, 20, 28, 29, 80–1, 84, 85, 91, 93, 94, 98, 104, 136, 147, 150, 170, 171, 267, 342, 355
  distortion 109, 110, 148, 314, 324, 325
  effect of clock rate 114
  feedback signal 104
  in terms of Normal spectra 100
  input 108
  long term double sided 161
  normalised 146
  of error signal 140
  output 107, 151
  rectangular 160
  sampled 103
Speech-reiteration delta modulation 214–19
Speech transmission 270, 289, 314, 334
Speech wave 214
Speech waveform 218
Speed control 368–71
Staircase function 26–7, 91
Statistical delta modulation 224
Step-size 4, 35, 112, 171, 214, 216, 217, 243, 249, 252, 259, 261
Step input response 230, 235, 240, 246, 251

Sub-optimal differential pulse code modulation 314
Syllabically Companded All Logic Encoder. *See* SCALE
Syllabically companded delta modulation 183–222, 289
  versus logarithmic pulse code modulation 288
Syllabically companded delta-sigma modulation 184–200, 296
  basic equations 189
  block diagram 185
  codec parameters 190–3
  dynamic range 193
  high order companding 195
  idling pattern 187
  n syllabic expanders 198
  operating principle 186
  overload characteristic 187
  performance 288, 296
  signal-to-noise ratio 187
  signal-to-quantization noise ratio 187, 188
  transmission errors 197
Syllabically companded two-bit pulse code modulation 330
  decoder output 334
Synchronization 296
Synchronization code 30–1

Telecommunication channels, imperfections of 1
Television transmission 262–5, 296, 329
  transmission errors 264
  weight values 264
Threshold control 237
Threshold signal 70
Time-division multiplexing 29–31
Time quantization 105
Time scaler 358–62
  applications 360
  as oscilloscope aid 362
  frequency limitations 360
Time sharing 297
Transfer function 51, 58, 64, 65, 129, 130, 131, 136, 161
  complex 139
  generalised 139
  of linear filter model 139
Transmission errors 158–63, 197, 226, 233, 295
  in delta-sigma modulation 163
  syllabically companded delta-sigma modulation 197
  television performance 264

Ultra violet recorders   360
'Up-pulses'   229, 230

Wattmeter, delta-sigma   365–8
Wave modulation, rectangular   178–81
Waveform tracking multi-level d.m.   304
Weight values, television performance   264
Weighter   303

Weighting coefficients   339, 342, 347
Weighting sequence   340, 347, 351
  binary   348

Zero-crossing   23, 39
  asynchronous delta modulation   166–8, 171–2
  asynchronous pulse delta modulation   174–6
z-transform   316, 339–40

HCZ

A ous
Booui i ki
 no Aciams

OB op 16 7/4 15